NEUROMETHODS

Series Editor
Wolfgang Walz
University of Saskatchewan
Saskatoon, SK, Canada

For further volumes:
http://www.springer.com/series/7657

Neuromethods publishes cutting-edge methods and protocols in all areas of neuroscience as well as translational neurological and mental research. Each volume in the series offers tested laboratory protocols, step-by-step methods for reproducible lab experiments and addresses methodological controversies and pitfalls in order to aid neuroscientists in experimentation. *Neuromethods* focuses on traditional and emerging topics with wide-ranging implications to brain function, such as electrophysiology, neuroimaging, behavioral analysis, genomics, neurodegeneration, translational research and clinical trials. *Neuromethods* provides investigators and trainees with highly useful compendiums of key strategies and approaches for successful research in animal and human brain function including translational "bench to bedside" approaches to mental and neurological diseases.

Animal Models of Reproductive Behavior

Edited by

Raúl G. Paredes

Instituto de Neurobiología, Universidad Nacional Autónoma de México, Querétaro, México; Escuela Nacional de Estudios Superiores, Unidad Juriquilla, Universidad Nacional Autónoma de México, Querétaro, México

Wendy Portillo

Instituto de Neurobiología, Universidad Nacional Autónoma de México, Querétaro, México

Marie Bedos

Escuela Nacional de Estudios Superiores, Unidad Juriquilla, Universidad Nacional Autónoma de México, Querétaro, México

Humana Press

Editors
Raúl G. Paredes
Instituto de Neurobiología
Universidad Nacional Autónoma de México
Querétaro, México

Escuela Nacional de Estudios Superiores
Unidad Juriquilla
Universidad Nacional
Autónoma de México
Querétaro, México

Wendy Portillo
Instituto de Neurobiología
Universidad Nacional Autónoma de México
Querétaro, México

Marie Bedos
Escuela Nacional de Estudios Superiores
Unidad Juriquilla
Universidad Nacional
Autónoma de México
Querétaro, México

ISSN 0893-2336 ISSN 1940-6045 (electronic)
Neuromethods
ISBN 978-1-0716-3233-8 ISBN 978-1-0716-3234-5 (eBook)
https://doi.org/10.1007/978-1-0716-3234-5

This Humana imprint is published by the registered company Springer Science+Business Media, LLC, part of Springer Nature.
The registered company address is: 1 New York Plaza, New York, NY 10004, U.S.A.

Preface to the Series

Experimental life sciences have two basic foundations: concepts and tools. The Neuromethods series focuses on the tools and techniques unique to the investigation of the nervous system and excitable cells. It will not, however, shortchange the concept side of things as care has been taken to integrate these tools within the context of the concepts and questions under investigation. In this way, the series is unique in that it not only collects protocols but also includes theoretical background information and critiques which led to the methods and their development. Thus, it gives the reader a better understanding of the origin of the techniques and their potential future development. The Neuromethods publishing program strikes a balance between recent and exciting developments like those concerning new animal models of disease, imaging, in vivo methods, and more established techniques, including immunocytochemistry and electrophysiological technologies. New trainees in neurosciences still need a sound footing in these older methods in order to apply a critical approach to their results.

Under the guidance of its founders, Alan Boulton and Glen Baker, the Neuromethods series has been a success since its first volume published through Humana Press in 1985. The series continues to flourish through many changes over the years. It is now published under the umbrella of Springer Protocols. While methods involving brain research have changed a lot since the series started, the publishing environment and technology have changed even more radically. Neuromethods has the distinct layout and style of the Springer Protocols program, designed specifically for readability and ease of reference in a laboratory setting.

The careful application of methods is potentially the most important step in the process of scientific inquiry. In the past, new methodologies led the way in developing new disciplines in the biological and medical sciences. For example, Physiology emerged out of Anatomy in the nineteenth century by harnessing new methods based on the newly discovered phenomenon of electricity. Nowadays, the relationships between disciplines and methods are more complex. Methods are now widely shared between disciplines and research areas. New developments in electronic publishing make it possible for scientists that encounter new methods to quickly find sources of information electronically. The design of individual volumes and chapters in this series takes this new access technology into account. Springer Protocols makes it possible to download single protocols separately. In addition, Springer makes its print-on-demand technology available globally. A print copy can therefore be acquired quickly and for a competitive price anywhere in the world.

Saskatoon, SK, Canada *Wolfgang Walz*

Preface

Different factors contribute to the optimal expression of reproductive behaviors which are crucial for the survival of the species. From the recognition of a possible mate to the optimal display of parental behavior, including the correct expression of sexual behavior, all the behaviors related to the different stages of reproductive function are precisely orchestrated. Current efforts are centered on understanding the bases for the normal expression of those behaviors and more importantly, the factors that can interfere with those processes, such as drugs, endocrine disruptors, or parental care. To this aim, a wide range of species have been used to study these topics including mice, rats, voles, rabbits, and zebra finch.

In this volume, we present classic and cutting-edge approaches to study several aspects of reproductive behavior in different species. Since reproductive behaviors are included in social behaviors and most often have shared cerebral circuits, several chapters that describe methods to investigate social behavior were included. A few of the chapters focused on sexual motivation and reward, both being fundamental for the initiation of sexual behavior and to guarantee that the behavior will be repeated, respectively. In many species, olfactory signaling is crucial and an important modulator of reproductive behaviors, so two chapters that deal with this topic were included: the first one about the complementary roles of olfaction and estradiol in the Bruce Effect (pregnancy block), and a second one about the use of olfactometer in awake behaving mice to record brain activity on olfactory tasks. Since sexual behavior is central to reproductive function, four chapters on this topic were included, covering different models and aspects of it. In particular, the chapter of Elisa Ventura-Aquino and Anders Agmo reviewed how sexual response can be studied in females. Finally, this volume also includes two chapters about parental behavior, one in rats and another one in rabbits, as well as the last chapter on how reproductive behaviors can induce permanent plastic changes, like neurogenesis in birds, rats, and sheep.

All the chapters provide step-by-step instructions with helpful notes as well as updated information on how these highlighted approaches are participating in increasing our understanding of reproductive behaviors. It is obvious that reproductive behaviors show sexual dimorphism, which is why it is very important to include as much as possible both sexes in those studies, and the reason why an emphasis was given to the importance of analyzing sex differences in this book.

We believe this book will be useful not only to other researchers but also to students who want to incorporate new methods into the study of reproductive behaviors.

Querétaro, México

Raúl G. Paredes
Wendy Portillo
Marie Bedos

Contents

Contributors

ANDERS ÅGMO • *Department of Psychology, University of Tromsø, Tromsø, Norway; University of Tromsø, Tromsø, Norway*

SARAEL ALCAUTER • *Instituto de Neurobiología, Universidad Nacional Autónoma de México (UNAM), Querétaro, Mexico*

NICOLE AREVALO • *Department of Cell and Developmental Biology, University of Colorado Anschutz Medical Campus, Aurora, CO, USA*

DARIO ASPESI • *Department of Psychology and Neuroscience Program, University of Guelph, Guelph, ON, Canada*

NOAH BASS • *Department of Psychology and Neuroscience Program, University of Guelph, Guelph, ON, Canada*

MARIE BEDOS • *Escuela Nacional de Estudios Superiores, Unidad Juriquilla, Universidad Nacional Autónoma de México, Querétaro, Mexico*

MARIO CABA • *Centro de Investigaciones Biomédicas, Universidad Veracruzana, Xalapa, Mexico*

FRANCISCO CASTELÁN • *Departamento de Biología Celular y Fisiología, Unidad Foránea Tlaxcala, Instituto de Investigaciones Biomédicas, Universidad Nacional Autónoma de Mé xico, Tlaxcala, Mexico; Centro Tlaxcala de Biología de la Conducta, Universidad Autónoma de Tlaxcala, Tlaxcala, Mexico*

ELENA CHOLERIS • *Department of Psychology and Neuroscience Program, University of Guelph, Guelph, ON, Canada*

XI CHU • *Chemosensory Lab, Department of Psychology, Norwegian University of Science and Technology, Trondheim, Norway*

REBECA CORONA • *Instituto de Neurobiología, Universidad Nacional Autónoma de México, Querétaro, Mexico*

DENYS DECATANZARO • *Department of Psychology, Neuroscience & Behaviour, McMaster University, Hamilton, ON, Canada*

ALONSO FERNÁNDEZ-GUASTI • *Departamento de Farmacobiología, CINVESTAV-Sede Sur, Mexico City, Mexico*

MARÍA REYNA FUENTES-MORALES • *Centro Tlaxcala de Biología de la Conducta, Universidad Autónoma de Tlaxcala, Tlaxcala, Mexico*

ARIANNA GENTILE-POLESE • *Department of Cell and Developmental Biology, University of Colorado Anschutz Medical Campus, Aurora, CO, USA*

IVAN E. GLADWYN-NG • *Taconic Biosciences GmbH, Leverkusen, Germany*

FAY A. GUARRACI • *Department of Psychology, Southwestern University, Georgetown, TX, USA*

MICHAEL HALL • *Machine Shop Core, University of Colorado Anschutz Medical Campus, Aurora, CO, USA*

ROY HEIJKOOP • *Department of Psychology, UiT the Arctic University of Norway, Tromsø, Norway*

CECILIA HERNÁNDEZ BONILLA • *Centro Tlaxcala de Biología de la Conducta, Universidad Autónoma de Tlaxcala, Tlaxcala, Mexico; Doctorado en Ciencias Biológicas, Universidad Autónoma de Tlaxcala, Tlaxcala, Mexico*

KURT L. HOFFMAN • *Centro de Investigación en Reproducción Animal, CINVESTAV-Laboratorio Tlaxcala, Universidad Autónoma de Tlaxcala, Tlaxcala, Mexico*
PATTY T. HUIJGENS • *Department of Psychology, UiT the Arctic University of Norway, Tromsø, Norway*
M. FERNANDA LÓPEZ-GUTIÉRREZ • *Instituto de Neurobiología, Universidad Nacional Autónoma de México (UNAM), Querétaro, Mexico*
JUSTIN LOSACCO • *Department of Cell and Developmental Biology, University of Colorado Anschutz Medical Campus, Aurora, CO, USA; Neuroscience Program, University of Colorado Anschutz Medical Campus, Aurora, CO, USA*
ROSA ANGÉLICA LUCIO • *Centro Tlaxcala de Biología de la Conducta, Universidad Autónoma de Tlaxcala, Tlaxcala, Mexico*
MING MA • *Department of Cell and Developmental Biology, University of Colorado Anschutz Medical Campus, Aurora, CO, USA*
MARGARITA MARTÍNEZ-GÓMEZ • *Departamento de Biología Celular y Fisiología, Unidad Foránea Tlaxcala, Instituto de Investigaciones Biomédicas, Universidad Nacional Autónoma de México, Tlaxcala, Mexico; Centro Tlaxcala de Biología de la Conducta, Universidad Autónoma de Tlaxcala, Tlaxcala, Mexico*
CONNOR MCCULLOUGH • *Department of Bioengineering, University of Colorado Anschutz Medical Campus, Aurora, CO, USA*
SARAH H. MEERTS • *Neuroscience Program and Department of Psychology, Carleton College, Northfield, MN, USA*
ANGEL I. MELO • *Centro de Investigación en Reproducción Animal, CINVESTAV, Laboratorio Tlaxcala, Universidad Autónoma de Tlaxcala, Tlaxcala, Mexico*
LAETITIA MERLE • *Department of Cell and Developmental Biology, University of Colorado Anschutz Medical Campus, Aurora, CO, USA*
ZACNITÉ MIER-QUESADA • *Instituto de Neurobiología, UNAM, Campus Juriquilla, Queré taro, Mexico*
ANDREW MORAN • *Department of Cell and Developmental Biology, University of Colorado Anschutz Medical Campus, Aurora, CO, USA*
JUAN J. ORTIZ • *Instituto de Neurobiología, Universidad Nacional Autónoma de México (UNAM), Querétaro, Mexico*
BARISH OZBAY • *Intelligent Imaging Innovations, Inc, Denver, CO, USA*
PIETRO PALETTA • *Department of Psychology and Neuroscience Program, University of Guelph, Guelph, ON, Canada*
RAÚL G. PAREDES • *Instituto de Neurobiología, UNAM, Campus Juriquilla, Querétaro, Mexico; Escuela Nacional de Estudios Superiores, Unidad Juriquilla, UNAM, Querétaro, Mexico*
ANDREW PARRA • *Department of Cell and Developmental Biology, University of Colorado Anschutz Medical Campus, Aurora, CO, USA*
WENDY PORTILLO • *Instituto de Neurobiología, Universidad Nacional Autónoma de México (UNAM), Querétaro, Mexico*
DANIEL RAMIREZ-GORDILLO • *Department of Cell and Developmental Biology, University of Colorado Anschutz Medical Campus, Aurora, CO, USA*
DIEGO RESTREPO • *Department of Cell and Developmental Biology, University of Colorado Anschutz Medical Campus, Aurora, CO, USA; Neuroscience Program, University of Colorado Anschutz Medical Campus, Aurora, CO, USA*
JOSE RIGUERO • *Department of Cell and Developmental Biology, University of Colorado Anschutz Medical Campus, Aurora, CO, USA*

NATALIA ROBLES • *Instituto de Neurobiología, UNAM, Campus Juriquilla, Querétaro, Mexico*

OLESYA T. SHEVCHOUK • *Institute of Neuroscience and Physiology, Department of Pharmacology, The Sahlgrenska Academy at the University of Gothenburg, Gothenburg, Sweden*

EELKE M. S. SNOEREN • *Department of Psychology, UiT the Arctic University of Norway, Tromsø, Norway*

FABIO SIMOES-DE SOUZA • *Department of Cell and Developmental Biology, University of Colorado Anschutz Medical Campus, Aurora, CO, USA*

KIRA STEINKE • *Department of Cell and Developmental Biology, University of Colorado Anschutz Medical Campus, Aurora, CO, USA*

GUILLERMO VALERA-MARÍN • *Instituto de Neurobiología, Universidad Nacional Autônoma de México (UNAM), Querétaro, Mexico*

ELISA VENTURA-AQUINO • *Escuela Nacional de Estudios Superiores, Unidad Juriquilla, Universidad Nacional Autónoma de México, Queretaro, Mexico*

RYAN WILLIAMSON • *IDEA Core, University of Colorado Anschutz Medical Campus, Aurora, CO, USA*

Chapter 1

The Study of Social Cognition: Social Recognition and Social Learning in Laboratory Rats and Mice

Pietro Paletta, Dario Aspesi, Noah Bass, and Elena Choleris

Abstract

Social cognition includes a variety of behaviors that are important for social life and the formation of relationships. Effective paradigms for these behaviors are needed in order for research to be conducted to understand the mechanisms underlying social cognition. In this chapter, we examine the experimental methodology to measure two of these behaviors, social recognition and social learning, in rodents. The materials required and the procedures for the paradigms that are used in our laboratory are highlighted. Finally, we discuss some possible issues that may occur in these paradigms and how they can be avoided, and we discuss some of the current knowledge base about the mechanisms that mediate social recognition and social learning.

Key words Social cognition, Social recognition, Social learning, Behavioral paradigms, Rats, Mice

1 Introduction

Social behavior is a broad term comprised of interactions among individuals of the same species and results in the formation of different kinds of relationships characterized by different natures, purposes, durations, and forms. The existence of sociality and cooperative behaviors in many species across taxa is an evolutionary puzzle, although solutions have been extensively analyzed over the last 60 years [1–5]. Ecological factors influence whether social aggregation may or may not be evolutionarily favored in a certain species because of the different weights of costs and benefits [6–8]. For example, advantages of sociality may be reduced risk of predation, easier access to resources such as food or breeding sites and sharing parental care or territory defense. On the other hand, the example of disadvantages of group living may relate to increased competition for resources such as mates or food and higher risks of the spread of parasites and diseases. The ratio between costs and benefits may prevent the evolution of sociality at all or can affect the

Raúl G. Paredes et al. (eds.), *Animal Models of Reproductive Behavior*, Neuromethods, vol. 200,
https://doi.org/10.1007/978-1-0716-3234-5_1,

kind of social interactions happening in a social species. Therefore, the different evolutionary pressures shape the form of social relationships and behaviors displayed by individuals of social species.

When social aggregation is adaptive, and thus favored in a species, different types and degrees of sociality evolve, with species ranging from low- to high-structured social organizations. An essential prerequisite of any form of social aggregation is the necessary interaction with other individuals, which requires both general and specialized skills [9]. The cognitive skills required for the emergence of social behavior include the recognition of others through properly processing the information emitted by those individuals, the ability of attending to and learning from others, and the display of appropriate behavioral and emotional response [10]. Among these social cognitive skills, the appropriate processing of information *about* (social recognition) and *from* others (social learning) is critical aspects of sociality. Social recognition is the ability to recognize previously investigated conspecifics, and it allows the information about those individuals to be used to adaptively adjust their behavior toward them. Social recognition can be based on various aspects of the conspecific, such as sex, reproductive and/or emotional state, status in the hierarchy, and familiarity [11]. There is also true individual recognition, which is the ability to identify specific individuals in the social group and remember the previous interaction with those individuals [11, 12]. Pair bond formation, hierarchal organizations, kin recognition, and several other aspects of group living depend on the ability of the individual to discriminate among previously encountered conspecifics and to learn from them.

In addition, living with others provide a source of information that can be obtained through social learning processes. Social learning is an extremely common and biologically adaptive form of learning and may be defined as "learning that occurs via the observation of, or interaction with, a conspecific or its products" [13–16]. This type of learning mitigates the costs of individual trial-and-error learning. These two essential social skills, social recognition and social learning, have been demonstrated in several species across different taxa including the two main species of rodents used in laboratory settings, rats (*Rattus norvegicus*) and mice (*Mus musculus*) [15, 17]. The outcome of social recognition and social learning can be essential for reproduction and survival. In laboratory settings, different behavioral paradigms have been developed to assess these social skills with a number of variables affecting the interpretation of the results. Therefore, it is essential to take into consideration the different components in the behavioral tests to understand whether they are effectively studying social recognition, which measures social memory and social information processing, or social learning, which measures the impact of social context on learning.

1.1 Social Recognition

In laboratory rodents, different paradigms can be used to measure social recognition in experimental animals. The original paradigm involves a first exposure to a conspecific (stimulus animal) and a second exposure to the same individual 30 min later [18]. A decreased time spent investigating the same stimulus animal the second time was considered an indication of recognition. However, it is difficult to establish whether the experimental animal is really recognizing the other individual because the time spent investigating the stimulus animal, or the time in-between exposures may be not enough or too long, respectively, to induce memory formation. For example, male rats fail to show a decline in the duration of investigation during the second exposure if the time in-between phases exceed 60 min [19, 20]. Other factors such as a lack of interest in social interaction or individual variability in investigation time may change the results of this test without specifically indicating changes in social recognition. A variation of this paradigm to assess social recognition is the habituation/dishabituation paradigm, in which the experimental animal is exposed to the same stimulus animal multiple times [10, 21]. During these repeated exposures, the time spent investigating the stimulus animal should decrease showing *habituation*. In the final trial, the experimental animal is presented with a novel stimulus animal replacing the previously investigated one. If the experimental animal can recognize that the stimulus is a novel animal, it typically spends more time investigating it (*dishabituation*). Finally, the social discrimination paradigm is also often used to investigate social recognition [10, 22]. In this case, the experimental animal, after single or repeated exposure to familiar stimuli, is simultaneously presented with a familiar and a novel conspecific. The experimental animal shows social recognition if it spends more time investigating the novel conspecific, demonstrating it can remember to have already investigated the familiar stimulus.

Recognition memory involves two distinct processes, often referred to as *familiarity* and *recollection*. The difference can be easily illustrated by the common experience of being able to recognize an individual as familiar without recollecting who the person is or any information about the previous encounter [23]. Interestingly, these processes seem to be mediated by different brain regions. Studies on both human amnesic patients and preclinical rat models suggested that the hippocampus may be critical to recollection, whereas the surrounding cortical areas to familiarity [24, 25]. However, this strict division of the brain regions is still controversial, and studies on animals have also failed to provide consensus [26–29].

It is important to underline here that recollection and familiarity are different. Familiar recognition is observed when an individual that receives a signal remembers the sender but is not able to identify the identity of that sender [30]. Familiar recognition is a type of class-level recognition that can also be considered a first step toward other class-level or true individual recognition because the individual starts creating a template of the class, or the individual based on the characteristic of that conspecific. Due to the higher complexity of true individual recognition, familiarity is more typically tested in the laboratory setting. Both the habituation/dishabituation and social discrimination paradigms test whether the experimental animal can recognize whether the novel stimulus animal is familiar to them or not. However, while both paradigms test familiarity, the social discrimination paradigm is more typically used because it was found to be a more sensitive paradigm allowing for the better detection of social recognition memory [22, 31]. The social recognition paradigms can be adapted to test true individual recognition by using stimulus animals that the experimental animal is equally familiar with but have had different experiences with, which will lead to different behavioral responses. For example, true individual recognition in golden hamsters has been shown, where the stimulus hamster has previously shown dominance over the experimental hamster, resulting in avoidance of the stimulus in the paradigm test [32]. Since the social discrimination paradigm is a more sensitive paradigm for testing familiarity, we will describe the experimental procedure required in the social discrimination paradigm below.

1.2 Social Learning

Many studies of social learning have been conducted outside of the laboratory in more natural settings, providing valuable information because they can address major ecological validity limitations and can include large-body or endangered species that would be difficult to study in laboratory settings. Further, it is critical to understand social learning in a natural context before trying to understand the interplay between biological and cultural aspects of social learning. As discussed below, numerous methodologies have been developed to study social learning in natural contexts including field observations, field experiments, theoretical and statistical models, group contrasts methods, ecological and genetic methodology, and phylogenetic models. For a more exhaustive collection of specific methodologies *see* Kendal et al. [33]. As new methodologies are developed, data from older field studies may also be analyzed by following the guidelines of the novel models.

Field Observations

In wild animals, social learning has been reported in numerous field observational studies. For example, this methodology has been used to demonstrate social learning in capuchin monkeys, cetaceans, orangutans, bonobos chimpanzees, and common chimpanzees [34–38]. Two basic approaches to studying social learning include trying to elucidate *how* information is transferred and *from who*. To understand how information is transferred between conspecifics, experimenters may observe an individual learn how to complete a task by observing the strategies used by a conspecific [39]. For example, during field observations, researchers could habituate wild troops of primates to the presence of investigators who then follow specific focal animals and record their behavior. The second approach involves investigating who the information is being transferred from under natural conditions within a stable social group [40]. The speed at which information is transferred leading to socially learned behaviors depends on spatial proximity, frequency of observed behavior, attention, environmental, demographic, and developmental/ontogenetic factors [41–43]. For example, stone handling (SH) is a nonadaptive object-play behavior that has been consistently observed in Japanese macaques [44, 45]. Notably, SH behaviors are typically acquired by all members of a group over the age of 5 typically from a mother [44, 45]. By studying SH behavior via field observation, it has been shown that social observation is necessary for the transmission and persistence of behaviors within a group [46]. Thus, field observations are a valuable way to study social learning in natural environments.

A limitation to studying social learning in wild animals in field studies is that it is extremely difficult to determine whether the transmission of information is truly due to social processes and not ecological and/or genetic processes [47]. Because of this, many other models of social learning have been developed including statistical, mathematical, ethnographic, and experimental models [33, 48–50]. When genetic (*see* below) and ecological explanations for social learning appear implausible, ethnographic models pool data from intensive and longitudinal field studies and draw causative conclusions based on the findings of those studies [50]. Field experiments are another way to test social learning in a way that allows for the study of culture and social learning without as many constraints or limitations related to ecological, genetic, or developmental factors as field observations.

1.2.1 Field Experimental Methods

The use of controlled experimental manipulations of free-living populations allows researchers to draw conclusions and confirmations about social learning. These studies provide the basis for future laboratory investigations to build on previously established social behaviors in different animal populations. Specifically, socially learned behaviors that animals would acquire naturally in the wild may later be taught and tested in a laboratory setting. In field experiments, socially learned information may be diffused to other members of the population and the pattern of diffusion is monitored by researchers. Diffusion in the wild may then be compared to laboratory studies to investigate the influence of factors such as social networks, social interaction patterns, nondemonstrator influences, or bystanders on social learning [51, 52]. For example, during a field experiment, social information about food acquisition tactics was observed to be transferred between white-throated magpie-jays (*Calocitta formosa*) [53]. Further, to test whether a bias of behavior copying by more dominant individuals in laboratory settings could be generalized to field settings, researchers utilized a field experiment and found that wild vervet monkeys (*Chlorocebus aethiops pygerythrus*) did not show a bias toward copying higher-ranked conspecifics [54]. Moreover, by utilizing field experiments, no learning bias was found between the sexes in a test of whether there is an interaction between payoff bias and sex. Specifically, researchers found that when payoff was equal between sexes, wild vervet monkeys copied females more than males, whereas when payoff was biased toward males, males were more likely to copy males than females were [55]. This methodology is also important because it could be used to target simplified research questions that may be overlooked or overcomplicated in laboratory experiments. In field experiments, different animal populations such as domesticated animals may be studied.

There are many ways field experiments are utilized and manipulated by researchers to test different aspects of social learning. This includes the following: (1) placing entire populations in a new unoccupied environment and observing which behaviors develop that persisted from the previous resident population. For example, this has been studied in Saddlebacks (*Philesturnus carunculatus*) and Bluehead wrasses (*Thalassoma bifasciatum*) [56, 57]. (2) Placing entire populations in an environment with native residents who may act as demonstrators "DEM"s and observing which social behaviors are adopted by the new population. For example, this has been studied in French grunt (*Haemulon flavolineatum*) [58]. (3) Placing entire populations in a new environment occupied by trained DEMs, as reported in honeybees (*Apis mellifera* and *Apis florea*), pigeons (*Columba livia*), white-throated magpie jays (*Calocitta formosa*), Florida scrub jays (*Aphelocoma coerulescens*), Keas (*Nestor notabilis*), and Meerkats (*Suricata suricatta*) [53, 59–68]. Similarly, (4) placing entire populations in a new environment occupied by trained DEMs with behaviors that were manipulated by researchers, as seen in Collared flycatcher (*Ficedula albicollis*), Pied flycatcher (*Ficedula hypoleuca*), and vervet monkeys (*Chlorocebus aethiops*) [69, 70]. (5) Placing entire populations in a new environment occupied by mechanical/robotic DEMs. For example, this was tested in honeybees (*Apis mellifera*) [71]. Finally, (6) placing entire populations in a new environment occupied by human DEMs. For example, this has been studied in Greylag geese (*Anser anser*) [72]. Thus, there are a number of experimental manipulations with wild animals in their natural environment that allow for the testing of different aspects of social learning in a variety of species. Collectively, studying the social behaviors of wild animals in field studies allows for the understanding of how human cultures and social behaviors likely developed.

1.2.2 Mate-Choice Copying

A type of social learning that is commonly studied and observed in multiple species is "mate choice copying" [73, 74]. With mate choice copying, social learning is used to select potential mates by observing the mating behavior of conspecifics. For example, early studies in Trinidadian guppies (*Poecilia reticulata)* showed that after observing another female mate, naïve female guppies preferred to mate with that male when given the choice between the same male or a second identical male that had not been observed mating with the other female [75]. In other field observations, female mate choice copying was reported early on in many other species including sage grouse, white-bearded manakin, and humans [76–78]. Experimentally, this was first reported with female guppies (*Poecilia reticulata*), and then in other species of fish and birds [75, 79–84]. Mate choice copying was later observed also in male fish, guppies, deer mice, laboratory mice, and humans [85–91].

Snouted pipefish (*Syngnathus typhle*) is a species where parental care is provided by the males. Notably, male, and not female, snouted pipefish demonstrate mate choice copying [92]. This sex difference reflects the reversed sex role patterns in snouted pipefish and suggests that in general, mate choice copying is more prevalent in the sex of the individual within the species that provide parental care. Indeed, in female Japanese quails, a species in which the females provide parental care, the preference for a mate lasted for up to 48 h after observing the mate choice of others, whereas shortly after observation, males avoided the previously mating female [93–95]. In this case, for males, mate choice copying may not be evolutionarily beneficial as it increases sperm competition. Thus, mate choice copying has been described in a variety of species and it appears to be more common in females than males. This sex difference may be due to the greater investment in parental care by the females of many species which in turn can lead to females being choosier than males when selecting a mate and rely more heavily on socially acquired information about mate choice [96]. Further research using this methodology is encouraged to expand the existing knowledge of the regulatory mechanisms of the social brain.

1.2.3 Theoretical/Statistical Methodologies

By utilizing theoretical or statistical methodologies, researchers can test the probability that a reported behavior was developed via asocial learning, and thus, social learning can be inferred [48, 97]. The multivariate/multifactorial analytical model is another important way to study social learning. It is utilized by compiling data and controlling for multiple factors via a multiple regression analysis to assess social behavior development over time. In doing so, researchers combine multiple factors such as sex, habitat, and specific behaviors into a single model as opposed to the more common approach of conducting multiple independent correlational analyses using multiple models. For example, this approach was used to compile 14 years of longitudinal and observational data collected from dolphin mothers and calves foraging tactics, which revealed evidence of the social transmission of social information and utilization of social learning strategies in addition to nonsocially learned behaviors [98]. This methodology is important because it provided evidence that social learning is adaptive in wild animals such as dolphins. Indeed, this methodology provides evidence for social learning in wild populations and emphasizes the value of including multiple factors of developmental data into a single model [99, 100]. Alternatively, using a single-factor model allows for the consideration of individual variability, but may not accurately assess social behaviors that are common throughout populations and may fail to control for developmental changes of social or ecological conditions. Thus, a multivariate analytical approach may be favorable when analyzing longitudinal developmental data, especially in field observations.

Group Contrasts Methodology and Discussions of Ecological and Genetic Models

Many of the social behaviors documented in long-term longitudinal field observations in chimpanzees (*Pan troglodytes*) have been shown to vary significantly between populations in a manner that suggested the existence of socially transmitted cultures [101–107]. To assess the bases of cultures in chimpanzees, new techniques were developed to study social behaviors that varied and could not be attributed to ecology or genetics and to study differences in behavioral repertoires within populations [38, 43, 108–110]. To begin, researchers scoped through the literature and compiled numerous behaviors that they believed to differ culturally within populations of chimpanzees finishing with 65 categories. Next researchers assigned each of those behaviors to 1 of 6 categories: (1) customary (occurring in most members of the culture within a specific age-sex group), (2) habitual (consistently socially transmitted culturally, but not customary), (3) present (neither customary nor habitual), (4) absent (unreported or unexplained behaviors), (5) ecological explanation (when absence (4) is not present), and (6) unknown (refers to a novel behavior regardless of whether it has been recorded) [38]. Using this methodology to understand behavior, researchers gathered and categorized over 151 years of observational data and concluded that cultural transmission of behavior in chimpanzees is common [38]. This method may be utilized with one or multiple groups, however, it has two major limitations [36, 43, 110]. The first being that excluding the possibility of the behavior being attributed to ecology or genetics essentially aims to prove the null hypothesis [111]. Unfortunately, it is common for researchers to fail to collect sufficient ecological or genetic data before excluding the possibility that ecology or genetics contribute to the behavior. Second, removing a behavior from its ecological context in an attempt to explain complex behavioral characteristics may conflict with many principles of ethology [112]. Further, it is likely that a single behavior may be related to numerous ecological or genetic factors making it extremely difficult to separate those factors from the behavior. As previously outlined, this method allows for the opportunity for researchers to estimate the frequency of social behaviors, but does not address possible genetic contributions [113–115]. Phylogenetic approaches that take genetic factors into consideration were developed that address these limitations [49, 116].

Studying social learning outside the laboratory has proven to be critical, as discussed above. However, laboratory experiments are also extremely valuable and have increased flexibility to address specific complex research questions. Social learning can be assessed by different behavioral tests. However, the social transmission of food preferences (STFP) paradigm test is commonly used to assess social learning and consists of three phases. In phase 1, a "demonstrator" individual (DEM) consumes a flavored diet. In phase 2, the DEM interacts with an "observer" conspecific (OBS). Finally, in phase 3, the OBS can choose between two novel diets, one of which was consumed by the DEM. Typically, the OBS shows a preference for the diet consumed by the DEM, suggesting they learned from the DEM.

The literature has revealed somewhat conflicting findings regarding the transmission of social information within social hierarchies and dominance rankings. For example, capuchin monkeys were more likely to spend time observing higher ranking and older conspecifics crack nuts [117]. Similarly, captive chimpanzees also preferred copying higher-ranked conspecifics during an object manipulation task [118]. Further, information about defensive responses to biting flies was socially transmitted from DEM to OBS deer mice which was reported to be modulated by kinship, familiarity, and dominance [119]. Moreover, in rats, a food preference was more effectively transmitted from a subordinate DEM to a dominant OBS [120, 121]. However, during an extractive foraging social learning task, capuchins did not show a dominance or rank bias, nor did female wild vervet monkeys [54, 122]. Thus, factors relevant to social hierarchies and dominance rankings between the OBSs and DEMs should be considered and reported when utilizing an STFP paradigm. Since the STFP paradigm can be manipulated in many ways, it is likely the most efficient and flexible way to study social learning in animals.

The results from behavioral studies performed in laboratory settings can be translated into naturalistic situations contributing to identify the causes and factors involved in the evolution of social behaviors and cooperation. For instance, social recognition is pivotal to reduce the risks associated with the enhanced parasitic exposure due to group living and specific neurohormonal mechanisms have been implicated in the evolution of this social skill [123]. Social learning is also essential for a species because, for example, learning from others which food to avoid and which to prefer decreases the chance to ingest toxic or poisonous food. Thus, an appropriate investigation of social recognition and social learning is essential to identify the elements that allowed the evolution of these social cognitive skills and, ultimately, of sociality itself.

2 Materials

2.1 Social Recognition

2.1.1 The Arena

The social discrimination paradigm can be performed in a neutral arena as well as in the home cage of the experimental animal. The home cage of the animal is often used to avoid the interference of anxiety-like behaviors in the display of social recognition caused by the novelty of the environment [124]. However, testing in the home cage can also introduce undesired pretenses due to various cues within the cage causing a preference for one part of the cage over another, possibly showing a preference not driven by the social stimuli. When using the home cage, there is the need to single house the experimental animal for at least 3 days before testing, necessary for the experimental animal to establish a home cage territory [124, 125]. For this reason, it is also important not to clean the cage for at least 3 days prior to testing. This can be a disadvantage of using the home cage since isolation can affect the animal's welfare and therefore affect their behavior [126–128]. In many studies, social recognition is tested in a neutral arena so that the animals do not need to be isolated and novel place-anxiety is reduced with pretesting habituation to the arena [129]. Neutral arenas used include a new clean cage and the three-chamber apparatus [129–131]. A transparent Plexiglas lid should be used when video recording is from above the cage. The experimental procedure is usually performed during the dark phase of the light cycle under red light illumination because it corresponds to the active phase of rodents. Since social recognition is cognitively demanding, performing the test when the animal is fully active and awake is preferable.

2.1.2 Stimulus Animals

The stimuli most often used in the social recognition paradigms are either juvenile animals or ovariectomized females [124]. Juveniles are often chosen because they are considered "neutral" social stimuli that would not elicit sexual or aggressive behavior from the experimental animal [132, 133]. While this is beneficial, it can be impractical to keep them as stimulus animals because of how quickly juveniles age [21]. Due to this, ovariectomized female animals are often used as stimuli as they can be kept and used for longer and it takes longer for male experimental animals to show aggressive behavior toward ovariectomized females compared to male stimuli [21]. It has also been suggested that using ovariectomized females will not elicit sexual behaviors from males, as shown in an experiment where scents from intact females elicited copulatory behaviors in males but scents from ovariectomized females did not [134].

Despite juveniles and ovariectomized females being used most consistently, the specific aspects of social recognition that are being targeted in the experiment need to be considered when choosing the stimuli for the paradigm. For example, a study showed that an oxytocin receptor antagonist administered to one brain region of male rats or mice (medial amygdala) impaired recognition of ovariectomized females but not juvenile males, whereas the administration of the antagonist to a different region (lateral septum) impaired social recognition for both stimuli [135]. This suggests that the different aspects of conspecifics being recognized, such as age or sex, can be processed through different mechanisms and/or systems. In addition, when the role of gonadal hormones in social recognition is investigated, the experimental mice are gonadectomized so the effects of the specific hormonal treatments can be identified [136–140]. Accordingly, in those studies, the stimulus mice are also gonadectomized, so the stimuli's hormonal status does not affect social investigation by the experimental animal. The stimulus animals are also age and sex matched to the experimental animals to similarly prevent differences in age or sexual motivation from affecting the social investigation by the experimental animal. In addition, the stimulus mice are housed in pairs, both of which are presented to the experimental animal in the sample phases. In the test phase of the paradigm, the novel stimulus mouse is taken from a different cage of stimulus mice so there is no scent of the previously presented mice on the novel mouse.

Ideally, the stimulus animals will be separated from the experimental animals in some way that prevents physical interaction without preventing the transmission of other sensory cues. One common way is that of containing the stimulus animals within enclosures. For example, a few labs use cylinders made out of Plexiglas that the stimulus mice can be put into and moved in and out of the experimental animal's cage [136, 138, 140, 141]. These cylinders have had holes drilled all around the lower part so that olfactory information can still be transmitted to the experimental animal allowing for high levels of investigation. Having this physical separation prevents the stimulus animals from initiating the social interactions with the experimental animal or the other social stimulus. In addition, the enclosure prevents the experimental animal from placing its own scent on the stimulus animals which could affect the amount of investigation upon subsequent exposures, thus ruling out that social recognition measured at the test is instead recognition of its own pheromones [142]. Finally, having the stimulus animals contained in enclosures with a solid bottom prevents them from leaving lingering odors into the experimental animal's home cage during inter-exposure intervals, thus controlling for the amount of exposure of the experimental animal to the stimulus animal and/or its cues. The stimulus animals should be habituated to the containers they will be placed in so that during the experiment, they do not manifest stress or anxiety that could affect the investigation by the experimental animal.

2.1.3 Video Recording Device

A video camera supported by a tripod is used to record the interaction of the experimental animal with the stimulus conspecifics. To avoid human disturbance and to be able to capture the complex behavioral interaction displayed by the animals in a social context toward the stimuli, live scoring is not advisable. Therefore, a video camera with sufficiently high resolution is an appropriate way to record a quality video that can be used later for behavioral scoring. Recording the various phases of the procedure on video also allows testing multiple animals at the same time. If there are excessive reflections on the Plexiglas lid (even when antiglare Plexiglas is used) because of the possible overhead illumination, a polarizing lens filter may be used with the camera.

2.1.4 Social Recognition Paradigm

The social discrimination paradigm to assess social recognition consists of a number of sample phases in which the experimental animal is repeatedly exposed to the same stimulus conspecifics followed by a test phase, during which one of the stimulus animals is replaced by a novel stimulus animal. The number of exposures to the sample phase, as well as the length of each phase, varies according to the experimental procedure and what is being tested. For instance, Phan et al. (2011) have developed two versions of the social discrimination paradigm: an *easy* and a *difficult* paradigm [136]. The easy version involves three sample phases, each 4 min in duration, followed by the 4 min test phase with 3 min break between phases. In this "easy" paradigm control, ovariectomized female mice show good social recognition, and this version of the paradigm can thus be used to assess impairing effects of treatment [136]. Conversely, the difficult paradigm consists of two sample phases, each 5 min in duration, and one test phase, also 5 min long, with 5 min breaks between phases, and it has been developed to assess enhancing effects on social recognition because control ovariectomized mice do not show social recognition [136]. The length of each phase can vary (1–10 min), but it is important to consider that the majority of social interactions typically happen immediately after the experimental animal has the opportunity to investigate the stimuli. For this reason, allowing for long interaction for too long (e.g., >10 min) does not necessarily lead to greater investigation. Instead, it may be counterproductive because other variables may ensue which interfere with social recognition (e.g., location preference). It is also important to note that these paradigms were designed to be used with ovariectomized CD-1 mice. Since gonadectomies have been shown to impair social recognition in both sexes, for gonadally intact animals the paradigms may need modifying in order for facilitating or impairing effects of treatments to be detected [143–145].

2.1.5 Behavioral Analysis

A trained observer blind to the treatments should perform the behavioral analysis with appropriate video analysis software. For a complete evaluation of the behavior emitted by the experimental animal toward the stimulus conspecifics, it is essential to score not only the time spent actively investigating the stimuli but to perform a complete ethological analysis. A list with brief descriptions of behaviors that are usually recorded during the social recognition paradigm is reported in Table 1. The social discrimination paradigm takes advantage of the fact that mice and rats will investigate a novel stimulus conspecific more than a previously encountered stimulus animal. An investigation ratio (IR) can be calculated as follows: $IR = N/(N + F)$, where N is the total time spent investigating the Novel stimulus animals (or during the sample phases, the

Table 1
List and description of the behaviors the experimental mouse performs during the social recognition paradigm [137]

Social recognition behaviors	
Behavior	**Description**
Social investigation	Sniffing of the stimulus mice
Nonsocial investigation	Sniffing of stimulus tubes above holes
Stretch approach	Stretching toward stimulus
Biting stimulus	Biting cylinder
Digging	Moving bedding backward with forepaws
Burying	Moving bedding forward with forepaws
Horizontal activity	Walking, exploring, and sniffing that does not fall into categories above
Vertical activity	Rearing/leaning on cage walls or stimulus mice tubes
Inactivity	Sitting, freezing, and sleeping
Grooming	Self-grooming and scratching
Stereotypies	Strange behaviors; spin-turns, repeated jumps, repeated lid chews, head shakes, etc.

stimulus mouse the changed at test) and *F* is the total time spent investigating the familiar stimulus conspecific. In addition to the IR, the social investigation duration of both stimuli combined can be used as an index of overall interest in social stimuli. A significant increase in the IR during the test phase (IR_{test}) compared to the average IR from the sample phases (IR_{sample}) demonstrates that on average the animals of the experimental group prefer, on average, to investigate the novel stimulus individual and, therefore, could recognize the previously investigated (familiar) conspecific. Assessing other behaviors such as locomotor activity, grooming, and stereotypes can determine whether the experimental animal is displaying a normal species-specific behavior.

2.2 Social Learning

2.2.1 Apparatus

Three days before experimentation, same-sex DEMs and OBSs are pair-housed in new, clean cages. If undergoing surgery, for example, stereotaxic surgery, we recommend the use of perforated aluminum cage dividers ($25.8 \times .2 \times 10.8$ cm^3; 0.3 cm diameter holes; 0.1 cm between each hole; approximately 1500 holes per divider) to eliminate the opportunity for mice to agitate the surgical areas (e.g., the head cap that secures the cannula) [146]. The perforations in the cage dividers allow optimal familiarization opportunities between mice through sensory contact while also avoiding the possibility of fighting. The day before experimentation, mice are moved from the colony room to the testing room while being kept on the same light cycle, and then food deprived 15 h prior to the beginning of the experiment to facilitate food consumption.

During the DEM feeding phase (phase 1) of the STFP, DEMs should be transferred to new, clean, and empty cages stocked with food jars containing one of the two flavored food diets. The empty cages reduce the possibility of foreign materials entering the food jars to ensure optimal accuracy when measuring food consumption. The glass jars used in our laboratory are 5 cm high and 7.5 cm wide (Dyets, Bethlehem, PA) with collared stainless-steel lids with a food access hole 2.5 cm in diameter. A perforated steel disk sits on top of the powdered food to minimize spillage.

During the DEM–OBS social interaction phase (phase 2) of the STFP, original DEM–OBS pairs are transferred back into the cage they were pair-housed in with nesting materials and cage dividers removed. For optimal video recording, clear plexiglass lids are placed on the cages during the social interaction.

During the choice test phase (phase 3) of the STFP, OBS is transferred to new larger cages immediately after the social interaction. The choice test cages are stocked with corncob bedding, stainless steel wired lids, new water bottles, and two openings that lead to two stainless-steel tunnels that protrude out the front of the cage that contain two different flavored food diets located in spill-proof feeding trays (*see* Fig. 1) that could be accessed through small holes (*see* Fig. 2, Tecniplast, Varese, Italy) [147].

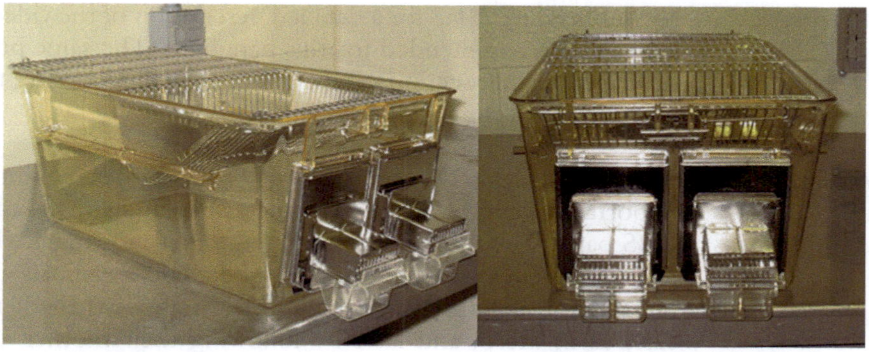

Fig. 1 Cage the observed mice are placed in for the choice phase (phase 3) of the STFP paradigm. The two protruding trays will contain the flavored food diets that the mice have free access to. These can be removed and weighed at the various time points of the choice test

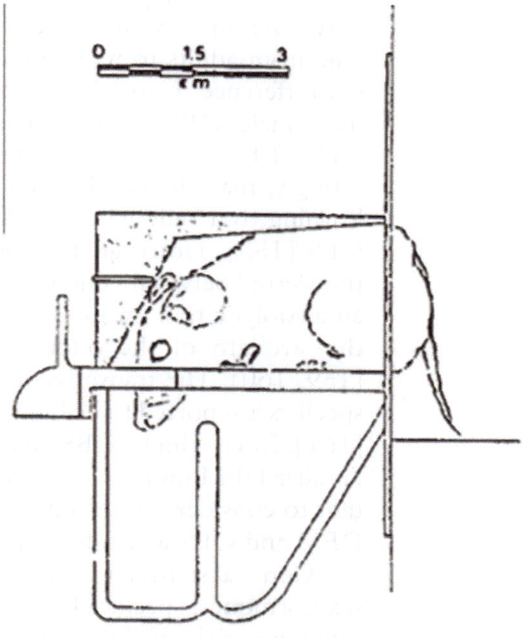

Fig. 2 This figure shows the feeding tray used during the STFP choice test. The tray is equipped with a section that collects food spillage for optimal consumption measurements [147]

2.2.2 Flavored Food Diets

Two commonly used flavored food diets for the STFP paradigm are a 1% cinnamon (CIN; McCormick Ground Cinnamon, McCormick Canada, London, Canada) and 2% cocoa (COC; Fry's Premium Cocoa, Cadbury Ltd., Mississauga, Canada) flavored food diet mixed with 14% ground rodent chow. These two flavored food diets are equally palatable and metabolically similar in CD1 mice (Charles Rivers, St. Constant, QC, Canada) [148, 149].

2.2.3 Video Recording Device

See Subheading 2.1.3 for a detailed description of the video recording device recommended for this paradigm. The same procedures described in that section can be applied when recording the social interactions of the STFP paradigm.

2.2.4 Demonstrators and Observers

When considering the methodology of an STFP paradigm, it is also important to consider age, relatedness, and familiarity of the DEM–OBS pairs. The mother is typically the first individual who learns food preference prior to food independence. Animals use social information from their mothers to learn about their diet from as early as in utero, and later by following their mother and other adult conspecifics to feeding locations [147, 150–152]. When this is the case, the young animal, typically weanling pups, acts as the OBSs, whereas the mother or other older adult acts as the DEM. Thus, young animals readily learn from older conspecifics. In adults, the socially acquired food preference may be transferred between same-sex rats, mice, and gerbils, and in mice, this remains true from adults to weanling offspring [153–156]. There was no sex difference in social learning ability between female–female or male–male OBS–DEM pairs in C57BL/6 mice, C57BL/6J × DBA/2Jhybrid mice, or Mongolian gerbils [155, 157, 158]. Interestingly, male–female DEM–OBS gerbil pairs also showed social learning, but only when the DEM was familiar or related to the OBS [155]. However, the socially acquired food preference was transferred between same-sex nonfamiliar gerbils when treated with an anxiolytic to a greater degree in males versus females, or when the strength of the odor was increased on the DEMs breath [159, 160]. The transmission of social information between conspecifics is reported to be intensified if animals are familiar or related [161]. Interestingly, OBS rats preferred foods consumed by a nonfamiliar DEM more than a familiar DEM [162]. Thus, it is important to consider specie and kinship and/or familiarity between the DEM and OBS when developing an STFP paradigm.

Gonadal status and sex hormones also influence STFP. Indeed, sex hormones, specifically estrogens, have consistently been implicated in social learning. For example, the estrus cycle regulates the STFP in mice [148]. Female mice in the proestrus phase of the estrus cycle, during a time of high circulating estrogens and progesterone, showed a prolonged duration of the expression of the socially acquired food preference in comparison to female mice in other phases of the estrus cycle and ovariectomized females [148, 163]. The socially acquired food preference was also prolonged following acute and chronic administration of estradiol benzoate and an estrogen receptor beta agonist [149, 164]. Thus, gonadectomizing DEMs is often done to eliminate the possibility of their sex hormones influencing the social interactions during the STFP paradigm. Further, gonadectomizing OBSs is often done to investigate the role of sex hormones. Thus, gonadal status and OBS–DEM sex should also be considered when developing a specific STFP paradigm.

2.2.5 Social Transmission of Food Preference Paradigm

Various aspects of the STFP paradigm can be adjusted to address the goals of the experiment. The social interaction (phase 2) of the STFP paradigm may be manipulated to test the enhancing or impairing effects of a drug or hormone on social learning. For example, by varying durations of phase 2 an "easy" or "difficult" paradigm can be created which can be used to test the impairing or enhancing effects of a drug, respectively. In laboratory mice, the duration of the social interaction during phase 2 of an "easy" STFP paradigm can range between 15 and 30 min, whereas in a "difficult" paradigm, OBSs would only get one sniff of the DEMs breath [165]. In an "easy" paradigm, mice show strong social learning. Thus, this allows for testing the impairing effects of a drug [148]. On the other hand, in a "difficult" paradigm, mice typically would not demonstrate social learning, so it allows for testing the enhancing effects of a drug [165]. In rats instead, OBS readily preferred the DEM diet following a 30 min social interaction, but not after a 15 min social interaction [166]. Thus, it is important to consider the length of the social interaction when developing a specific STFP paradigm in different species of laboratory rodents.

The STFP paradigm can also be used to investigate various types of memory such as short-term and/or long-term memory of a socially acquired food preference and/or the rapid or genomic effects of a drug or hormone. For example, by manipulating the length of the choice test (phase 3), one can test short-term memory and/or the rapid effects of a drug or hormone by measuring food intake shortly after phase 2 (e.g., every 5, 15, 30 min), or long-term memory by measuring food intake at longer intervals after the initial social interaction (phase 2) [153, 165]. Whether studying short-term- or long-term-memory testing, the food preference repeatedly over several hours (e.g., 1, 2, 4, 6, 8–24 h) allows for the assessment of the temporal expression of the socially acquired food preference [149]. Extending the length of phase 3 can also provide information on whether the researchers' manipulation, for example, drug administration, extends or reduces the duration of the expression of a socially acquired food preference.

2.2.6 Behavioral Analysis

A trained observer blind to the treatments should perform the behavioral analysis with appropriate video analysis software. *See* Table 2 for full lists and descriptions [146, 149, 167]. For the STFP paradigm, single and grouped social and nonsocial behaviors are scored. Behavioral data gathered while scoring the social interaction videos may be analyzed using mixed-design ANOVAs for the duration, latency, and frequency data of individual and grouped social and nonsocial behaviors (*see* Table 2).

To analyze the data, both a cinnamon (CIN) and DEM preference ratio can be calculated. The CIN preference ratio can be calculated to test social learning between each experimental group by using the following formula: total CIN consumption by

Table 2
Single social and nonsocial behaviors, and grouped behaviors produced by the observers during the 30 min social interactions phase of the STFP paradigm [149, 168]

Behavior	Description
Social behaviors	
Following DEM	The OBS actively follows or pursues and chases the DEM; reciprocal to avoid
Dominant behavior	OBS engages in controlling behavior over DEM; includes pinning, aggressive grooming, mounting attempt, and crawling over or on top
Attack delivered	OBS attacks DEM; includes physical attacks, dorsal/ventral bites. Measured by frequency of attacks
Boxing	Physical attacks; includes box/wrestle, offensive and defensive postures, lateral sideways threats and tail rattles
Open aggression	Physical attacks with a locked fight; includes tumbling, kick-away, and counterattack where the attacker cannot be identified
Avoidance of DEM	OBS withdraws and runs from DEM while DEM follows
Submissive behavior	DEM is in control; includes crawling under, supine posture (ventral side exposed), prolonged crouch, any other time the DEM is dominant (e.g., see dominant behavior description)
Attack received	Physical attacks; includes bites to dorsal/ventral regions. Measured by frequency of attacks
Social inactivity	OBS and DEM sit/lie/sleep together
Oronasal investigation	OBS actively sniffs DEM's oronasal area
Body investigation	OBS actively sniffs DEM's body
Anogenital investigation	OBS actively sniffs DEM's anogenital region
Stretched approach	Risk assessment behavior; back feet do not move and front feet approach demonstrator. Measured by frequency of stretched approaches
Approaching and/or attending to the DEM	OBS attention is focused on DEM, head tilted toward DEM and movements toward DEM often from across the cage. This becomes "following the DEM" once along the tail or sniff
Nonsocial behaviors	
Horizontal exploration	Movement around cage; includes active sniffing of air and ground
Vertical exploration	Movement to investigate upward with both front feet on the ground; includes sniffing, wall leans, and lid chews (less than 3)
Digging	Rapid stereotypical movement of forepaws in the bedding
Abnormal stereotypies	"Strange" behaviors; includes spin-turns, repeated jumps/lid chews/ head shake (more than 3)
Solitary inactivity	No movement; includes sitting, lie down, and sleep
Self-grooming	Rapid movement of forepays over facial area and along body

(continued)

Table 2
(continued)

Behavior	Description
Grouped behaviors	
Total activity	All behaviors involve activity, both social and nonsocial. Excluded from this group are "Solitary Inactivity," "Social Inactivity," and "Self-grooming"
Total social behavior	"Following the DEM," "Dominant Behavior," "Attack Delivered," "Boxing," "Open Aggression," "Avoidance of DEM," "Submissive Behavior," "Attack Received," "Attack Delivered," "Social Inactivity," "Oronasal Investigation," "Body Investigation," "Anogenital Investigation," "stretched approach," and "Approaching and/or Attending to the DEM." This composite behavior does not indicate whether the social interactions are affiliative or agonistic
Agonistic behavior delivered	"Following the DEM," "Dominant Behavior," and "Attack Delivered"
Agonistic behavior received	"Avoidance of DEM," "Submissive Behavior," and "Attack Received"
Total agonistic behaviors	"Agonistic Behavior Delivered" and "Agonistic Behavior Received" plus "Open Aggression," and "Boxing." This composite behavior does not indicate the direction of the agonistic behavior (e.g., whether agonistic behavior is directed toward OBS or toward DEM)
Dominance score	Total agonistic behavior delivered minus total agonistic behavior received. A negative score indicates that the OBS was the submissive animal in the pair, while a positive score signifies that the OBS was the dominant animal
Social investigation	"Oronasal Investigation," "Body Investigation," "Anogenital Investigation," "Stretched Approach," and "Approaching and/or Attending to the DEM"
Nonsocial behaviors	"Horizontal Exploration," "Vertical Exploration," "Digging," "Abnormal Stereotypies," "Solitary Inactivity," and "Self-grooming"
Nonsocial locomotor behaviors	"Horizontal Exploration," "Vertical Exploration," and "Digging"
Nonsocial nonlocomotor behaviors	"Solitary Inactivity" and "Self-grooming"

OBS/(CIN + COC consumption). A ratio of 1 would indicate that the OBS completely preferred the CIN diet, whereas 0 would indicate the OBS completely preferred the COC diet. The DEM preference ratio can be used to test the strength of the socially acquired food preference within experimental groups by using the following formula: DEM food consumed by OBS/ (CIN + COC consumption). A ratio of 1 would indicate that the OBS completely preferred the DEM diet. These analyses can be calculated at each time point during the OBS choice test.

3 Methods: Behavioral Procedure

3.1 Social Recognition

Before starting to test social recognition, place the animals (experimental and stimulus) in the testing room for a period of acclimation (generally at least 1 h). Avoid having other animals in the same room for other tests at the same time because olfactory and ultrasonic cues can affect their behavior.

Place the video camera in the right position to obtain the highest possible resolution of the animal and start video recording before placing the stimulus animals in the cage/arena with the experimental individual to avoid losing any social interaction between the animals. Hold the stimulus animals by the tails and gently place them into the cylinders, and then place the cylinders into the home cage/arena of the experimental animal (if the cylinders or any other enclosures are not used, just place the stimulus conspecifics directly in the home cage/arena) (*see* Fig. 3). Start the timer as soon as the stimulus animals are in the cage with the experimental animal because the social interaction typically begins immediately.

When completely leaving the room is not possible, during the sample and test phases, the experimenter should maintain a reasonable distance and avoid making any noise or movement that may interfere with the animal's behavior. At the end of the phase, stop video recording and immediately remove the stimulus animals from the cage/arena. At this moment, a break phase begins. For the subsequent phase (sample or test), repeat the same procedure. Between experimental animals, it is essential to properly clean the arena with odorless detergent and allow it to dry completely. When cylinders or other enclosures are used for the stimulus animals, they need to be washed between exposures with odorless detergent and dried. This ensures stimulus investigation is not driven by odor cues left on the cylinders from previous experimental phases.

Fig. 3 Social recognition paradigm. This version of the paradigm takes place in the experimental mouse's home cage with the stimulus mice placed in Plexiglas cylinders. The video camera is placed above the cage to observe the entire cage and record the paradigm phases for later analysis

The overall experimental design should reflect proper counterbalancing of the variables specific to the experiment. For example, the position of stimulus animals in the home cage or arena has to be consistent for each animal but should be counterbalanced for different experimental animals to avoid possible confounding variables such as preference for specific locations within the cage/arena. Similarly, which stimulus conspecific is replaced by the novel conspecific in the test phase should also be counterbalanced across experimental animals.

3.2 Social Learning

As discussed above, same-sex DEM–OBS pairs are pair-housed and separated by perforated cage dividers 3 days before the STFP experiment. The day before the experiment, the DEM–OBS pairs are moved to the experimental room to habituate and are food deprived 15 h prior to the start of the experiment. The experiment begins with the DEM feeding (phase 1) at the start of the dark cycle (8:00 AM) where DEMs are moved to their feeding cages stocked with either 1% CIN or 2% COC flavored food diets and are free to consume the food for 1 h. Food measurements are recorded before and after the DEM feeding. If mice do not consume at least 0.10 grams of food, they will not proceed to the social interaction phase of the STFP paradigm.

Next, original DEM–OBS pairs are placed back into the cage they were pair-housed in (without the cage dividers or nesting materials) for a 30 min social interaction (phase 2). Here, information about the recently consumed food may be transmitted from the DEM to the OBS via odor cues [147, 169]. The social interaction is video recorded for a later behavioral analysis.

Finally, immediately following the social interaction, during the OBS choice test (phase 3), OBSs are placed into individual choice cages that are stocked with both the 1% CIN and 2% COC flavored food diets for 8 h. Food intake is measured at 1, 2, 4, 6, and 8 h. If social learning occurs, the OBS prefers the flavored food diet that their respective DEM consumed before their social interaction. The preference for a food that was previously consumed by a conspecific may be rooted in the evolutionary need to avoid poisonous foods. Hence, if an individual knows a conspecific consumed a certain food and was safe afterward, then they may assume that food is safe to consume themselves [169]. Social learning may be studied in different ways by manipulating the social interaction (phase 2), the intervals between phase 2 and the choice test (phase 3), the length of phase 3, and various factors such as relatedness, sex, and familiarity of the DEM–OBS pair, dominance ranking, and gonadal status to test different aspects of social learning or effects of treatment.

4 Notes

4.1 Possible Paradigm Issues

These behavioral paradigms are common and effective methods, as well as relatively simple ways to test social behaviors in laboratory animals; however, there may be minor issues that researchers can encounter when conducting experiments with these paradigms. For example, in the social recognition paradigm, an issue that may occur is the stimulus mice's activity causing the cylinders they are in during the paradigm to move around the cage. This can draw the attention of the experimental animal, causing investigation levels to be affected by factors other than recognition. This issue can be detected during the period of habituating the stimulus mice to the cylinders. When well habituated, the stimulus mice show little resistance to being put into the cylinders and once in the cylinders show reduced behaviors of reaching toward the tops of the cylinders/cage lids. To address this, more sessions can be added to the habituation period, so the stimulus mice are less active in the cylinders. Another issue that may occur is scents interfering with the experimental animal's investigation. Scents of the stimulus animals can be left in the cylinders after they are removed and can affect social recognition in subsequent phases of the paradigm. This effect would not be easy to detect so to address this, new and clean cylinders should be used in each phase of the paradigm and the cylinders should be cleaned with soap and baking soda and allowed to air dry completely in order to remove any odors that may have been left behind.

Similarly, issues can occur in the STFP paradigm, such as the DEM consuming less than 0.10 gram of their food diet during the DEM feeding. This could result in the food odor not being detectable by the OBS during the social interaction. Therefore, those mice should be excluded from the study. It is also important to keep the corn cob bedding as far back in the OBS choice cage as possible. This would reduce the possibility of OBSs transferring bedding into their food containers which could interfere with the food weighing. Another issue a researcher may run into during the STFP paradigm could be if the food containers of the choice test are not filled high enough for the OBS to reach the food. Since the containers have two compartments: one for the food diet and one to contain spillage, it is important to only fill the food diet compartment (the one furthest from the choice cage) and leave the spillage compartment empty (e.g., instead of filling each halfway).

4.2 Underlying Mechanisms for these Social Behaviors

Social recognition and social learning are essential social skills, evolutionarily conserved in most vertebrates. Investigating why and how animals use social information is necessary to understand social behavior and sociality [170, 171]. Social information processing is a perquisite for social recognition and the facilitation of social interactions and group living. The social context also provides individuals the opportunity to obtain and use information from the behavior of other conspecifics. This socially acquired information can be useful for a wide variety of scopes ranging from learning about resources such as food to obtaining and choosing mates [10]. A thorough analysis of the behaviors displayed by individuals of social species toward other conspecifics is critical to understand social information processing and the evolution of complex mammalian societies [170].

An increasing number of studies are investigating the neurobiological underpinnings of social cognition, by effectively employing the described approaches to study social recognition and social learning, with several neurotransmitter and neuroendocrine systems being targeted [10, 172]. In this context, it is fundamental to highlight that the genes and neuropeptides involved in social recognition and social learning are often sexually differentiated, reflecting the sex differences observed in the emission of social behaviors in several species [173]. This evidence may reveal a key role played by sex hormones in the neurobiological processes underlying and modulating in social cognition. Sex hormones, including androgens, estrogens, and progesterone, not only mediate the activation of genes involved in social behaviors, but they can also interact with different neurochemical systems in various brain regions to ultimately modulate social recognition and social learning. Among the neurotransmitters known to be involved in these social skills, the nonapeptides oxytocin (OT) and arginine-vasopressin (AVP) have been extensively shown to play a prominent role, as both are involved in prosocial (affiliation) and antisocial (aggression) behaviors. OT and AVP are considered key factors for the evolution and expression of different types of social systems. Interestingly, both OT and AVP are under the control of sex hormones, which can influence the organization and the activation of these systems in mice and rats [10, 172, 174, 175]. The effects of gonadal hormones on these systems may explain well-known sex differences in social behaviors such as aggression or social information processing. The investigation of the neurobiological mechanisms underlying social behaviors can help understand the evolutionary mechanisms involved in sociality. For example, it has shown that various types of aggression in mice, which are expressed to fulfill different evolutionary needs, can be differentially affected by psychoactive drugs [176]. Although an intriguing picture of the regulation of neurobiological mechanisms essential for social information processing by the sex hormones is emerging, the neuroendocrine mechanisms as well as the brain areas involved in social cognitive processes are yet to be fully understood.

Establishing the neuroendocrinological underpinning of social cognition is fundamental for the understanding of sexually different conditions characterized by abnormal social behaviors such as autism spectrum disorder (ASD) and schizophrenia. Individuals with ASD show impairment in the recognition of faces, and a study has shown that AVP or OT can improve social emotion recognition and reduce autistic-typical repetitive stereotypical behaviors [177]. Borie et al. (2022) have also recently reviewed the research investigating OT and AVP as biomarkers for ASD and targeting the OT and AVP systems in alleviating the symptoms of ASD [178]. Intriguingly, the incidence of ASD is highly sexually differentiated affecting approximately four males for every affected female, which points at a role for gonadal hormones [179]. One theory suggests that testosterone facilitates, and estrogens may protect from, the worst symptoms of the disease [179]. Taken together, the evidence presented here highlights the relevance of an appropriate investigation of social information processing in rodents not only to understand the evolutionary establishment of societies but also to gain more knowledge about functional and dysfunctional human social behavior.

References

1. Axelrod R, Hamilton WD (1981) The evolution of cooperation. Science (New York, NY) 211(4489):1390–1396. https://doi.org/10.1126/SCIENCE.7466396
2. Fletcher JA, Doebeli M (2009) A simple and general explanation for the evolution of altruism. Proc R Soc B Biol Sci 276(1654):13–19. https://doi.org/10.1098/RSPB.2008.0829
3. Hamilton WD (1964) The genetical evolution of social behaviour. J Theor Biol 7(1):1–16. https://doi.org/10.1016/0022-5193(64)90038-4
4. Nowak MA, Sigmund K (2005) Evolution of indirect reciprocity. Nature 437(7063):1291–1298. https://doi.org/10.1038/NATURE04131
5. Smead R, Forber P (2016) The coevolution of recognition and social behavior. Sci Rep 6. https://doi.org/10.1038/SREP25813
6. Alexander RD (1974) The evolution of social behavior. Annu Rev Ecol Syst 5(1):325–383. https://doi.org/10.1146/ANNUREV.ES.05.110174.001545
7. Blumstein DT, Ebensperger LA, Hayes LD, Vásquez RA, Ahern TH, Burger JR, Dolezal AG, Dosmann A, González-Mariscal G, Harris BN, Herrera EA, Lacey EA, Mateo J, McGraw LA, Olazábal D, Ramenofsky M, Rubenstein DR, Sakhai SA, Saltzman W et al (2010) Toward an integrative understanding of social behavior: new models and new opportunities. Front Behav Neurosci 4:34. https://doi.org/10.3389/FNBEH.2010.00034/BIBTEX
8. Silk JB (2007) The adaptive value of sociality in mammalian groups. Philos Trans R Soc B Biol Sci 362(1480):539–559. https://doi.org/10.1098/RSTB.2006.1994
9. Sherman P, Reeve H, Pfenning D (1997) Recognition systems. In: Krebs J, Davies N (eds) Behavioural ecology: an evolutionary approach, 4th edn. Blackwell Science, pp 69–97
10. Choleris E, Clipperton-Allen AE, Phan A, Kavaliers M (2009) Neuroendocrinology of social information processing in rats and mice. Front Neuroendocrinol 30:442–459. https://doi.org/10.1016/j.yfrne.2009.05.003
11. Choleris E, Clipperton-Allen A, Phan A, Valsecchi P, Kavaliers M (2012) Estrogenic involvement in social learning, social recognition and pathogen avoidance. Front Neuroendocrinol 33:140–159
12. Gheusi G, Bluthe R, Goodall G, Dantzer R (1994) Social and individual recognition in rodents: methodological aspects and neurobiological bases. Behav Process 33:59–88
13. Heyes CM (1994) Social learning in animals: categories and mechanism. Biol Rev 69:207–231
14. Galef BG (1988) Imitation in animals: history, definition and interpretation of data from the psychological laboratory. In: Zentall TR, Galef BG (eds) Social learning: psychological and biological perspectives. Erlbaum, Hillsdale, pp 3–28

15. Galef BG, Laland KN (2005) Social learning in animals: empirical studies and theoretical models. Bioscience 55(6):489–499. https://doi.org/10.1641/0006-3568(2005)055[0489:SLIAES]2.0.CO;2
16. Box HO (1984) Primate behavior and social ecology. Chapman & Hall, London
17. Ferguson JN, Young LJ, Insel TR (2002) The neuroendocrine basis of social recognition. Front Neuroendocrinol 23(2):200–224. https://doi.org/10.1006/FRNE.2002.0229
18. Thor DH, Holloway WR (1981) Persistence of social investigatory behavior in the male rat: evidence for long-term memory of initial copulatory experience. Anim Learn Behav 9(4):561–565. https://doi.org/10.3758/BF03209791
19. Dantzer R, Bluthe RM, Koob GF, le Moal M (1987) Modulation of social memory in male rats by neurohypophyseal peptides. Psychopharmacology 91(3):363–368. https://doi.org/10.1007/BF00518192
20. Sekiguchi R, Wolterink G, van Ree JM (1991) Short duration of retroactive facilitation of social recognition in rats. Physiol Behav 50(6):1253–1256. https://doi.org/10.1016/0031-9384(91)90591-B
21. Winslow JT, Camacho F (1995) Cholinergic modulation of a decrement in social investigation following repeated contacts between mice. Psychopharmacology 121(2):164–172. https://doi.org/10.1007/BF02245626
22. Engelmann M, Wotjak CT, Landgraf R (1995) Social discrimination procedure: an alternative method to investigate juvenile recognition abilities in rats. Physiol Behav 58(2):315–321. https://doi.org/10.1016/0031-9384(95)00053-L
23. Yonelinas AP (2002) The nature of recollection and familiarity: a review of 30 years of research. J Mem Lang 46(3):441–517. https://doi.org/10.1006/JMLA.2002.2864
24. Fortin NJ, Wright SP, Eichenbaum H (2004) Recollection-like memory retrieval in rats is dependent on the hippocampus. Nature 431(7005):188–191. https://doi.org/10.1038/nature02853
25. Yonelinas AP, Kroll NEA, Quamme JR, Lazzara MM, Sauvé MJ, Widaman KF, Knight RT (2002) Effects of extensive temporal lobe damage or mild hypoxia on recollection and familiarity. Nat Neurosci 5(11):1236–1241. https://doi.org/10.1038/nn961
26. Manns JR, Hopkins RO, Reed JM, Kitchener EG, Squire LR (2003) Recognition memory and the human hippocampus. Neuron 37(1):171–180. https://doi.org/10.1016/S0896-6273(02)01147-9
27. Mumby DG (2001) Perspectives on object-recognition memory following hippocampal damage: lessons from studies in rats. Behav Brain Res 127(1–2):159–181. https://doi.org/10.1016/S0166-4328(01)00367-9
28. Mumby DG, Gaskin S, Glenn MJ, Schramek TE, Lehmann H (2002) Hippocampal damage and exploratory preferences in rats: memory for objects, places, and contexts. Learn Mem 9(2):49–57. https://doi.org/10.1101/LM.41302
29. Nemanic S, Alvarado MC, Bachevalier J (2004) The hippocampal/parahippocampal regions and recognition memory: insights from visual paired comparison versus object-delayed nonmatching in monkeys. J Neurosci 24(8):2013–2026. https://doi.org/10.1523/JNEUROSCI.3763-03.2004
30. Gherardi F, Aquiloni L, Tricarico E (2012) Revisiting social recognition systems in invertebrates. Anim Cogn 15(5):745–762. https://doi.org/10.1007/S10071-012-0513-Y
31. Choleris E, Ogawa S, Kavaliers M, Gustafsson J-A, Korach KS, Muglia LJ, Pfaff DW (2006) Involvement of estrogen receptor a, b and oxytocin in social discrimination: a detailed behavioral analysis with knockout female mice. Genes Brain Behav 5:528–539
32. Lai W-S, Johnston RE (2002) Individual recognition after fighting by golden hamsters: a new method. Physiol Behav 76:225–239
33. Kendal RL, Custance DM, Kendal JR, Vale G, Stoinski TS, Rakotomalala NL, Rasamimanana H (2010) Evidence for social learning in wild lemurs (Lemur catta). Learn Behav 38:220–234
34. Perry S, Baker M, Fedigan L, Gros-Louis J, Jack K, MacKinnon K, Manson J, Panger M, Pyle K, Rose L (2003) Social conventions in wild white-faced capuchin monkeys: evidence for traditions in a neotropical primate. Curr Anthropol 44:241–268
35. Rendell LE, Whitehead H (2001) Culture in whales and dolphins. Behav Brain Sci 24:309–382
36. van Schaik CP, Ancrenaz M, Borgen G, Galdikas B, Knott CD, Singleton I, Suzuki A, Utami SS, Merrill M (2003) Orangutan cultures and the evolution of material culture. Science 299:102–105
37. Hohmann G, Fruth B (2003) Culture in bonobos? Between species and within-species variation in behavior. Curr Anthropol 44:563–571

38. Whiten A, Goodall J, McGrew WC, Nishida T, Reynolds V, Sugiyama Y, Tutin CEG, Wrangham RW, Boesch C (1999) Culture in chimpanzees. Nature 399:682–685
39. Custance D, Whiten A, Fredman T (1999) Social learning of an artificial fruit task in capuchin monkeys (Cebus apella). J Comp Psychol 113:13–23
40. Biro D, Inoue-Nakamura N, Yamakoshi G, Sousa C, Matsuzawa T (2003) Cultural innovation and transmission of tool use in wild chimpanzees: evidence from field experiments. Anim Cogn 6:213–223
41. Coussi-Korbel S, Fragaszy DM (1995) On the relationship between social dynamics and social learning. Anim Behav 50:1441–1453
42. Huffman MA, Hirata S (2003) Biological and ecological foundations of primate behavioral tradition. In: Fragaszy DM, Perry S (eds) The biology of traditions: models and evidence. Cambridge University Press, Cambridge, UK, pp 267–296
43. van Schaik CP, Fox EA, Fechtman LT (2003) Individual variation in the rate of use of tree-hole tools among wild orang-utans: implications for hominin evolution. J Hum Evol 44: 11–23
44. Huffman MA (1984) Stone play of Macaca fuscata in Arashiyama B troop: transmission of a non-adaptive behavior. J Hum Evol 13: 725–735
45. Huffman MA (1996) Acquisition of innovative cultural behaviors in nonhuman primates: a case study of SH, a socially transmitted behavior in Japanese macaques. In: Galef BG Jr, Heyes C (eds) Social learning in animals: roots of culture. Academic Press, San Diego, pp 267–289
46. Huffman MA, Nahallage CAD, Leca J-B (2008) Cultured monkeys: social learning cast in stones. Cult Dir Psychol Sci 17(6):410–414
47. West MJ, King AP, White DJ (2003) Discovering culture in birds: the role of learning and development. In: de Waal FBM, Tyack PL (eds) Animal social complexity: intelligence, culture, and individualized societies. Harvard University Press, Cambridge, MA, pp 470–492
48. Hoppitt W, Kandler A, Kendal JR, Laland KN (2010) The effect of task structure on diffusion dynamics: implications for diffusion curve and network-based analyses. Learn Behav 38:243–251
49. Lycett SJ (2010) The importance of history in definitions of culture: implications from phylogenetic approaches to the study of social learning in chimpanzees. Learn Behav 38: 252–264
50. Lonsdorf EV, Bonnie KE (2010) Opportunities and constraints when studying social learning: developmental approaches and social factors. Learn Behav 38(3):195–205
51. Reader SM, Biro D (2010) Experimental identification of social learning in wild animals. Learn Behav 38(3):265–2283
52. Lefebvre L, Giraldeau L-A (1994) Cultural transmission in pigeons is affected by the number of tutors and bystanders present. Anim Behav 47:331–337
53. Langen TA (1996) Social learning of a novel foraging skill by white throated magpie jays (Calocitta formosa, Corvidae): a field experiment. Ethology 102:157–166
54. Botting J, Whiten A, Grampp M, van de Waal E (2018) Field experiments with wild primates reveal no consistent dominance-based bias in social learning. Anim Behav 136:1–12
55. Bono AE, Whiten A, van Schaik C, Krützen M, Eichenberger F, Schnider A, van de Waal E (2018) Payoff-and sex-biased social learning interact in a wild primate population. Curr Biol 28(17):2800–2805
56. Jenkins PF (1978) Cultural transmission of song patterns and dialect development in a free-living bird population. Anim Behav 25: 50–78
57. Warner RR (1990) Male versus female influences on mating-site determination in a coral-reef fish. Anim Behav 39:540–548
58. Helfman GS, Schultz ET (1984) Social transmission of behavioural traditions in a coral reef fish. Anim Behav 32:379–384
59. Gould JL (1974) Honey bee communication. Nature 252:300–301
60. Gould JL (1975) Communication of distance information by honeybees. J Comp Physiol 104:161–173
61. Gould JL (1975) Honey bee recruitment: the dance-language controversy. Science 189: 685–693
62. Esch HE, Zhang S, Srinivasan MV, Tautz J (2001) Honeybee dances communicate distances measured by optic flow. Nature 411: 581–583
63. Sherman G, Visscher PK (2002) Honeybee colonies achieve fitness through dancing. Nature 419:920–922
64. Lefebvre L (1986) Cultural diffusion of a novel food-finding behaviour in urban pigeons: an experimental field test. Ethology 71:295–304

65. Banks AN, Guilford T (2000) Accurate route demonstration by experienced homing pigeons does not improve subsequent homing performance in naive conspecifics. Proc R Soc B 267:2301–2306
66. Midford PE, Hailman JP, Woolfenden GE (2000) Social learning of a novel foraging patch in families of free-living Florida scrub-jays. Anim Behav 59(6):1199–1207. https://doi.org/10.1006/anbe.1999.1419
67. Gajdon GK, Fijn N, Huber L (2004) Testing social learning in a wild mountain parrot, the kea (Nestor notabilis). Learn Behav 32(1):62–71. https://doi.org/10.3758/BF03196007
68. Thornton A, Malapert A (2009) The rise and fall of an arbitrary tradition: an experiment with wild meerkats. Proc R Soc B 276:1269–1276
69. Seppänen J-T, Forsman JT (2007) Interspecific social learning: novel preference can be acquired from a competing species. Curr Biol 17:1248–1252
70. van de Waal E, Renevey N, Favre CM, Bshary R (2010) Selective attention to philopatric models causes directed social learning in wild vervet monkeys. Proc R Soc B 277:2105–2111
71. Michelsen A, Andersen BB, Storm J, Kirchner WH, Lindaner M (1992) How honeybees perceive communication dances, studied by means of a mechanical model. Behav Ecol Sociobiol 30:143–150
72. Fritz J, Bisenberger A, Kotrschal K (2000) Stimulus enhancement in graylag geese: socially mediated learning of an operant task. Anim Behav 59:1119–1125
73. Miller GF, Todd PM (1998) Mate choice turns cognitive. Trends Cogn Sci 2:190–198
74. Verzijden MN, ten Cate C, Servedio MR, Kozak GM, Boughman JW, Svensson EI (2012) The impact of learning on sexual selection and speciation. Trends Ecol Evol 27: 511–519
75. Dugatkin LA (1992) Sexual selection and imitation: females copy the mate choice of others. Am Nat 139:1384–1389
76. Wiley RH (1973) Territoriality and non-random mating in sage grouse Centrocercus urophasianus. Anim Behav Monogr 6: 85–169
77. Lill A (1974) Sexual behavior of the lek-forming white-bearded Manakin (Manacus manacus trinitatis Hartert). Z Tierpsychol 36:1–36
78. Sigall H, Landy D (1973) Radiating beauty—effects of having a physically attractive partner on person perception. J Pers Soc Psychol 28: 218–224
79. Alonzo SH (2008) Female mate choice copying affects sexual selection in wild populations of ocellated wrasse. Anim Behav 75:1715–1723
80. Godin JG, Hair KPE (2009) Mate-choice copying in free-ranging Trinidadian guppies, Poecilia reticulata. Behaviour 146:1443–1461
81. Goulet D, Goulet TL (2006) Nonindependent mating in a coral reef damselfish: evidence for mate choice copying in the wild. Behav Ecol 17:998–1003
82. Witte K, Ryan MJ (2002) Mate choice copying in the sailfish molly Poecilia latipinna in the wild. Anim Behav 63:943–949
83. Grant JWA, Green LD (1996) Mate copying versus preference for actively courting males by female Japanese medaka Oryzias latipes. Behav Ecol 7:165–167
84. Hoglund J, Alatalo RV, Gibson RM, Lundberg A (1995) Mate choice copying in black grouse. Anim Behav 49:1627–1633
85. Schlupp I, Ryan MJ (1997) Male sailfish mollies, Poecilia latipinna, copy the mate choice of other males. Behav Ecol 8:104–107
86. Schlupp I, Marler C, Ryan MJ (1994) Benefit to male sailfin mollies of mating with heterospecific females. Science 263:373–374
87. Witte K, Ryan MJ (1998) Male body length influences mate-choice copying in the sailfin molly Poecilia latipinna. Behav Ecol 9:534–539
88. Auld H, Godin J (2015) Sexual voyeurs and copiers: social copying and the audience effect on male mate choice in the guppy. Behav Ecol Sociobiol 69:1795–1807
89. Kavaliers M, Bishnoi IR, Ossenkopp KP, Choleris E (2022) Odor-based mate choice copying in deer mice is not affected by familiarity or kinship. Anim Cogn 25(2):241–248
90. Kavaliers M, Choleris E, Ågmo A, Braun WJ, Colwell DD, Muglia LJ, Ogawa S, Pfaff DW (2006) Inadvertent social information and the avoidance of parasitized male mice: a novel role for oxytocin. Proc Natl Acad Sci USA 103(11):4293–4298
91. Yorzinski LJ, Platt ML (2010) Same-sex gaze attraction influences mate-choice copying in humans. PLoS One 5(2):e9115
92. Widemo MS (2006) Male but not female pipefish copy mate choice. Behav Ecol 17(2):255–259
93. White DK, Galef BG Jr (1999) Mate choice copying and conspecific cueing in Japanese

quail, Coturnix coturnix japonica. Anim Behav 57:465–473

94. White DK, Galef BG Jr (1999) Social effects on mate choice of male Japanese quail, Coturnix japonica. Anim Behav 57:1005–10012
95. White DK, Galef BG Jr (2000) Differences between the sexes in the direction and duration of response to seeing a potential sex partner mate with another. Anim Behav 59:1235–1240
96. Kavaliers M, Choleris E (2017) Social cognition and the neurobiology of rodent mate choice. Integr Comp Biol 57(4):846–856
97. Franz M, Nunn CL (2010) Investigating the impact of observation errors on the statistical performance of network-based diffusion analysis. Learn Behav 38:235–242
98. Sargeant BL, Mann J (2009) Developmental evidence for foraging traditions in wild bottlenose dolphins. Anim Behav 78(3):715–721
99. Lonsdorf V, Eberly LE, Pusey AE (2004) Sex differences in learning in chimpanzees. Nature 428:715–716
100. Thornton A, McAuliffe K (2006) Teaching in wild meerkats. Science 313:227–229
101. McGrew WC, Tutin CEG (1978) Evidence for a social custom in wild chimpanzees? Man 13:234–251
102. Goodall J (1986) The chimpanzees of Gombe: patterns of behavior. Harvard University Press, Cambridge, MA
103. Nishida T (1990) The chimpanzees of the Mahale mountains: sexual and life history strategies. Tokyo University Press, Tokyo
104. McGrew WC (1992) Chimpanzee material culture: implications for human evolution. Cambridge University Press, Cambridge
105. Sugiyama Y (1993) In: Berthelet A, Chavaillon J (eds) The use of tools by human and non-human primates. Clarendon, Oxford, pp 175–187
106. Wrangham RW, McGrew WC, de Waal FBM, Heltne PG (eds) (1994) Chimpanzee cultures. Harvard University Press, Cambridge, MA
107. Boesch C (1996) The emergence of cultures among wild chimpanzees. Proc Br Acad 88: 251–268
108. Boesch C, Marchesi P, Marchesi N, Fruth B, Joulian F (1994) Is nut cracking in wild chimpanzees a cultural behaviour? J Hum Evol 26(4):325–338
109. McGrew WC, Marchant LF (1997) On the other hand: current issues in and meta-analysis of the behavioral laterality of hand function in nonhuman primates. Am J Phys Anthropol 104(S25):201–232
110. Krützen M, Mann J, Heithaus MR, Connor RC, Bejder L, Sherwin WB (2005) Cultural transmission of tool use in bottlenose dolphins. Proc Natl Acad Sci USA 102:8939–8943
111. Fragaszy DM, Perry S (2003) The biology of traditions. Cambridge University Press, Cambridge
112. Tinbergen N (1963) On aims and methods of ethology. Z Tierpsychol 20(4):410–433
113. Galef BG (2009) Culture in animals? In: Laland KN, Galef BG (eds) The question of animal culture. Harvard University Press, Cambridge, MA, pp 222–246
114. Laland KN, Hoppitt W (2003) Do animals have culture? Evol Anthropol 12:150–159
115. Laland KN, Kendal JR, Kendal RL (2009) Animal culture: problems and solutions. In: Laland KN, Galef BG (eds) The question of animal culture. Harvard University Press, Cambridge, MA, pp 174–197
116. Tehrani J, Collard M (2002) Investigating cultural evolution through biological phylogenetic analysis of Turkmen textiles. J Anthropol Archaeol 21:443–463
117. Canteloup C, Hoppitt W, van de Waal E (2020) Wild primates copy higher-ranked individuals in a social transmission experiment. Nat Commun 11:459
118. Kendal R, Hopper LM, Whiten A, Brosnan SF, Lambeth SP, Schapiro SJ, Hoppitt W (2015) Chimpanzees copy dominant and knowledgeable individuals: implications for cultural diversity. Evol Hum Behav 36(1):65–72
119. Kavaliers M, Colwell DD, Choleris E (2005) Kinship, familiarity and social status modulate social learning about 'micropredators' (biting flies) in deer mice. Behav Ecol Sociobiol 58(1):60–71
120. Williamson CM, Lee W, Curley JP (2016) Temporal dynamics of social hierarchy formation and maintenance in male mice. Anim Behav 115:259–272
121. Awazu S, Fujita K (1998) Influence of social relationships on food transmission in rats. Jpn J Anim Psychol 48(2):183–190
122. Dindo M, Leimgruber KL, Ahmed R, Whiten A, de Waal F (2011) Observer choices during experimental foraging tasks in brown capuchin monkeys (Cebus apella). Am J Primatol 73(9):920–927
123. Kavaliers M, Choleris E, Ågmo A, Muglia LJ, Ogawa S, Pfaff DW (2005) Involvement of the oxytocin gene in the recognition and avoidance of parasitized males by female mice. Anim Behav 70(3):693–702. https://

doi.org/10.1016/J.ANBEHAV.2004.12.016
124. Winslow J (2003) Mouse social recognition and preference. Curr Protoc Neurosci 22(1):8.16.1–8.16.16
125. Choleris E, Gustafsson J, Korach KS, Muglia LJ, Pfaff DW, Ogawa S (2003) An estrogen-dependent four-gene micronet regulating social recognition: a study with oxytocin and estrogen receptor-a and -b knockout mice. Proc Natl Acad Sci USA 100(10):6192–6197
126. Koike H, Ibi D, Mizoguchi H, Nagai T, Nitta A, Takuma K, Nabeshima T, Yoneda Y, Yamada K (2009) Behavioral abnormality and pharmacologic response in social isolation-reared mice. Behav Brain Res 202:114–121
127. Okada R, Fujiwara H, Mizuki D, Araki R, Yabe T, Matsumoto K (2015) Involvement of dopaminergic and cholinergic systems in social isolation-induced deficits in social affiliation and conditional fear memory in mice. Neuroscience 299:134–145
128. Amiri S, Haj-Mirzaian A, Rahimi-Balaei M, Razmi A, Kordjazy N, Shirzadian A, Mehr SE, Sianati H, Dehpour AR (2015) Co-occurrence of anxiety and depressive-like behaviors following adolescent social isolation in male mice; possible role of nitrergic system. Physiol Behav 145:38–44
129. Macbeth AH, Edds JS, Young WS (2009) Housing conditions and stimulus females: a robust social discrimination task for studying male rodent social recognition. Nat Protoc 4(11):1574–1581
130. Karlsson SA, Haziri K, Hansson E, Kettunen P, Westberg L (2015) Effects of sex and gonadectomy on social investigation and social recognition in mice. BMC Neurosci 16(1):83
131. Yu W, Yen YC, Lee YH, Tan S, Xiao Y, Lokman H, Khoo ATT, Ganegala H, Kwon T, Ho W, Je HS (2019) Prenatal selective serotonin reuptake inhibitor (SSRI) exposure induces working memory and social recognition deficits by disrupting inhibitory synaptic networks in male mice. Mol Brain 12(1):29
132. Thor DH (1980) Testosterone and persistence of social investigation in laboratory rats. J Comp Physiol Psychol 94(5):970–976
133. Taylor G (1982) Urinary odors and size protect juvenile laboratory mice from adult male attack. Dev Psychobiol 15(2):171–186
134. Ingersoll D, Weinhold L (1987) Modulation of male mouse sniff, attack, and mount behaviors by estrous cycle-dependent urinary cues. Behav Neural Biol 48:24–42
135. Lukas M, Toth I, Veenema A, Neumann I (2013) Oxytocin mediates rodent social memory within the lateral septum and the medial amygdala depending on the relevance of the social stimulus: male juvenile versus female adult conspecifics. Psychoneuroendocrinology 38:916–926
136. Phan A, Lancaster KE, Armstrong JN, MacLusky NJ, Choleris E (2011) Rapid effects of estrogen receptor α and β selective agonists on learning and dendritic spines in female mice. Endocrinology 152(4):1492–1502. https://doi.org/10.1210/en.2010-1273
137. Phan A, Gabor CS, Favaro KJ, Kaschack S, Armstrong JN, MacLusky NJ, Choleris E (2012) Low doses of 17b-estradiol rapidly improve learning and increase hippocampal dendritic spines. Neuropharmacology 37: 2299–2309
138. Phan A, Suschkov S, Molinaro L, Reynolds K, Lymer JM, Bailey CDC, Kow L, MacLusky NJ, Pfaff DW, Choleris E (2015) Rapid increases in immature synapses parallel estrogen-induced hippocampal learning enhancements. Proc Natl Acad Sci USA 112(52):16018–16023
139. Lymer JM, Robinson A, Winters BD, Choleris E (2017) Rapid effects of dorsal hippocampal G-protein coupled estrogen receptor on learning in female mice. Psychoneuroendocrinology 77:131–140
140. Lymer JM, Sheppard PAS, Kuun T, Blackman A, Jani N, Mahbub S, Choleris E (2018) Estrogens and their receptors in the medial amygdala rapidly facilitate social recognition in female mice. Psychoneuroendocrinology 89:30–38
141. Kudryavtseva NN, Bondar NP, Avgustinovich DF (2002) Association between experience of aggression and anxiety in male mice. Behav Brain Res 133:83–93
142. Insel T, Fernard R (2004) How the brain processes social information: searching or the social brain. Annu Rev Neurosci 27:697–722
143. Sawyer TF, Hengehold AK, Perez WA (1984) Chemosensory and hormonal mediation of social memory in male rats. Behav Neurosci 98(5):908–913
144. Bluthe RM, Dantzer R (1993) Role of the vomeronasal system in vasopressinergic modulation of social recognition in rats. Brain Res 604:205–210
145. Hlinak Z (1993) Social recognition in ovariectomized and estradiol-treated female rats. Horm Behav 27:159–166

146. Matta R, Underwood EA, Leach ZK, Vertes AC, Atabakhsh V, da Silva MB, Choleris E (2018) Dopamine D2-type receptors in the dorsal hippocampus mediate social learning in female but not male mice. Soc Neurosci Abstr 639:16–18
147. Valsecchi P, Mainardi M, Sgoifo A, Taticchi A (1989) Maternal influences on food preferences in weanling mice Mus domesticus. Behav Process 1-3:155–166
148. Choleris E, Clipperton-Allen AE, Gray DG, Diaz-Gonzalez S, Welsman RG (2011) Differential effects of dopamine receptor D1-type and D2-type antagonists and phase of the estrous cycle on social learning of food preferences, feeding, and social interactions in mice. Neuropsychopharmacology 36:1689–1702
149. Clipperton AE, Spinato JM, Chernets C, Pfaff DW, Choleris E (2008) Differential effects of estrogen receptor alpha and beta specific agonists on social learning of food preferences in female mice. Neuropsychopharmacology 33: 2362–2375
150. Hepper PG (1988) Adaptive fetal learning: prenatal exposure to garlic affects postnatal preferences. Anim Behav 36:935–936
151. Galef BG (1980) Development of olfactory control of feeding-site selection in rat pups. J Comp Physiol Psychol 4:615–622
152. Galef BG, Clark MM (1971) Parent–offspring interactions determine time and place of first ingestion of solid food by wild rat pups. Psychon Sci 25:15–16
153. Galef BG, Kennett DJ, Wigmore SW (1984) Transfer of information concerning distant food in rats: a robust phenomenon. Anim Learn Behav 12:292–296
154. Valsecchi P, Galef BG (1998) Social influences on the food preferences of house mice (Mus musculus). Int J Comp Psychol 2:245–256
155. Valsecchi P, Choleris E, Moles A, Guo C, Mainardi M (1996) Kinship and familiarity as factors affecting social transfer of food preferences in adult Mongolian gerbils (Meriones unguiculatus). J Comp Psychol 110: 243–251
156. Chloeris E, Guo C, Mainardi M, Valsecchi P (1997) The effect of demonstrator age and number on duration of socially-induced food preferences in house mouse (Mus domesticus). Behav Process 41:69–77
157. Kogan JH, Frankland PW, Blendy JA, Coblentz J, Marowitz Z, Schutz G, Silva JA (1996) Spaced training induces normal long-term memory in CREB mutant mice. Curr Biol 7:1–11
158. Wainwright PE, Huang Y-S, Coscina DV, Levesque S, McCutcheon B (1994) Brain and behavioral effects of dietary n-3 deficiency in mice: a three generational study. Dev Psychobiol 27:467–487
159. Choleris E, Valsecchi P, Wang Y, Ferrari P, Kavaliers M, Mainardi M (1998) Social learning of food preference in male and female Mongolian gerbils is facilitated by the anxiolytic chlordiazepoxide. Pharmacol Biochem Behav 60:575–584
160. Galef BG Jr, Rudolf B, Whiskin EE, Choleris E, Mainardi M, Valsecchi P (1998) Familiarity and relatedness: effects on social learning about foods by Norway rats and Mongolian gerbils. Anim Learn Behav 26: 448–454
161. Monfils MH, Afee LA (2019) Insights from social transmission of information in rodents. Genes Brain Behav 18:1–13
162. Panksepp J, Panksepp JB (2013) Toward a cross-species understanding of empathy. Trends Neurosci 36(8):489–496
163. Sánchez-Andrade G, James BM, Kendrick KM (2005) Neural encoding of olfactory recognition memory. J Reprod Dev 51(5):547–558
164. Clipperton-Allen AE, Flaxey I, Webster HK, Choleris E (2009) Pre-acquisition administration of estradiol benzoate prolongs the preference for a socially transmitted food preference in ovariectomized CD1 mice. In: 39th annual society for neuroscience conference, Chicago, IL, United States, p 681.14
165. Ervin KSJ, Lymer JM, Matta R, Clipperton-Allen AE, Kavaliers M, Choleris E (2015) Estrogen involvement in social behaviors in rodents: rapid and long-term actions. Horm Behav 74:53–76
166. Galef BG, Whiskin EE (2008) Effectiveness of familiar kin and unfamiliarnonkin demonstrator rats in altering food choices of their observers. Anim Behav 76:1381–1388
167. Blanchard RJ, Yudko EB, Rodgers RJ, Blanchard DC (1993) Defense system psychopharmacology: an ethological approach to the pharmacology of fear and anxiety. Behav Brain Res 58:155–165
168. Grant EC, Mackintosh JH (1963) A comparison of the social postures of some common laboratory rodents. Behavior 21:246–259
169. Munger SD, Leinders-Zufall T, McDougall LM, Cockerman RE, Schmid A, Wandernoth P, Wennemuth G, Biel M, Zufall F, Kelliher KR (2010) An olfactory subsystem that detects carbon disulfide and

mediates food-related social learning. Curr Biol 20:1438–1444

170. Bonnie KE, Earley RL (2007) Expanding the scope for social information use. Anim Behav 74(2):171–181. https://doi.org/10.1016/J.ANBEHAV.2006.12.009

171. Danchin É, Giraldeau LA, Valone TJ, Wagner RH (2004) Public information: from nosy neighbors to cultural evolution. Science 305(5683):487–491. https://doi.org/10.1126/SCIENCE.1098254/ASSET/FA8399AE-D8D8-4479-B243-689E495D2B51/ASSETS/GRAPHIC/ZSE0280427060003.JPEG

172. Aspesi D, Choleris E (2022) Neuroendocrine underpinning of social recognition in males and females. J Neuroendocrinol 34(2):e13070. https://doi.org/10.1111/JNE.13070

173. Wolff JO, Sherman PW (2007) In: Wolff JO, Sherman PW (eds) Rodent societies, an ecological and evolutionary perspective. University of Chicago Press

174. Bredewold R, Veenema AH (2018) Sex differences in the regulation of social and anxiety-related behaviors: insights from vasopressin and oxytocin brain systems. Curr Opin Neurobiol 49:132. https://doi.org/10.1016/j.conb.2018.02.011

175. Dumais KM, Veenema AH (2016) Vasopressin and oxytocin receptor systems in the brain: sex differences and sex-specific regulation of social behavior. Front Neuroendocrinol 40:1. https://doi.org/10.1016/j.yfrne.2015.04.003

176. Parmigiani S, Ferrari PF, Palanza P (1998) An evolutionary approach to behavioral pharmacology: using drugs to understand proximate and ultimate mechanisms of different forms of aggression in mice. Neurosci Biobehav Rev 23(2):143–153

177. Carter CS (2007) Sex differences in oxytocin and vasopressin: implications for autism spectrum disorders? Behav Brain Res 176:170. https://doi.org/10.1016/j.bbr.2006.08.025

178. Borie AM, Theofanopoulou C, Andari E (2022) The promiscuity of the oxytocin-vasopressin systems and their involvement in autism spectrum disorder. Handb Clin Neurol 182:121–140. https://doi.org/10.1016/B978-0-12-819973-2.00009-5

179. Krishnan MC (2018) Sex differences in autism spectrum disorder. In: The complexity of autism spectrum disorders. https://doi.org/10.4324/9780429454646

Chapter 2

Social and Sexual Preference in Socially Monogamous Species: Prairie Voles (*Microtus ochrogaster*) and Zebra Finch (*Taeniopygia guttata*)

M. Fernanda López-Gutiérrez, Guillermo Valera-Marín, Sarael Alcauter, and Wendy Portillo

Abstract

Social monogamy is a reproductive strategy characterized by long-lasting pair bonds, showing a selective preference for the sexual partner, mate guarding, and biparental caregiving of offspring. This chapter focuses on the standard methodology to evaluate pair bonding in the prairie vole (*Microtus ochrogaster*), a rodent from the *Cricetidae* family, and the zebra finch (*Taeniopygia guttata*), a passerine monogamous bird.

In the prairie vole, partner preference is the standard test to determine affiliative behavior for the sexual partner. Resident-intruder tests allow the evaluation of aggressive behavior directed toward other conspecifics to maintain the pair bond, and the conditioned place preference test has been extensively used to assess the reinforced properties of drugs and behaviors in rodents. In the zebra finch, a standard test to determine mate and sexual partner choice is the partner preference test performed in a communal cage, since it achieves a closer resemblance to the finch's behavior in natural settings. This chapter focuses on the detailed methodology, data analysis, and test variations.

Key words Social monogamy, Partner preference test, Resident-intruder test, Prairie vole, Zebra finch

1 Introduction

Monogamy is the reproductive strategy that characterizes around 3–10% of mammals. It is defined by selective attraction and cohabitation with a specific conspecific, and it is commonly associated with the shared caregiving of the offspring [1]. Although this kind of strategy is relatively common in some canid species, it is considered to be rare in most mammal taxons. For example, jackals (*Canis mesomelas*) usually form long-lasting pair bonds and older offspring contribute with caring for the new litters [2]. On the other hand, monogamy is common in birds, as both sexes can raise the progeny without major differences [2]. Sexual monogamy, also called genetic monogamy, is rare. Most monogamous animals are socially

Raúl G. Paredes et al. (eds.), *Animal Models of Reproductive Behavior*, Neuromethods, vol. 200,
https://doi.org/10.1007/978-1-0716-3234-5_2,

monogamous, which means that they exhibit a clear social preference for the partner, but they may also show extra-partner copulation [3].

Pair bond formation involves several steps. First, a memory of the partner-specific cues shall be formed, and it must evoke a positive hedonic response or attraction [4, 5]. At the same time, to maintain the pair bond the pair must display rejection or aggressive behavior toward other conspecifics [4, 6]. These steps have been assessed in laboratory conditions. Usually, partner preference tests evaluate attraction to the partner and rejection to the strange animal; this is a reliable way to assess pair bond formation [7]. However, there are tests that evaluate these aspects of the pair bond formation separately. The positive qualities of cohabitation with the partner may be assessed with a conditioned place preference (CPP) test, and the rejection toward strange animals may be assessed with a selective aggression test. In this chapter, we describe the partner preference test, conditioned place preference, and resident-intruder test in the prairie vole (*Microtus ochrogaster*), and the methodological strategies to evaluate pair bonding in zebra finch (*Taeniopygia guttata*).

2 Prairie Vole

2.1 Partner Preference Test

2.1.1 Introduction

The partner preference test (PPT) is the gold standard for pair bonding evaluation in the prairie vole (*Microtus ochrogaster*). The test is based on evidence that prairie voles are highly social and prone to affiliative behavior in contrast to other rodents [8] and even to other species of voles [9], usually displaying a selective preference for social contact with their partner rather than with a stranger conspecific. With the intention of studying the behavioral and neurobiological mechanisms underlying pair bonding, Sue Carter's research group first described and implemented the partner preference test [10]. The basic objective of the test is to assess the pair bonding of one test vole by a manner of choice, that is, by comparing the amount of time it spends in social contact with its partner or with a stranger vole within a specific timeframe. Obtaining both time measures allow researchers to calculate the level of preference toward the partner, commonly expressed as a preference index. The test has proven effective in evaluating pair bonding in both sexes and has also been used to assess monogamous behavior between different species of voles [9].

2.1.2 Materials

Subjects

Since the test evaluates pair bond formation, the minimum requirement is to have a male and female pair that has had some form of previous social interaction. These voles must be sexually mature adults (i.e., approximately over 45 days of age) [11] to be pair bonded, since this is a form of sexual behavior and voles use sex-related cues to recognize conspecifics [12]. Social interaction between a pair is enabled through cohabitation, usually in a home cage under laboratory conditions (*see* **Note 1**). Once pair bonded, the female–male pair is typically housed in its own cage.

Second, a stimulus "stranger" vole is required as an alternate choice for the tested subject, and therefore must be of the opposite sex to the experimental subject and of the same sex as the partner. "Stranger" voles should not be family related to any other vole in the test and should preferably be in the same hormonal condition as the partner vole, since familiar odors and behavioral sexual receptivity, respectively, may be confounding factors during the test. Control of the hormonal condition is particularly relevant in female voles, and different approaches are possible (*see* **Note 2**). It is also recommended for male "stranger" voles to be sexually experienced, which can be done by previously screening them with a sexually receptive female but housing them alone prior to testing.

Apparatus

The most common apparatus for the PPT is composed of three interconnected chambers, in which a central chamber is connected to two opposite chambers (Fig. 1). The central chamber is designated as the neutral or solo chamber, while one opposite chamber is occupied by the partner vole and the other by the stranger vole (stimulus voles). The tested subject is allowed to roam freely between the chambers so time spent with each stimulus vole can be measured separately. The apparatus in this method has a linear or straight configuration, in which the three chambers are open from the top and are separated by short tunnels (*see* **Note 3**).

In the classical configuration of PPT, physical social interaction between the subject and the stimulus voles is desired, so restraining the stimulus voles to their respective chambers is necessary (*see*

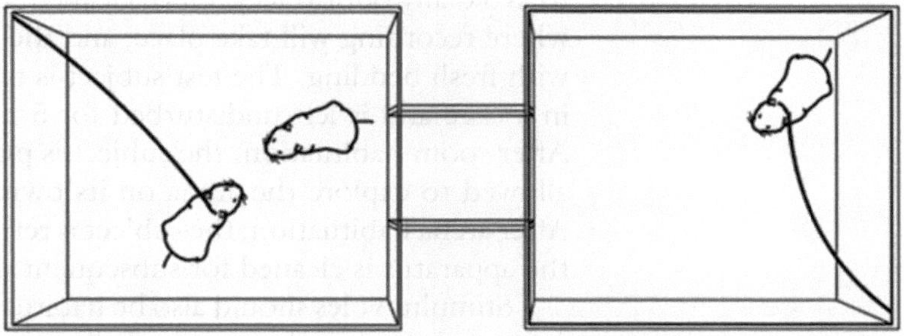

Fig. 1 Partner preference test (PPT) apparatus. The arena is a linear configuration of interconnected chambers, in which a central, short tunnel connects two larger opposite chambers; each securely housing the stimulus voles through a tether. The test vole can roam freely among the chambers and through the tunnel to interact with the stimulus voles

Note 4). This is usually done with a collar and a tether that allow free movement within the chamber but prevent any physical contact with the opposite stimulus vole (*see* **Note 5**).

Tools for Test Setup and Analysis

The test should preferably be made in a separate, noise-free room where subjects can remain undisturbed. Video recording of the test is necessary for subsequent analysis; therefore, room lighting should be dim enough to allow adequate recording but not too intense to disturb or stress the animals. Fresh bedding, masking tape, paper towels, odor-free soap (Extran, Millipore), or ethanol (70%) should be readily available for cleaning. A piece of paper or index card is useful to register test-related data, including the ID of the test subject and stimulus voles. A timer or stopwatch is also handy to keep track of the test time.

While using a human observer to watch the video and perform manual analysis is possible, it may be time-consuming and impractical considering that the test lasts 3 h (*see* **Note 6**) and several PPTs are usually analyzed in a single study. In addition, manual scoring should consider blind analyses and inter-rater reliability between multiple experimenters. Therefore, computer-automated data analysis is highly recommended. There are several alternatives designed for such purpose, and free or licensed software is available. The method described below is based on the use of the open-source software *UMATracker*, a quantitative analysis toolkit for animal experiments [13].

2.1.3 Methods

Arena Preparation and Habituation

Habituation of subjects to the apparatus is necessary to avoid the expression of other behaviors that may interfere in any sociosexual interactions expected in the test. It is recommended to habituate subjects at least once, 1 day before the test. Before and between tests, the apparatus must be properly cleaned with soap or ethanol to avoid any odor traces. Once cleaned, the apparatus is positioned where recording will take place, and the apparatus floor is covered with fresh bedding. The test subject is taken into the testing room in a cage and is left undisturbed for 5 min for room habituation. After room habituation, the subject is placed on the apparatus and allowed to explore the arena on its own for 30 min (*see* **Note 7**). After arena habituation, the subject is returned to its home cage and the apparatus is cleaned for subsequent use.

Stimulus voles should also be habituated, particularly if they are going to be tethered, since the feeling of restraint might be stressful for the animals at the beginning. Stimulus voles may be habituated simultaneously for 30 min in the same apparatus if they remain securely restrained within their respective chamber. Typical data from the test are provided in Fig. 2.

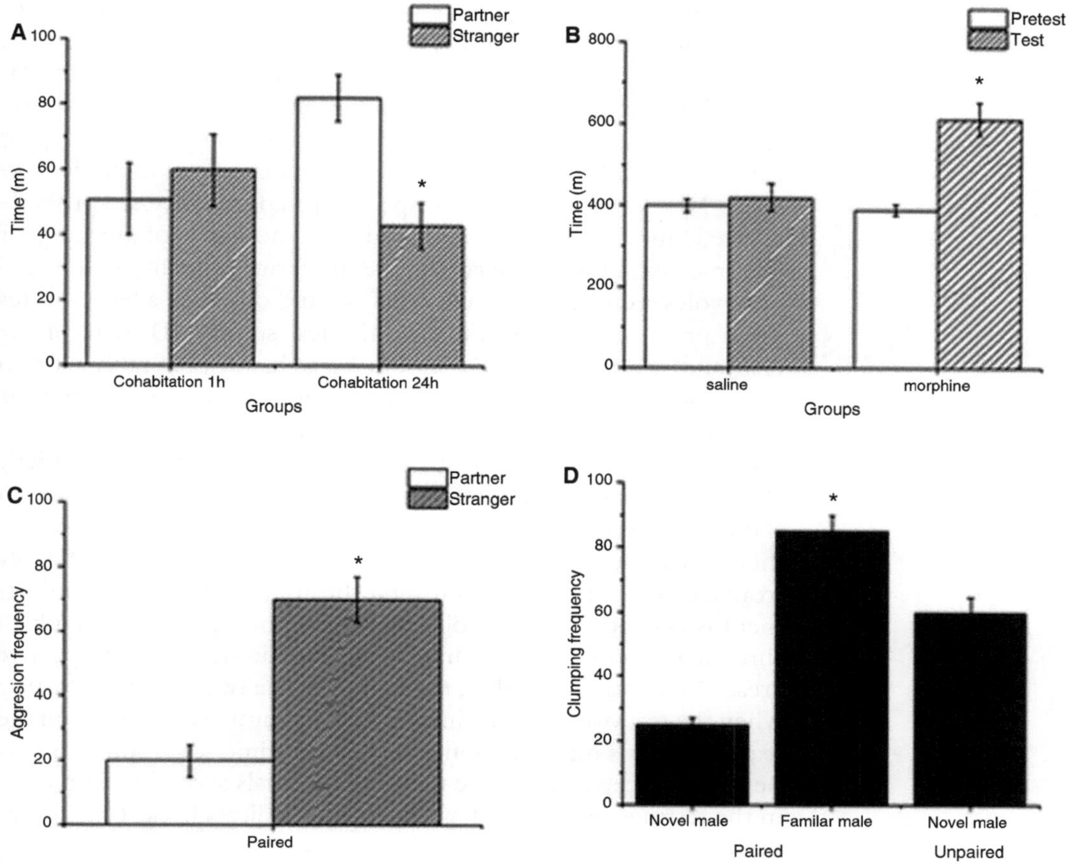

Fig. 2 Typical data from the behavioral tests. (**a**) Partner preference test in prairie voles. Usually, 1 h of cohabitation with mating is not enough to induce a preference for the sexual partner. A strong preference for the sexual partner will be observed after 24 h of cohabitation. (**b**) Conditional place preference in prairie voles. Morphine administration induces a reward state but not a saline injection. (**c**) Resident-intruder test. Pair-bonded prairie voles display higher aggression frequency toward a stranger than their sexual partner. (**d**) Female pairing behavior in the social interaction test. Paired female zebra finches displayed higher levels of clumping toward their mates than novel males. Paired females spend more time clumping their mates than unpaired females clumping novel males
* Different from the partner in the partner preference and resident intruder test. $p < 0.05$
* Different from the pretest in the conditional place preference test. $p < 0.05$
* Different from the novel male in paired and unpaired female zebra finches. $p < 0.05$

Partner Preference Test

On the day of the test, the cleaned apparatus is positioned where recording will take place and its floor is covered lightly with fresh bedding (*see* **Note 8**). At this moment, it is advisable to check that any necessary electronic or camera setup for recording is working properly (batteries, available memory storage, etc.). The minimum recording time should be 3 h, which is the actual duration of the test. For this configuration, the camera is mounted on the ceiling for a top view of the arena, and the camera field of view is adjusted to include the entire apparatus (*see* **Note 9**). Once the apparatus and recording setup are ready, voles are taken into the testing room in separate cages and left undisturbed for 5 min for room habituation. After room habituation, each stimulus vole is placed and tethered in its respective chamber, ensuring that its chamber and ID (partner or "stranger") are properly matched and identified (*see* **Note 10**). Voles are visually inspected to ensure they are properly secured, moving comfortably, and with no signs of freezing or extreme agitation. Otherwise, the tethering system is adjusted until voles seem comfortable, which should only take a few minutes given prior habituation. Further, the test subject ID, date of the test, and stimulus vole IDs should be at hand, preferably written down or printed legibly on a card or piece of paper that will be shown at the beginning of the video. Once recording has started, the subject ID or the card mentioned previously is shown briefly (but clearly) to the camera, and the test vole is carefully placed in the central or neutral chamber. It is recommended for the experimenter to observe for the first few minutes of the test, preferably from a distance to avoid disturbing the animals, in case any adjustment is needed. The test vole should be moving freely through all three chambers with the stimulus voles remaining in their respective areas. As mentioned earlier, the length of the test is of 3 h, starting when the test vole is placed in the neutral chamber, that is, when the experimenter's manipulation has ended. A timer or stopwatch may be used to keep track of the time. The animals should be left alone in the testing room and a few discrete periodical checks (e.g., every hour or every 45 min) are advised to ensure that the test is running smoothly for both the recording and the animal's status. While defensive behaviors (chasing and squawking) may happen between the test vole and the stimulus voles, if frequent direct aggression such as lunging or biting is observed, the experimenter should evaluate if the test should continue to prevent unnecessary harm to the animals.

When the test is over, the recording is stopped and each vole is removed from the apparatus separately (usually beginning with the ones that are tethered) and taken from the test room, also making sure their IDs match their home cage. The bedding from the apparatus is discarded, and the arena is cleaned with soap or ethanol. At this moment, it is suggested to check if the video was recorded properly and, if applicable, transfer it to the device where analysis will be performed.

Data Preparation

The basic data analyzed in the PPT is the amount of social contact the test vole has with each of the stimulus voles (i.e., the partner and the stranger voles). Most researchers score the cumulative amount of huddling, which is described as side-to-side bodily contact between individuals (*see* **Note 11**). This is because huddling is considered a prosocial, affiliative behavior in the prairie vole, and a widely proven indicator of social preference [9, 14, 15]. However, other behaviors may also be scored according to the experimenter's own protocol and objectives. The method described below scores the amount of huddling and is based on the use of the open-source software UMATracker [13].

Before using the software tools, the video must be properly edited to be prepared for tracking analysis. First, the video must be identified and labeled to match the tested vole's ID; keeping the original video and making a copy for analysis is highly recommended. Second, the video used for the analysis is trimmed with a video editor from the moment the test vole is placed inside the neutral chamber (and the experimenter has ceased manipulation) up to the 3-h mark. Third, using UMATracker's *Filter Generator* software, the input video goes through various modifications to generate a filter (*see* **Note 12**). The resulting filter should enable proper vole detection and tracking by the software. Once the filter is saved and ready to use, UMATracker's *Tracking* software uses the filter data to generate tracking nodes, each node corresponding to each vole. By employing the algorithm *Group Tracker GMM* for three objects (*see* **Note 13**), the video tracking is accomplished (which should normally take hours of processing), and position data are obtained and saved. After assessing the tracking result manually, the app *Tracking Corrector* may be used if the software had errors or swapped tracking IDs during the process. Finally, *Area51* software simply uses the position data obtained from tracking each vole, that is in every frame of the video, to calculate the amount of interaction between prairie voles and score social contact (*see* **Note 14**). Huddling will be scored efficiently if the radius of the tracking nodes of two voles overlaps. The results graph will show a small network of interacting nodes, each vole representing a node, and data containing the number of frames corresponding to such interactions is saved. It is important to correctly identify and match each node to the respective test and stimulus voles.

Data Analysis

The number of frames calculated by the software for both stimulus voles is equivalent to the amount of time the tested subject had social interaction with each stimulus vole (*see* **Note 15**), which allows the obtention of the partner preference index by the following formula:

$$\text{Partner preference index} = \frac{\text{Time with partner vole}}{\text{Time with partner vole} + \text{time with stranger vole}}$$

This index considers specifically the amount of social contact with the stimulus voles and the proportion of such contact during the entirety of the test; ranging from 0 to 1, being 0 a null preference for the partner, an index of 0.5 interpreted as an equal preference toward any of the stimulus voles, and 1 being an absolute preference for the partner vole (*see* **Note 16**). Though the quantitative index indicates partner preference at an individual level [16], partner preference should be properly assessed at a sample or group level to determine a statistically significant preference in time spent with the partner compared with time spent with the "stranger" vole. For the analysis, a normality test (e.g., Shapiro–Wilk test) should be made a priori to evaluate if a nonparametric test is suitable given the data distribution. Usually, a standard Student *t*-test or a nonparametric Wilcoxon signed-rank test or Mann–Whitney rank sum test are adequate to determine a group-wise statistical significance in the partner preference over a "stranger vole."

Even though prairie voles are known to display selective partner preference as a result of pair bonding, a considerable amount of literature has shown that several factors influence the development and maintenance of the latter [7, 17, 18]; and even individual variations in partner preference have been observed in the same controlled conditions [15]. Though this may be most advantageous in further exploring the mechanisms of pair bonding, hence the importance of relying on a proper procedure and statistical analysis to interpret the results obtained in the PPT.

While studies using the PPT and variations in its methods are extensive, this may be a comprehensive guide in its most basic form. It is important to consider that prairie voles are live animals, and all experimental procedures must be carried out in accordance with bioethics local guidelines and protocols for the use of laboratory animals.

2.1.4 Notes

1. In laboratory conditions, voles are reported to pair bond by at least 6 h of continuous cohabitation if full physical interaction is allowed [10]. This is because mating has been observed as an important enhancer in the formation of the bond, especially in males [16, 19]. However, pair bonding is possible with restricted physical interaction (i.e., without mating but with olfactory and visual interaction enabled) if cohabitation is extended for a 24-h period [10].
2. If females are to remain intact, estrus induction by daily estradiol benzoate (EB) administration or estrus induction by exposure to soiled bedding of a sexually mature male are both common options usually done at least 4 days before the day of the test. On the other hand, bilateral ovariectomy with or without hormonal replacement (EB administration) is also applicable. Male voles can also be vasectomized if it is suitable for the experimental protocol.
3. Another common apparatus variant is with the chambers arranged as an inverted "Y" shape (branched) connected

through the tubing. Customized polycarbonate home cages connected by plastic tubing are regularly used, but for custom-built chambers, acrylic plastic is the material of choice. Another linear configuration consists of three chambers divided by acrylic separators. Materials with washable and non-porous surfaces are recommended, as they are easier to clean and do not retain odors.

4. Alternately, the interaction between the subject and stimulus voles can be restricted by barriers that allow visual, auditory, and olfactory social contact, but not physical or mating behavior. Since prosocial behaviors such as huddling require physical contact and are commonly scored to evaluate preference, this variant should rely instead on measuring time spent in the proximal area related to each of the stimulus voles to determine partner preference.

5. Plastic zip ties have been found very useful as collars in the tethering system since they are easily adjustable and durable against gnawing or chewing. Zip ties may be placed before the day of the test (just before habituation), ensuring they are loose enough to enable tethering but tight enough to prevent them from slipping through the vole's skull. Also, zip ties may be easily removed with a clipper if voles have finished behavioral tests. Small-sized fishing swivels may also be useful to avoid tether twisting during the test.

6. Different partner preference test durations (e.g., from 10 min to 8 and 24 h) have been used in previous reports; however, evidence suggests that selective social preference is only evident after more extended testing time. In general, the 3-h test length has become a reliable standard in determining partner preference in the prairie vole in several research groups [8, 10].

7. Protective gloves are recommended for prairie vole handling since they have long incisors, and they may bite. It is strongly advised to use a different set of gloves for handling each sex to avoid sex-specific odor mixing.

8. For efficiency, many research groups test more than one subject simultaneously, requiring more than one apparatus (and sometimes more than one camera). No visual contact between voles of different tests is recommended to avoid distractions.

9. It is advisable to check and fix for any evident flares from reflective surfaces in the camera view and disable any auto-focus features of the camera that may be troublesome for subsequent video analyses. A camera top view over an open-top apparatus diminishes lighting or flare issues in comparison to a side view.

10. For a blind analysis, the stimulus voles' ID may be replaced by other labeling as long as it is appropriately matched and

registered. Additionally, if the PPT is to be repeated a second time or more for the same test subject or its partner, alternating or switching positions of stimulus voles in the opposite chambers of the apparatus is recommended to avoid any spatial or context bias over the test vole that may be confounding. Changing the "stranger vole" for subsequent PPTs is also necessary to avoid familiarity-based bias.

11. While voles during the test may display other types of sociosexual behavior bouts, such as anogenital sniffing, licking, grooming, or mating behavior like lordosis in females or mounts in males, these behaviors are usually brief and may not necessarily indicate a selective social preference. Nevertheless, depending on the experimenter's objectives, they may be useful behaviors to consider.

12. The filter parameters shown useful for this configuration are the following: *Distance* (set to dark brown), *BGRToGray*, *Threshold* (low numbers), and a *Rectangle selection* covering only the area of interest; then, the morphology operators *Erosion kernel*, *Dilation* and *Closing kernel* (in that order) are applied to the video. The ideal filter is a clean, binarized version of the video in which each vole is transformed into a clearly defined white shape (blob) over a black background. For more details on how to use the software, documentation for the entire software package is available on the developer's website (https://umatracker.github.io/UMATracker-manual-en/).

13. The recommended set of parameters after opening the app is *# of objects* set to 3 and *Radius* set to 10; then, the filter is opened with the software and it should automatically detect the three voles, otherwise the filter should be modified until vole detection.

14. The *Area51* app will calculate the amount of interaction by processing the position data obtained from the *Tracking* app by simply clicking *Run* and *Calculate.* In the app, the *Radius* parameter of each detected vole (node) can be increased or decreased accordingly to adjust the detection of closeness between voles and score social contact. However, this app also enables the use of multiple regions of interest, which may be useful for other PPT variants.

15. The *Tracking* software generates an "info" text file that specifies the video's number of frames per second (FPS). By dividing the number of total frames by FPS, the tracking data are converted to time units (seconds), which may be more practical for subsequent analyses.

16. If convenient, the index can also be analyzed and presented in percentage form (0–100%). While data (time huddling or time in close proximity) can be analyzed and compared statistically

without an index calculation, the index is a convenient way to analyze partner preference between different groups or treatments. If multiple groups with different treatments are to be compared, a two-way analysis of variance (ANOVA) is a common choice.

3 Conditioned Place Preference Test

3.1 Introduction

The conditioned place preference (CPP) test has been extensively used to assess the reinforcing properties of several drugs or the reinforcing properties of several situations or behaviors in rodents [5, 20–23].

The CPP procedure is based on Pavlovian conditioning in which an unconditioned stimulus, the rewarding stimulus, is associated with a conditioned stimulus, the chamber, and its contextual cues [20]. A reinforcing event, such as a morphine injection, induces a positive hedonic state. If the animal is placed in a chamber with contextual cues, it will learn that the contextual cues are associated with a positive hedonic state [24, 25]. Some social experiences may have the same effect; an animal may experience a positive hedonic state during several minutes after a socially reinforcing event. For instance, if a rat mother is feeding her pups, the reinforcing properties of the experience will last for a long time. If the mother is moved to a chamber with contextual cues, she will associate the positive experience with those cues [22]. On the other hand, social interaction with a same-sex conspecific has also been demonstrated to have rewarding properties in mice [21]. Regarding pair bonding, this kind of test has been used to assess not only the reinforcing properties of mating in prairie voles [5], but also those of the whole experience of cohabitation and mating during the pair bond formation [18]. The CPP test may also be used to compare the reinforcing properties of pair bond formation in animals that pair bond during the cohabitation with mating in comparison with animals that do not pair bond [18]. It has also been used to assess the reinforcing properties of cohabitating with a long-lasting partner [23].

Therefore, the CPP test is an established tool to assess the positive hedonic cues from the partner in pair bond formation. The protocol for CPP evaluation in other rodents is described in this chapter assessment of sexual reward with the conditioned place preference paradigm in this book.

3.2 Methods

3.2.1 Subjects

To assess the reinforcing properties of cohabitating and/or mating with a conspecific, the experimental animals should be adults (2–-3 months old). The stimulus prairie voles should have previous sexual experienced (*see* **Note 1**). To avoid pregnancy, it is advisable

Fig. 3 Diagram of a CPP cage with three compartments. In the CPP test, the white and black chambers (23 × 37 × 32 cm) are communicated by a smaller gray chamber (23 × 24 × 32 cm). All chambers are separated by removable guillotine doors. The CPP procedure consists of a pretest, three nonreinforced sessions, three reinforced sessions, and a test. During the pretest (day 1), prairie voles are allowed to explore the three chambers. During the conditioning, the guillotine doors are closed, and the voles are placed in the preferred (days 2, 4, and 6) or nonpreferred or reinforced chamber (days 3, 5, and 7). During the test, the voles are allowed to move freely between the three chambers

to use vasectomized males and ovariectomized females treated with estradiol to induce sexual receptivity (*see* **Note 2**).

On the other hand, several controls may be used, for instance, to compare the reinforced effects of cohabitation with an opposite-sex sexually active stimulus animal and social cohabitation with a same-sex partner or a gonadectomized partner. It is also possible to compare the reinforcement in prairie voles with that of a similar nonbonding polygamous species like the meadow vole [23].

3.2.2 Apparatus and Preparation

The CPP test is performed in a cage with two or three compartments. The animal can discriminate between these compartments, as they have distinct contextual cues (Fig. 3). Contextual cues can be different colors, odors, textures, or figures. The compartments should be separated by removable doors to restrict the animals or allow them to explore at will. CPP cages can be made with black, gray, and white acrylic, instead of other materials like wood, to ensure that the cages can be properly cleaned between animals. Many prairie voles try to escape from the CPP arena; therefore, the cages must have high walls or be enclosed, in which case proper ventilation must be ensured. Cleaning the cages between procedures with odor-free soap followed by ethanol (10–70%) is mandatory.

3.3 Conditional Place Preference Test

The CPP consists of eight sessions: a pretest, three nonreinforced sessions alternated with three reinforced sessions, and a test (*see* **Note 3**). During the pretest, prairie voles are placed in the CPP cage and allowed to explore all the compartments for 30 min (day 1). This procedure allows us to establish a baseline and determine the preferred compartment for each animal (*see* **Note 4**).

The compartment in which the animal spends most of the time is considered the preferred compartment (nonreinforced). The compartment where the animal spends less time is considered the nonpreferred compartment (reinforced).

The day after the pretest (day 2), without any stimulation the animal is placed for 40 min in the preferred compartment. On day 3, voles are exposed to an event that will reinforce, such as mating with a sexual partner; then, they are gently placed in the reinforced chamber for 40 min. After six alternated sessions [three reinforced (days 2, 4, and 6) and three nonreinforced (days 3, 5, and 7)], the preference for each compartment is tested again (test) under the same conditions as the pretest (day 8).

A CPP is defined by a significant increase of time spent in the reinforced chamber during the test compared to the pretest.

3.4 Statistical Analysis

Since each vole is evaluated before (pretest) and after conditioning (test), each animal can be used as their own control. The Kolmogorov–Smirnoff and Levine tests can be used to evaluate normality and homoscedasticity. If data have a normal distribution, it can be analyzed with a repeated measures t-test. If data are not normally distributed, it can be analyzed with a Wilcoxon test. A $p < 0.05$ will be set as a threshold for statistical significance.

3.5 Variations

There are several relevant variations in the CPP test that shall be considered. Our research group uses a three-chamber CPP cage. The small compartment in the middle allows the animal to stay in a neutral nonassociated chamber and decide whether to explore the two main chambers (*see* **Note 5**). However, other groups have used protocols with only two chambers. The use of a forced-choice apparatus may increase the sensibility of the test but will also bias for the initial chamber [23, 26].

The number of nonreinforced and reinforced sessions is subject to variation as well. Some behaviors, such as paced mating (females are allowed to control the rate of sexual stimulation) in naturally cycling female rats, may show reinforcing properties in one session [27]. It is recommended to standardize your protocol with 1–3 reinforcing and nonreinforcing sessions. Prus and coworkers (2009) recommend as well to establish a baseline during several days [26].

Other protocols choose one of the chambers as the reinforced compartment, which is called an unbiased design; however, it is possible that the reinforcing effects are hidden because the animals initially prefer the reinforced chamber [28].

The contextual cues used shall be relevant to the behavior analyzed. For instance, color, figures, or textures can be used.

The olfactory cue may facilitate contextual learning. In prairie voles, the CPP test works well only using the contextual color of the chambers (*see* **Note 6**).

An interesting protocol was used by Goodwin and coworkers [23], used as a contextual cue for the type of bedding that was present in the animal home cage. In this condition, the experience of cohabitation and mating with a long-lasting partner is associated with the characteristics of the nest.

The procedure can be used to induce a Conditioned Place Aversion by using an aversive stimulus instead of a rewarding one. This protocol may be used to demonstrate if encountering an opposite-sex conspecific is aversive once the pair bond is formed. Typical data from the test are provided in Fig. 2.

3.6 Notes

1. To acquire sexual experience, vasectomized male prairie voles can mate with a sexually receptive female in three behavioral tests of 1 h each. The experimenter should only select males that mount and intromit in all mating sessions. For stimulus females that performed three sexual behavior tests of 1 h, the experimenter should select those who display lordosis posture and do not show rejection behaviors.
2. Vasectomized stimulus males did not require hormonal replacement to induce sexual receptivity. Ovariectomized females should be treated with EB to induce sexual receptivity [5] or a silastic cannula with estradiol can be implanted [15]. Female voles can be ovariectomized under deep anesthesia with a mixture of ketamine (60 mg/kg) and xylazine (4 mg/kg) should have at least 2 weeks to recover after surgery. To induce sexual receptivity, females can be injected daily with estradiol benzoate (EB) (0.5 mg/vole) for 4 consecutive days before the behavioral test [29]. This species does not need progesterone to induce sexual receptivity [30].
3. The CPP test is a robust analysis that allows to compare differences in the rewarding characteristics of the partner. However, it is sensitive to variations in the procedure; for instance:
   - Manipulation stress may suppress the rewarding properties of the reinforcing stimulus or have a negative effect on the nonreinforcing session. It is strongly recommended to manipulate the animals several weeks before the experiment to habituate them to the experimenter.
   - As some prairie voles are biters, it is advisable to always manipulate them with protective gloves, not just for personal protection but to be as gentle as possible.
   - Animal handling should be done with the whole hand. The animal is grasped gently around the body, placing its head

between the thumb and forefinger. In this position, the animal will not try to bite or escape.

4. Note that if a prairie vole does not explore one of the lateral chambers or spends less than 1 min in any of the lateral chambers, this animal should be removed from the experiment. For prairie voles, all behavioral tests should be performed during the light phase of the light/dark (14:10) cycle.
5. It is important to note that prairie voles will prefer the black chamber.
6. It is particularly important to choose the best protocol, apparatus, and time-lapse for the experiment [26]. For example, our group failed to reinforce females with the CPP test during pair bond formation, but we did so with males, but it is possible with males. However, reinforcing females can be done once the pair bond is strongly established using bedding as a contextual cue [23].

4 Resident-Intruder Test

4.1 Introduction

A crucial component of the pair bond is the ability to maintain it once it has been formed. This happens when both sexes display mutual interest and energy in maintaining the bond. Pair-bonded animals inhabit and defend their territory to access food and other resources. To maintain the bond, the pair must respond appropriately to extra-pair individuals. Selective aggression is important in pair maintenance (*see* **Note 1**). Monogamous male song sparrows (*Melospiza melodia*) actively defend territories during the breeding and nonbreeding seasons [31], and monogamous California mice (*Peromyscus californicus*) also show aggression behavior to maintain their bonding [32, 33]. Sexually naïve prairie voles are highly social with other conspecifics. Male prairie voles that mate exhibit more aggressive behavior toward a male intruder than males that cohabitate with a female but do not mate or are sexually naïve [34, 35]. After mating prairie voles develop pair bonds, show rejection and aggressive behaviors to other potential mates in their territory, and display mate guardian behavior [36]. In addition, partner separation is important for pair-bond maintenance, as it generates a stress response and the need to seek out the sexual partner [37].

The most common test to evaluate rejection and aggressive behaviors toward other potential mates is the resident-intruder test [19, 38, 39].

Several studies have demonstrated that rejection/attack behavior toward stranger prairie voles in the resident-intruder test is a good measure to evaluate the maintenance and strength of a pair bond [19, 34, 40, 41]. In the field and under laboratory conditions, pair-bond maintenance depends on selective aggression toward novel conspecifics as well as selective affiliation with the mating partner. Male and female paired voles exhibit high levels of aggressive behavior toward same-sex stranger voles. Rejection behaviors are less frequent toward opposite-sex animals [34, 35, 40] (*see* **Note 2**).

4.2 Resident-Intruder Test

The resident-intruder test has two components: habituation and the test.

4.2.1 Habituation

If the resident-intruder test is going to be performed for both members of the pair, the member that will not be tested first is removed from its home cage and placed in a clean cage for no more than 30 min [36]. The resident vole is habituated to the room conditions for at least 10 min in its home cage.

4.2.2 Test

The test starts when an intruder (*see* **Note 3**) is placed in the resident prairie vole home cage. The intruder can be of the same or opposite sex. Same-sex intruder and resident voles should be age and weight matched. Opposite-sex intruders should be age, weight, and under the same hormone condition as the sexual partner.

The test usually lasts between 6 and 10 min [19, 38–40]. Video recording of the test is advisable to avoid distracting the animals. The following behaviors will be analyzed during the test: frequency of aggressive interactions, such as lunges, pushing, chasing, offensive rears toward the strange prairie vole, jumping and biting, and supine posture definitions in [42]. During the resident-intruder test, it is also necessary to evaluate the duration of the affiliative behavior, such as olfactory investigation, anogenital sniffing, and side-by-side contact [19] (Table 1). The test must be ended if the experimenter considers that the prairie voles can be seriously injured.

Some researchers analyze each of the aggressive and affiliative behaviors individually [19]. However, also it is common to analyze them as a group. For example, the frequency of aggressive interactions includes the number of lunges, bites, chases, and offensive rears, and the frequency of affiliative behaviors grouping olfactory investigation, anogenital sniff, and side-by-side contact [36, 41].

4.3 Statistical Analysis

Determine if data have a normal distribution, by running normality and equal variance test. If data are normally distributed, and two groups are analyzed (e.g., control and experimental), each parameter will be analyzed with a Student *t*-test. If data are not normally distributed, then a Mann–Whitney test will be used.

Table 1
Ethogram of the aggressive and affiliative behaviors in the resident-intruder test

	Definition
Aggressive interactions	
Lunges	Sudden thrust or pass
Pushing	The attacker applies force to move the intruder
Chasing	The resident runs after the intruder
Offensive rears	The defender rears up onto its hind legs and faces the opponent
Biting	Lunging at the opponent with the mouth open and can be classified according to the targeted body area (head, back, flanks, and rump)
Jumping	The attacker leaps at the defender with it four limbs breaking contact with the ground
Supine posture	The resident lies on its back and faces the attacker
Affiliative behaviors	
Olfactory investigation	The resident's nose touches the intruder's body
Anogenital sniffing	Active olfactory investigation takes place beneath the tail
Side-by-side contact	The two animals are directly next to each other

Behavioral definitions were taken from Winslow et al. [19] and Pellis et al. [42]

4.4 Notes

1. Resendez and coworkers demonstrated that males attack more frequently than females [36], and this behavior persists for at least 2 weeks after mating and endures in the absence of constant exposure to a sexual partner [16].
2. In male voles, aggression also depends on the physiological state of their sexual partner. Males are more aggressive if their sexual partner is pregnant. The gestational state of the females does not influence their aggressivity [43].
3. To easily identify the intruder, a small (3 × 3 cm) patch of fur on the back should be shaved. This will be done 1–2 weeks before the test to avoid stress.

5 The Zebra Finch

5.1 Introduction

The Zebra Finch (*Taeniopygia guttata*) is a passerine bird of the estrildid family naturally inhabiting Central Australia that has become increasingly relevant as a model organism for animal behavior. Easily kept and bred in captivity, the zebra finch has been studied in several fields that range sexual selection, pair bonding, vocal signaling, song learning, genomics, and genetics [44]. Like many bird species, Zebra finches are colonial, sexually dimorphic, and socially monogamous, and both males and females tend to care for offspring [45]. Hence, Zebra finches form pair bonds, a mating relationship as much as an affiliative social relationship that is frequently formed at the onset of reproductive maturity; the pair remains in close physical proximity all year even when no reproduction is occurring and lasts even until death. These social monogamous pairs show a characteristic behavioral repertoire of close physical contacts, such as clumping (perching with side-to-side contact), mutual preening (allopreening), and nest sharing [46]. Once pair bonded, females behaviorally discriminate between mate and stranger calls [47] and become aggressive toward new males, avoiding close contact with them [48]. Previous work describing and implementing protocols for the study of sociosexual behaviors in this finch species have focused on the mechanisms involving mate choices and pair bonding, such as its neuroendocrine regulation [49, 50], song learning and perception [47, 51], and the selection of phenotypic and genetic attributes [52].

Zebra finches kept and raised in laboratory conditions are generally maintained in mixed-sex aviaries (communal flight cages) made of wire of a minimum mesh of 12.7 × 15.2 mm, at temperatures between 20 and 25 °C, 40% of humidity, and on a day-night cycle ranging from 12:12 to 10:14 h [53] (*see* **Note 1**). The cage should be provided with ad libitum water, grit, and seeds; their diet sometimes enriched with cuttlebone and other kinds of fresh food (leafy greens and cooked egg) (*see* **Note 2**). The cage floor should be lined with bedding that must be changed every 3 days, while water must be changed daily and contained in water bottles sterilized at least twice a week [53]. When reaching adulthood, males and females are usually separated into unisex aviaries with acoustic and visual isolation, which is particularly important for studies involving pair bonding and sexual selection [54].

5.1.1 Group Aviary Test

In the zebra finch, the evaluation of pair bonding has been relatively straightforward. Pairing has been conceptually divided into three stages: a brief courtship phase, a short pair formation phase, and a pair maintenance phase of indefinite duration. Typically, pair bond formation can take up to 2 weeks, but studies have shown it happens in the order of hours to days [55, 56]. The process of pair bonding in these birds starts within the first minutes of introduction to potential mates, in which relevant affiliative behaviors, such as directed song and allopreening (preening or grooming the skin or feathers of another bird, Merriam Webster), are carefully quantified. The consistency of such behaviors during cohabitation is compared to evaluate pair bond formation, which is considered completely established 2 weeks after initial pairing [45]. Earlier reports focusing on mate choice paradigms used what is usually called forced pairing, in which the experimenter, randomly, or purposely, chooses which male or female should be paired. In this protocol, a sexually naïve male and female bird were placed in a cage and introduced to each other, and affiliative behaviors were scored to evaluate pair bond formation. However, researchers have favored a more natural method of pair bonding in which birds are allowed to choose a partner from an assorted few in a mixed-sex aviary [57]. The protocol described below will focus on the group aviary test [50] since the latest research on pair bonding in zebra finches has employed this simple method or similar variants.

5.2 Materials

5.2.1 Subjects

Subjects must be male and female naïve (no prior sexual or pairing experience), adult zebra finches (4 months old or older), previously isolated from visual contact from the opposite sex, that is, housed in same-sex aviaries, at least a week before any trial is ran (*see* **Note 3**).

As mentioned earlier, sexual selection is present in the zebra finch and its mechanisms have been widely studied, thus control or consideration of variables related to mate choice is important for the interpretation of social bonding studies. Both sexes are involved in partner choice [56] and show phenotypic bias toward characteristics that denote overall health conditions such as weight, size, and beak color [58]. Also, mate preference in female zebra finches has shown that song rate is an important trait for male choice [51]. Male song rate has been reported to be heritable and to reflect body health (*see* **Note 4**). Therefore, it is highly recommended to use healthy birds of similar physical features, breed, or plumage pattern (e.g., wild-type, white, etc.), to avoid additional confounds related to mate choice. If subjects are to be physically labeled, plastic leg bands that avoid bias in attractiveness or uniquely marked metal rings are also suggested (do not use black, red, or dark green leg bands or tags, but *see* ref. [59]).

5.2.2 Apparatus

Tools for Test Setup and Analysis

The apparatus consists of a large communal cage with the capability of housing at least ten birds (e.g., of dimensions 180 × 180 × 100 cm), placed in a sound-proof room where the animals will not be frequently disturbed for the duration of the trial. The cage should be provided with all the amenities mentioned previously for zebra finch housing (*see* **Note 5**). A scale is also necessary for a bird weighing before the test. Smaller individual cages with proper covering (i.e., specialized cage covers or clean blankets) are needed for bird transportation into the testing cage (35.6 × 35.6 × 45.7 cm) and testing room. A transport cart is recommended for handling several cages (Fig. 4).

The trial has a duration of 5 days, from which daily 15-min intervals are randomly chosen in selected time ranges for experimental observation. This real-time manual quantification of behaviors by trained observers is recommended behind a one-way viewing screen to avoid disturbing the birds (*see* **Note 6**), alongside

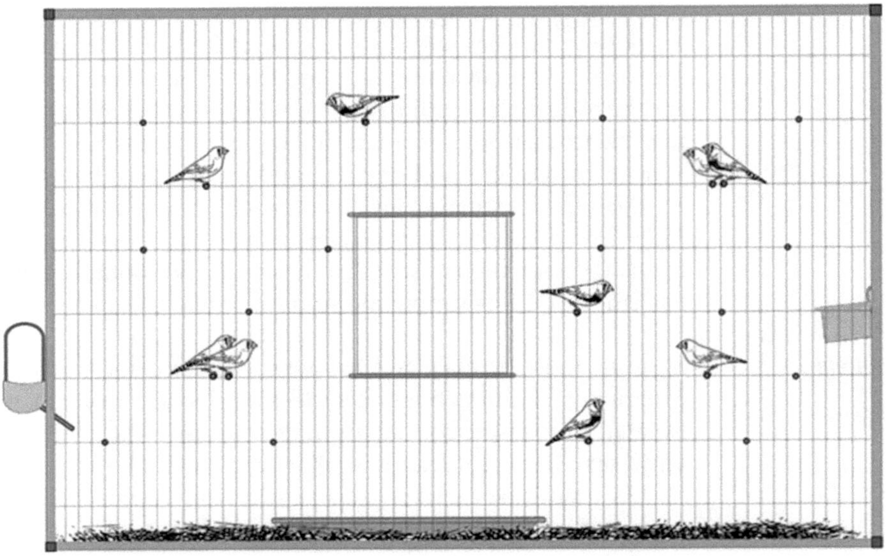

Fig. 4 Example schematic of a communal cage for the zebra finch Group aviary test protocol for pair bond formation and evaluation. Perches should be placed at different heights and distances to enable interaction. The cage should provide ad libitum water, grit, and seeds

video recording of each of the observation intervals (*see* **Note 7**). Manual scoring should consider inter-rater reliability between multiple experimenters. A premade check sheet for behavioral scoring for each observer is quite useful for quick and comparable quantifications along with a stopwatch or timer for tracking the time of each observation interval (*see* **Note 8**).

5.3 Methods

5.3.1 Preparation

Since the test cage will house mixed-sex birds, it is suggested for flocks in each test to be composed of biased sex ratios (e.g., five males with four females or vice versa), which has been reported to facilitate pair bonding [60] (*see* **Note 9**). Finches selected for the test should be previously identified by their ID labels and match as much as possible in relevant phenotypic characteristics and registered accordingly, especially in weight and plumage (*see* **Note 10**). To minimize stress in handling the birds and allow proper display of social behavior, it is recommended for each bird taken into the testing room to be allowed to enter the test cage on their own by joining the transport cage to the communal test cage momentarily.

5.3.2 Behavioral Scoring

Several sociosexual behaviors have been extensively characterized in the zebra finch during pair introduction, courting, mating, pair bonding, breeding, and pair bond maintenance (*see* **Note 11**). While diverse social bonding stages and behavior expressions may be recorded during the trial, several studies have reported clumping as one of the most relevant behaviors for pair bond assessment.

Clumping can be defined as two birds perching with physical contact and facing the same direction [48, 56, 61, 62]. Thus, the percentage of time spent clumping is the behavior assessed in this method, considering its common occurrence and the relative easiness in its visual detection by observers.

Premade check sheets for each observer are prepared to facilitate scoring according to the number of subjects used in the study and the number of intervals chosen. Zebra finches are mostly diurnal, and observations may be restricted to a specific time range (e.g., from 800 to 1900). Each subject is to be observed for 15 min intervals per day for each of the 5 days of the trial, recording the number and duration of clumping bouts with each particular bird in that period and accurately registering their identities (e.g., on day 1, female-3 clumped 3 min with male-1, and so on). The observation order of subjects should be preferably randomized in each interval. Since clumping necessarily involves two birds, it is always a directed behavior, though video playbacks of recorded intervals are useful to verify bird identity and behavior occurrence.

5.3.3 Trial

Before the start of the trial, it is advisable to check that all electronic or camera setups are working properly (batteries, available memory storage, etc.). The test cage must be properly cleaned before the trial (*see* **Note 12**). Every subject selected for the test should be identified, caught from its home cage, weighed, and taken into a smaller transport cage that should be covered during the process to avoid unnecessary stimuli and stress (*see* **Note 13**). Each bird is then taken into the testing room and allowed to enter the test cage on its own. Once all birds are inside the cage, recording of the first interval is begun, and the occurrence and duration of clumping should be scored in real-time with aid of the premade check sheet. The first minutes of the trial comprise the introduction stage and are particularly active because males attack each other, establishing dominance between them, and the first directed songs toward females occur [56]. Video recording is stopped each time an observation interval ends, and daily random observation intervals are then selected in the consecutive days up to the fifth day of the trial. As the trial elapses, the aggressiveness between males tends to diminish and male–female pair bonding will be visually evident due to the display of selective affiliative behaviors, in which an increased incidence of clumping with a specific bird is observed; the bond typically established by the fourth day of the test (*see* **Note 14**). Once the last observation interval of the fifth day has passed, each bird is removed from the testing room and housed with their mates (*see* **Note 15**), ensuring their IDs are properly matched (*see* **Note 16**).

5.3.4 Data Analysis

Pair bonding is assessed by the amount of expression of clumping with each finch of the opposite sex, expressed in percentage. Since pair bonding implies mutual preference, the index may be calculated only for females, for males, or both, depending on the objective of the experiment.

In this method, female preference will be evaluated by obtaining the clumping percentage per day of the test:

$$\text{Clumping preference female 1 day 1} = \frac{\text{Time clumping with male 1}}{\begin{array}{c}\text{Time with male 1} + \text{time with male 2} + \text{time with male 3}\\ +\text{time with male 4} + \text{time with male 5}\end{array}} \times 100$$

The formula considers the amount of clumping with all the opposite-sex finches and the proportion of such contact during each daily interval. Different clumping bouts registered in the same day interval with a specific male are summed to obtain a total clumping percentage per male per day. The preference may range from 0% to 100%, being 0 a null preference for a particular bird, an index of 50% an equal preference, and 100% being an absolute preference for a male finch. When clumping preference is greater than 50%, a pair bond toward a finch may be inferred, while less than 50% is considered unpaired. This index is calculated for each subject and for each testing day. Considering proportion estimates may not be approximately normally distributed and sample sizes may be small, an arcsine-based transformation or similar alternatives may be useful for data normalization [61]. To compare indexes across days, paired-sample *t*-tests are recommended.

In conclusion, the group aviary test is useful to compare different treatments and their influence on pair bond formation, in which hormones [50, 61], neurotransmitters [46], and song [51], have all been found to play a role in the behavior of this particularly fascinating bird, and much more is yet to be understood. This may be a basic guide in how to study pair bonding in the zebra finch. All experimental procedures must be carried out in accordance with local guidelines and protocols for the use of laboratory animals. Typical data from the test are provided in Fig. 2.

5.4 Notes

1. Suggested minimum cage dimensions for a communal flight cage is of 244 × 122 × 213 cm, while for a single breeding pair is of 35.6 × 35.6 × 45.7 cm [53]. Placing perches at different heights and positions is highly recommended, avoiding placing them over food or water.
2. Placing natural fibers on the cage floor (grass or bark) and a shallow dish of water for bathing are highly recommended for environmental enrichment [53].

3. To identify male from female zebra finches, several criteria may be used. First, only male birds sing. Second, wild-type (plumage) females have a tawny-gray color that fades into a white breast. Wild-type young males have black and white stripes along the throat and chestnut markings on the flank and develop an orange cheek patch approximately 35 days after hatching [45].
4. While it has been reported that the ideal housing for the zebra finch is in mixed-sex aviaries, overall health in male juveniles as well as interaction with other adult males will influence song learning and pattern, a variable which should be considered for certain experiments.
5. While not strictly necessary, many researchers allow pair-bonded finches to breed during these types of studies. For nesting material, hay or coconut husks are suggested together with nest boxes; eggs are laid once per day over 2–7 days and require a 14–15-day incubation period [53]. Note that nest construction might occlude the assessment of behaviors happening inside it. If breeding is not to be allowed, do not provide nest boxes or nesting material.
6. If different treatments and their effects over pair bonding are going to be evaluated in a study, a blind-to-treatment approach is highly recommended for experimenters when scoring behavior.
7. Video recording trial will allow verification of certain behaviors or subsequent analyses. The entirety of the test may be recorded, but consider that the memory capacity of the device used has to be much greater. The camera is usually mounted for a front view of the cage covering the entire interaction zone. Some researchers have found easier scoring by dictating observations vocally into a digital recorder and transcribing the data for analyses.
8. Several reports have shown that song rate is important for mate preference in zebra finches [60]. Though it is not strictly essential for the evaluation of pair bond formation, scoring or recording song in males may be useful in studying mechanisms of pair bond maintenance and mate selection [56].
9. Fewer birds may be used in each trial (e.g., 2–3 male/female ratio or vice versa) depending on the goal of the study or the available cage size. However, a larger flock will behave closer to what is seen in natural settings.
10. Finches should be weighed since heavier males usually win fights, become dominant, and may bond sooner [56], and if not registered might produce a confound when analyzing the data. A kitchen scale is sufficient for zebra finch weighing. Birds may be placed headfirst inside an empty toilet paper tube and

turned upside down on the scale to be weighed. Placing them inside a medium-sized plastic that is briefly covered (with a hand or lid) during the time of weighing is another alternative.

11. Other pairing behaviors commonly observed and scored are male-directed singing, allopreening, coordinated preening, time spent together on nest (co-nesting), mate guarding, spatial proximity, nest-building, tail quivers, male aggressiveness, copulations, beak wiping, and among others [45]. Kruskal–Wallis tests may be adequate in comparing more than one pairing behavior, using Mann–Whitney *U* tests as follow-up, and treatment effects on pairing status can be analyzed using chi-square analyses. For more details, *see* refs. [51, 61].
12. Cage maintenance, such as cleaning or food and water replacement, should preferably be scheduled to not interfere during experimental observations.
13. For catching birds inside a cage, a low-stress method is turning off the room lights and using a cloth bag to catch the bird, using a flashlight to aid in illumination. The *bander's grip* is adequate for handling finches, which can be done by holding the bird's back against the palm of the experimenter and placing the bird's head between the index and the middle fingers. Transporting subjects to the testing room should be done following the same method for each bird and should take a similar amount of time to avoid confounds related to stress due to handling.
14. At this point, the pair bond will be established for most birds, but since the flock in the test has a sex-biased ratio, at least one finch will not pair bond. Zebra finches will select a pair bond rather than remain single [45]; thus, birds that have no clear evidence of bonding when the test ends can be paired in subsequent trials (either by another group aviary test or by forced pairing). However, these birds, while they should be paired with unfamiliar birds, will have prior sociosexual experience, which should be taken into account when analyzing and interpreting the data. Another variant is testing finches with an equal male–female ratio, so all birds have a mate available for pair bonding.
15. It should be considered that if a pair-bonded zebra finch male–female dyad is separated (for at least 5 min), behaviors associated with greeting and pair maintenance will manifest significantly upon reunion [48]. To avoid separation-related stress, male–female dyads may also be transported in the same cage back to their housing.
16. Pairs may be subsequently observed in their home aviaries for pair-bond maintenance assessment and breeding and may be housed separately as a male–female dyad or in mixed-sex aviaries depending on the objective of the experiment.

Acknowledgments

We thank Jessica Norris for her editing. This work was supported by grants UNAM-DGAPA-PAPIIT IN-212219 to Sarael Alcauter and IN-208221 to Wendy Portillo, CF-2023-G-206 to Wendy Portillo. María Fernanda López-Gutiérrez is a doctoral student from the Programa de Doctorado en Ciencias Biomédicas, Universidad Nacional Autónoma de México (UNAM), and has received CONACyT fellowship (2020-00002602NACF-17340, cvu600922).

References

1. Carter CS, DeVries AC, Getz LL (1995) Physiological substrates of mammalian monogamy: the prairie vole model. Neurosci Biobehav Rev 19(2):303–314
2. Mock DW, Fujioka M (1990) Monogamy and long-term pair bonding in vertebrates. Trends Ecol Evol 5(2):39–43
3. Carter CS, Perkeybile AM (2018) The monogamy paradox: what do love and sex have to do with it? Front Ecol Evol 6:202
4. Numan M, Young LJ (2016) Neural mechanisms of mother-infant bonding and pair bonding: similarities, differences, and broader implications. Horm Behav 77:98–112
5. Ulloa M et al (2018) Mating and social exposure induces an opioid-dependent conditioned place preference in male but not in female prairie voles (Microtus ochrogaster). Horm Behav 97:47–55
6. Numan M (2014) Neurobiology of social behavior: toward an understanding of the prosocial and antisocial brain. Elsevier, London
7. Ahern TH, Young LJ (2009) The impact of early life family structure on adult social attachment, alloparental behavior, and the neuropeptide systems regulating affiliative behaviors in the monogamous prairie vole (microtus ochrogaster). Front Behav Neurosci 3:17
8. Beery AK et al (2018) Specificity in sociality: mice and prairie voles exhibit different patterns of peer affiliation. Front Behav Neurosci 12:50
9. Salo AL, Shapiro LE, Dewsbury DA (1993) Affiliative behavior in different species of voles (Microtus). Psychol Rep 72(1):316–318
10. Williams JR, Catania KC, Carter CS (1992) Development of partner preferences in female prairie voles (Microtus ochrogaster): the role of social and sexual experience. Horm Behav 26(3):339–349
11. Mateo JM et al (1994) Sexual maturation in male prairie voles: effects of the social environment. Physiol Behav 56(2):299–304
12. Zheng DJ et al (2013) Social recognition is context dependent in single male prairie voles. Anim Behav 86(5):1085
13. Yamanaka O, Takeuchi R (2018) UMATracker: an intuitive image-based tracking platform. J Exp Biol 221(Pt 16):jeb182469
14. Amadei EA, Johnson ZV, Kwon YJ, Shpiner AC, Saravanan V, Mays WD, Ryan SJ, Walum H, Rainnie DG, Young LJ, Liu RC (2017) Dynamic corticostriatal activity biases social bond formation in female prairie voles. Nature 546:297
15. Lopez-Gutierrez MF et al (2021) Brain functional networks associated with social bonding in monogamous voles. eLife 10:e55081
16. Insel TR, Preston S, Winslow JT (1995) Mating in the monogamous male: behavioral consequences. Physiol Behav 57(4):615–627
17. Ophir AG, Wolff JO, Phelps SM (2008) Variation in neural V1aR predicts sexual fidelity and space use among male prairie voles in seminatural settings. Proc Natl Acad Sci U S A 105(4):1249–1254
18. Valera-Marin G et al (2021) Raised without a father: monoparental care effects over development, sexual behavior, sexual reward, and pair bonding in prairie voles. Behav Brain Res 408: 113264
19. Winslow JT et al (1993) A role for central vasopressin in pair bonding in monogamous prairie voles. Nature 365(6446):545–548
20. Bardo MT, Bevins RA (2000) Conditioned place preference: what does it add to our preclinical understanding of drug reward? Psychopharmacology 153(1):31–43
21. Panksepp JB, Lahvis GP (2007) Social reward among juvenile mice. Genes Brain Behav 6(7): 661–671
22. Mattson BJ et al (2001) Comparison of two positive reinforcing stimuli: pups and cocaine throughout the postpartum period. Behav Neurosci 115(3):683–694

23. Goodwin NL et al (2019) Comparative role of reward in long-term peer and mate relationships in voles. Horm Behav 111:70–77
24. Tzschentke TM (2007) Measuring reward with the conditioned place preference (CPP) paradigm: update of the last decade. Addict Biol 12(3–4):227–462
25. Tzschentke TM (1998) Measuring reward with the conditioned place preference paradigm: a comprehensive review of drug effects, recent progress and new issues. Prog Neurobiol 56(6):613–672
26. Prus AJ, James JR, Rosecrans JA (2009) Conditioned place preference. In: Buccafusco JJ (ed) Methods of behavior analysis in neuroscience. CRC Press, Boca Raton
27. Camacho FJ, Garcia-Horsman P, Paredes RG (2009) Hormonal and testing conditions for the induction of conditioned place preference by paced mating. Horm Behav 56(4):410–415
28. Calcagnetti DJ, Schechter MD (1994) Nicotine place preference using the biased method of conditioning. Prog Neuro-Psychopharmacol Biol Psychiatry 18(5): 925–933
29. Roberts RL, Cushing BS, Carter CS (1998) Intraspecific variation in the induction of female sexual receptivity in prairie voles. Physiol Behav 64(2):209–212
30. Dluzen DE, Carter CS (1979) Ovarian hormones regulating sexual and social behaviors in female prairie voles, *Microtus ochrogaster*. Physiol Behav 23(3):597–600
31. Heimovics SA, Ferris JK, Soma KK (2015) Non-invasive administration of 17beta-estradiol rapidly increases aggressive behavior in non-breeding, but not breeding, male song sparrows. Horm Behav 69:31–38
32. Kowalczyk AS, Davila RF, Trainor BC (2018) Effects of social defeat on paternal behavior and pair bonding behavior in male California mice (Peromyscus californicus). Horm Behav 98: 88–95
33. Rieger NS, Marler CA (2018) The function of ultrasonic vocalizations during territorial defence by pair-bonded male and female California mice. Anim Behav 135:97–108
34. Gobrogge KL, Wang ZW (2011) Genetics of aggression in voles. Adv Genet 75:121–150
35. Wang Z, Hulihan TJ, Insel TR (1997) Sexual and social experience is associated with different patterns of behavior and neural activation in male prairie voles. Brain Res 767(2): 321–332
36. Resendez SL et al (2012) kappa-Opioid receptors within the nucleus accumbens shell mediate pair bond maintenance. J Neurosci 32(20):6771–6784
37. Eisenberger NI (2012) The neural bases of social pain: evidence for shared representations with physical pain. Psychosom Med 74(2): 126–135
38. Gobrogge KL et al (2007) Anterior hypothalamic neural activation and neurochemical associations with aggression in pair-bonded male prairie voles. J Comp Neurol 502(6): 1109–1122
39. Gobrogge KL et al (2009) Anterior hypothalamic vasopressin regulates pair-bonding and drug-induced aggression in a monogamous rodent. Proc Natl Acad Sci U S A 106(45): 19144–19149
40. Aragona BJ et al (2006) Nucleus accumbens dopamine differentially mediates the formation and maintenance of monogamous pair bonds. Nat Neurosci 9(1):133–139
41. Walcott AT, Ryabinin AE (2017) Alcohol's effects on pair-bond maintenance in male prairie voles. Front Psych 8:226
42. Pellis SM, Pellis VC (1988) Play-fighting in the Syrian golden hamster Mesocricetus auratus Waterhouse, and its relationship to serious fighting during postweaning development. Dev Psychobiol 21(4):323–337
43. Resendez SL et al (2016) Dopamine and opioid systems interact within the nucleus accumbens to maintain monogamous pair bonds. eLife 5:e15325
44. Swaddle JP (2010) Zebra finches. In Encyclopedia of Animal Behavior, edited by Michael D. Breed and Janice Moore, 629–632. Oxford: Academic Press. https://doi.org/10.1016/B978-0-08-045337-8.00048-6
45. Zann RA (1996) The zebra finch: a synthesis of field and laboratory studies. Oxford University Press
46. Adkins-Regan E (2011) Neuroendocrine contributions to sexual partner preference in birds. Front Neuroendocrinol 32(2):155–163
47. Vignal C, Mathevon N, Mottin S (2008) Mate recognition by female zebra finch: analysis of individuality in male call and first investigations on female decoding process. Behav Process 77(2):191–198
48. Silcox AP, Evans SM (1982) Factors affecting the formation and maintenance of pair bonds in the zebra finch, Taeniopygia guttata. Anim Behav 30:1237–1243
49. Mansukhani V, Adkins-Regan E, Yang S (1996) Sexual partner preference in female zebra finches: the role of early hormones and social environment. Horm Behav 30(4): 506–513

50. Adkins-Regan E (1999) Testosterone increases singing and aggression but not male-typical sexual partner preference in early estrogen treated female zebra finches. Horm Behav 35(1): 63–70
51. Tomaszycki ML, Adkins-Regan E (2005) Experimental alteration of male song quality and output affects female mate choice and pair bond formation in zebra finches. Anim Behav 70:785–794
52. Houtman AM (1992) Female zebra finches choose extra-pair copulations with genetically attractive males. Proc R Soc Lond B 249:3–6
53. Olson CR, Wirthlin M, Lovell PV, Mello CV (2014) Proper care, husbandry, and breeding guidelines for the zebra finch, Taeniopygia guttata. Cold Spring Harb Protoc 12:1243–1248
54. Svec LA, Licht KM, Wade J (2009) Pair bonding in the female zebra finch: a potential role for the nucleus taeniae. Neuroscience 160(2): 275–283
55. Prior NH et al (2020) Monogamy in a moment: how do brief social interactions change over time in pair-bonded zebra finches (Taeniopygia guttata)? Integr Org Biol 2(1): obaa034
56. Ikebuchi M, Okanoya K (2006) Growth of pair bonding in zebra finches: physical and social factors. Ornithol Sci 5:65–75
57. Griffith SC (2017) Variation in reproductive success across captive populations: methodological differences, potential biases and opportunities. Ethology 123:1–29
58. Wynn SE, Price T (1993) Male and female choice in zebra finches. Auk 110:635–638
59. Burley N, Krantzberg G, Radman P (1982) Influence of colour-banding on the conspecific preferences of zebra finches. Anim Behav 30: 444–455
60. D'Amelio PB, Trost L, Ter Maat A (2017) Vocal exchanges during pair formation and maintenance in the zebra finch (Taeniopygia guttata). Front Zool 14:1–12
61. Smiley KO, Vahaba DM, Tomaszycki ML (2012) Behavioral effects of progesterone on pair bonding and partner preference in the female zebra finch (Taeniopygia guttata). Behav Process 90(2):210–216
62. Prior NH et al (2016) Sex steroid profiles and pair-maintenance behavior of captive wild-caught zebra finches (Taeniopygia guttata). J Comp Physiol A Neuroethol Sens Neural Behav Physiol 202(1):35–44

Chapter 3

Basic Protocols for the Study of Maternal Behavior in Rabbits

Mario Caba, Margarita Martínez-Gómez, Cecilia Hernández Bonilla, Kurt L. Hoffman, and Angel I. Melo

Abstract

Maternal behavior in rabbits is unusual among mammals. During pregnancy, the doe builds a nest and after parturition the only maternal behavior displayed is crouching over the litter to nurse the pups. More remarkably is that the nursing visit of the mother occurs just once a day with a periodicity of slightly less than 24 h. This visit lasts less than 5 min and during that time the doe does not lick or retrieve her pups. Immediately after nursing, the mother jumps out of the nest and does not return for another 24 h. In this chapter, we will briefly describe the different components of maternal behavior in rabbits and some basic protocols for studying both nest-building behavior and nursing under laboratory conditions. We describe protocols comprising the following: (1) nest-building behaviors and systemic and central administration of hormones to experimentally induce them; (2) two different protocols for recording and analyzing maternal behavior during parturition and during the postpartum period; and (3) induction of maternal responsiveness by anosmia, induced by lesioning the accessory olfactory bulb and intranasal infusion of zinc sulfate.

Key words Digging, Straw carrying, Nursing, Olfaction, Maternal responsiveness, Nest building

1 Introduction

As for all altricial mammals, the objective of maternal behavior (MB) in rabbits is to ensure the pup's growth, development, and survival by providing them with nutrients, protection, warmth, and social stimuli during the postnatal preweaning period. In contrast to most mammals, rabbits and other lagomorphs (hares, pikas) show very minimal maternal responsiveness: they do not transport or retrieve their pups, they do not lick their pups, and the time spent in close contact with them is minimal. Moreover, during the prepartum nest-building stage, the pregnant rabbit suffers hair loosening, allowing her to pluck tufts of her own hair from her body, which she uses to line the maternal nest. But the most striking and important characteristic that distinguishes rabbit MB from that

Raúl G. Paredes et al. (eds.), *Animal Models of Reproductive Behavior*, Neuromethods, vol. 200,
https://doi.org/10.1007/978-1-0716-3234-5_3,

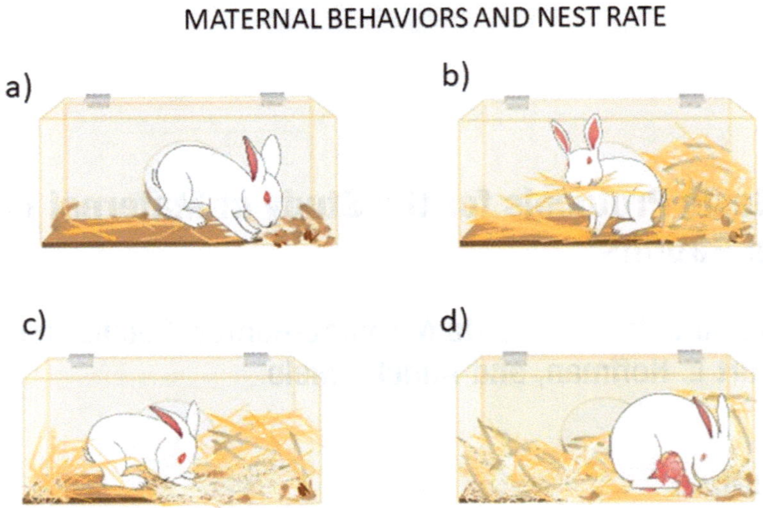

Fig. 1 Schematic representation of the different compounds of maternal behavior in rabbit and nest rate. (**a**) Digging behavior/Nest = 1. (**b**) Straw carrying behavior/Nest = 2 or "Straw-Nest." (**c**) Hairpulling behavior/Nest = 3 or "Good maternal nest." (**d**) Nursing behavior/Nest = 4 or "Excellent maternal nest"

of other mammals is the unusual strategy of nursing: in contrast to rodents, canines, and felines that nurse the pups many times through the day, rabbits nurse their litter only once per day and for only a few minutes [1–4].

Under natural, farm, or laboratory conditions, prepartum nest building is the first MB displayed by the rabbit. Nest building in the rabbit involves three behavioral components: digging a burrow, straw carrying, and hair-plucking (Fig. 1). In the laboratory, increased digging is observed across the first 3 weeks of pregnancy (gestation is 30–32 days) and can be observed experimentally with the aid of a sheet of compressed cardboard placed inside the maternal nest box (Fig. 1a). In nature, the pregnant rabbit constructs an underground burrow during this time. During the 2–4 days prior to parturition, the pregnant rabbit collects nest material, such as dry grass, straw, hay, or other similar material in her mouth and carries it back to the nest site: a behavior known as straw carrying (Fig. 1b). There is also a notable decline in food intake across this period. Finally, when parturition is imminent, she displays hairpulling (Fig. 1c), in which she plucks out hair from her ventrum and thighs, mixes it with the collected nest material, and uses it to line the completed nest [4–8].

The temporal expression of each behavioral component of nest-building behavior during pregnancy coincides with specific circulating hormonal patterns. Thus, the expression of digging coincides with high plasma levels of estradiol (E) and progesterone (P), straw carrying with a high concentration of E and a steady decline of P, and hairpulling (preceded by hair loosening) with low levels of P and increased levels of circulating prolactin (PRL) and testosterone (T) [8]. The expression of each of these nest-building behaviors can be reproduced experimentally under laboratory conditions in ovariectomized (Ovx) females by the systemic administration of specific combinations of these hormones (*See* Subheading 3; [2, 9–11]).

During parturition, the rabbit adopts an upright crouching posture (ventroflexion and arching back) while lowering her head and rump, and then starts licking the newborns (the only moment that a rabbit mother displays this behavior). At this time, she devours the amniotic membranes and placenta (placentophagia) [12, 13]. After parturition, the mother leaves the nest, covers the burrow or nest chamber with additional nest material to protect her offspring, and returns once per day at around the same time for a few minutes, when she enters the nest and crouches over the litter for nursing [4, 14, 15].

Due to their high reproductive rate and worldwide distribution, along with their highly organized hormonal, behavioral, physiological, and chronobiological patterns, the rabbit is an interesting and important animal model of MB. In the present chapter, we will review basic protocols for studying MB in the rabbit under laboratory conditions before, during, and after parturition. First, we describe methods for analyzing nest-building behaviors and their hormonal induction by systemic and central administration of steroid hormones (Subheadings 2, 3, and 4). Next, we describe methods for recording MB during parturition (Subheading 5). Then we explore nursing behavior: the short, once-a-day visit of the doe to the litter (Subheading 6). In addition, considering that under seminatural [6] and laboratory [4] conditions nursing occurs only once a day with a circadian periodicity slightly shorter than 24 h (thereby advancing slightly each day), we briefly describe the method used to assess this periodicity (Subheading 7). Furthermore, we review methods for inducing maternal responsiveness in virgin and Ovx female rabbits by lesioning the accessory olfactory bulbs (AOB) or the main olfactory epithelium by infusing zinc sulfate ($ZnSO_4$) and using a "sensitization" paradigm similar to that used in rats [16] (Subheading 8).

2 Protocol 1. Nest-Building Behavior

Rabbit nest-building behaviors have been studied since the 1950s, due to the correlation of the quality of the maternal nest with the survival of the offspring and the reproductive success of female rabbits [4, 14, 17–19]. In nature, the doe excavates a nest burrow, collects materials (e.g., dry grass) to form a nest, and finally she plucks hair from her ventrum and legs (in a vertical head-bobbing motion), which she mixes with the nest materials and uses to line the finished nest. Although it has been reported that under natural and seminatural conditions the construction of the nest starts during the last third of pregnancy [14, 15], González-Mariscal and collaborators (1994) found that pregnant rabbits in laboratory conditions begin to display increased digging as soon as at post-mating day 7, with this behavior reaching its maximum expression at 25–27 after mating (8–6 days prepartum). Thereafter, digging follows a gradual decline to a low level by prepartum day 2. This decline coincides with an increase in the expression of straw carrying behavior (expressed across the final 4 days of pregnancy). Straw carrying is maximally expressed at prepartum day 1 or 2, which coincides with the onset of hairpulling behavior, which persists until 3–4 days postpartum [7].

Although it has been reported that the quality of the nest improves through successive pregnancies [4, 5], there is some disagreement on this point, probably due to inconsistencies among studies with regard to specific housing conditions (e.g., laboratory versus seminatural conditions) and how nest quality was scored. For example, one group of investigators used a scale of 0–6, based on the degree of the hollow in the center of the nest, the arrangement of the nest materials, and the final nest lining with hair, as follows: absence of the nest (score 0), poor (score 1), fair (score 3), good (score 4), very good (score 5), and excellent (score 6) [4, 20]. Other researchers, who described nest building under seminatural conditions, reduced the scoring to poor, fair, good, and excellent: poor (slight depression of the ground and pups are not covered), fair (larger depression than the previous score, lined with some straw and pups covered partially), good (an open burrow or a deep hole covered), and excellent (a deep nest chamber far away from the entrance of the burrow, lined with straw, and pups covered with hair) [5]. Another study proposed a classification according to the presence of nest material mixed with hair, with score 1 being the total absence of nest material and hair, and score 6 being a lot of hair mixed with nest material; between these extremes the quantity of straw and hair showing an incremental increase [9]. Finally, other researchers made a distinction between a "straw nest," in which the nest site contains a fully formed straw nest and the rabbit is no longer collecting nest material, and a "maternal nest," in which the "straw nest" is lined with a considerable amount of hair [21–23].

Taking into consideration the above reports, here we propose four scores for registering nest quality under laboratory conditions (Fig. 1): Score 0, the absence of digging into the cardboard, and the absence of straw inside the nest box; Score 1, the compressed cardboard in the nest box is destroyed for at least a third of its size (Fig. 1a); Score 2, a "straw nest," a considerable amount of straw inside the nest box, but not taking the form of a nest, and not containing hair (Fig. 1b); Score 3, a "good" maternal nest, a completed straw nest that has a fair amount of tufts of hair scattered around inside the nest box and not mixed or only slightly mixed with the straw (Fig. 1c); and Score 4, an "excellent" maternal nest being a straw nest with a lot of hair mixed with the straw completely covering the pups (Fig. 1d).

2.1 Materials

(1) Sexually active males. (2) Pregnant rabbits of 6–12 months of age (3.5–4.5 kg body weight). (3) Individual wire mesh cages (52 cm long × 42 cm wide × 41 cm high). (4) Maternal wire mesh cages (90 cm long × 60 cm wide × 40 cm high). (5) Wooden or acrylic nest box (50 cm long × 30 cm wide × 32 cm high), with a round (24 cm diameter) or rectangular (20 cm × 22 cm) opening in one side. (6) Compressed cardboard cut to the size of the floor of the wooden or acrylic nest box. (7) Dry straw. (8) Balance that measures grams to weigh compressed cardboard and straw.

2.2 Procedure

Although all nest-building behaviors are assessed at the same time during the test procedure, we will describe each method in turn. The test begins when the doe is mated with a sexually active male rabbit and this day is considered as pregnancy day 0. We suggest recording behavioral tests on pregnancy days 7, 14, 21, 25, 27, 29, and 30 and onward until postpartum days 3–4, as was performed by González-Mariscal and collaborators [7].

Digging behavior (Fig. 2a): (1) Introduce a wooden or acrylic nest box that contains a sheet of compressed cardboard (previously weighed, and the weight recorded) into the wire mesh cage. The compressed cardboard is cut to the exact size necessary to completely cover the floor of the nest box and serves as the substrate for the rabbit's digging. (2) 24 h later, remove and weigh the cardboard that remains intact, discarding the small pieces that the rabbit had dislodged from it. The daily difference in weight is used as a quantitative measure of digging behavior.

Straw carrying behavior (Fig. 2b): (1) Introduce a wooden or acrylic box that contains a sheet of compressed cardboard into the wire mesh cage (*see* previous section). Weigh 100 g of straw and place it inside the cage, but outside of the nest box. (2) 24 h later, remove the straw inside the nest box, weigh it, and discard it. This procedure is repeated on each of the days that straw carrying is quantified (days 7, 14, 21, 25, 27, 29, 30, until postpartum days 3–4).

Hairpulling behavior (Fig. 2c): Due to the relative difficulty of observing this discretely expressed behavior, the presence of tufts of hair inside the cage and/or the nest box is considered to be an indication of its expression. Thus, every morning from pregnancy day 14 onward, the nest box and cage should be checked for the presence of tufts of hair. Hairpulling is therefore registered daily as a categorical (yes, no) event.

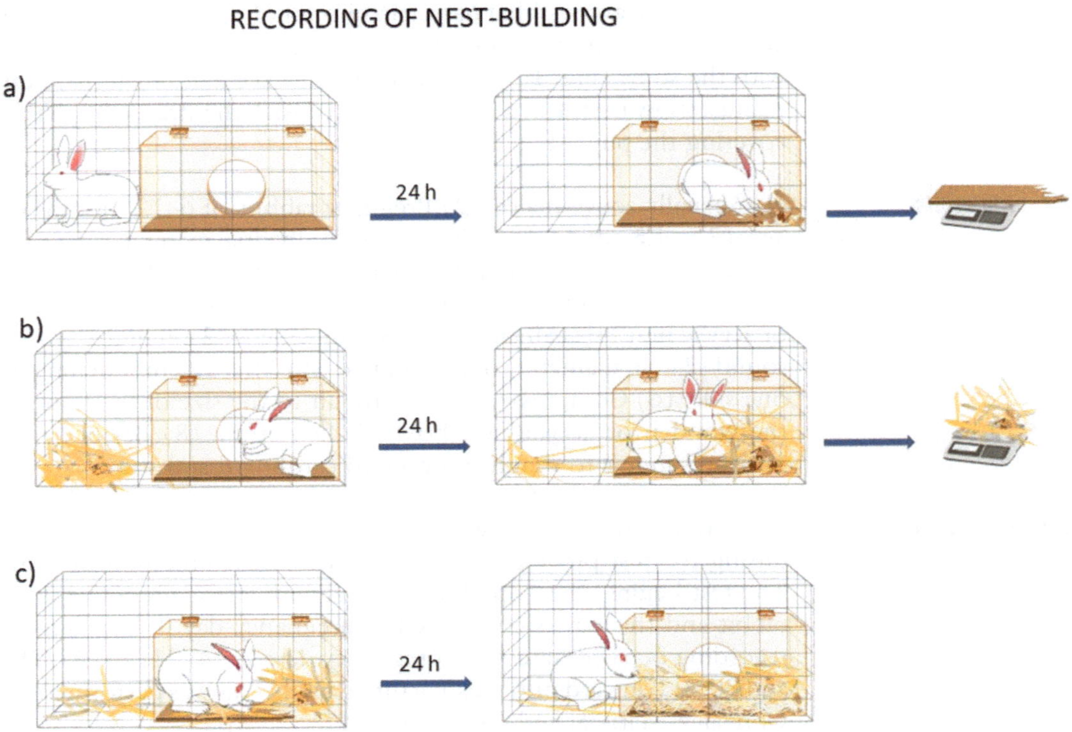

Fig. 2 Schematic representation of the recording of digging (**a**), straw carrying (**b**), and hairpulling (**c**)

An important physiological event that could be considered during the analysis of nest building is hair loosening, which occurs prior to hairpulling. This process is promoted by the decline of circulating P along with increased conversion of T to 5 alpha-dihydrotestosterone (DHT). To test the rabbit for hair loosening, the pregnant rabbit is placed on her back and the experimenter grasps a tuft of hair between the thumb and middle/index fingers, and gently pulls by flexing the wrist in a backward direction. This is done for three locations on the rabbit's ventrum: the upper chest, the middle abdomen, and the lower abdomen [2]. Ideally, this procedure would be performed on the same days in which the presence/absence of hair inside the cage/nest box is registered. Hair loosening is expressed as the percentage of does for which at least one tuft of hair was dislodged during this procedure.

Notes (1) Before weighing the cardboard for the digging test, it is necessary to allow it to air-dry to eliminate any moisture (e.g., urine) that would cause inaccurate weight measures. (2) Hairpulling and hair loosening are reported quantitatively as the percent of females that displayed each of these characteristics on each day of measurement.

3 Protocol 2. Induction of Nest Building by Systemic Administration of Hormones

Nest-building behaviors in rabbits coincide with specific changes in the levels of circulating E, P, PRL and T [23]. In Ovx, (with or without previous nest building or maternal experience), experimental hormone treatments that mimic these natural hormonal changes can reproduce the corresponding nest-building behaviors. Briefly, digging behavior begins after 4 days of treatment with estradiol benzoate (EB)+P, plateaus at day 17 (one day after the last injection of P, according to the protocol described below), and then declines. Straw carrying is observed beginning at approximately day 18 of this treatment, reaches a peak at day 20, and then decreases. Finally, hairpulling is observed beginning on days 20–21 [9] (*see* Hormonal Treatment 1).

Considering that elevated plasma levels of PRL coincide with hairpulling, a study was carried out to test whether PRL is a hormonal trigger for this behavior. Ovx, EB/P-treated female rabbits with bromocriptine (Br; a dopamine antagonist that inhibits the synthesis of PRL) displayed digging but not straw carrying or hairpulling, consistent with a role for this hormone in triggering these behaviors. In order to confirm that PRL is necessary for the expression of these behaviors, the second experiment was done in which Ovx, EB/P-treated female rabbits received Br + ovine PRL (oPRL) [9] (*see* Hormonal Treatment 2).

Furthermore, given that levels of circulating T also increase during the expression of hairpulling [7], experiments were done to test the possible role of T for the expression of these behaviors. It was observed that the combination of T with P triggered digging, and either T or 5 alpha-dihydrotestosterone propionate (DHTP) induced hair loosening in 71% and 100% of females, respectively [2] (*see* Hormonal Treatment 3). Here, we describe the hormonal protocols to induce these behaviors.

3.1 Materials

(1) Multiparous adult female rabbits (3.0–3.5 kg body weight). (2) Individual and maternal wire mesh cages. (3) Wooden or transparent acrylic nest boxes with a round or rectangular opening on one side for the construction of the nest. (4) Compressed cardboard and dry straw (*see* Subheading 2.1). (5) EB, P, T, DHT, oPRL, and Br (all hormones and drugs are obtained from Sigma, St Louis, MO, USA). (6) Vegetable oil. (7) Ketamine and rompun for ovariectomy. (8) Penicillin. (9) Balance in grams to weigh the cardboard and straw. (For details on specific dimensions of cages and nest boxes, *see* Subheading 2.1).

3.2 Procedure

Ovariectomize female rabbits under anesthesia with ketamine (25 mg/kg) and rompun (8 mg/kg) (for details on the surgical techniques for ovariectomy [24]). After surgery, 200,000 units of penicillin are administered s.c., and the animal is allowed 4 weeks to recover before beginning the experiment.

Hormonal Treatment 1 Females are transferred to a maternal wire mesh cage that contains a wooden or acrylic nest box (day 0). The nest box contains a sheet of compressed cardboard for quantifying digging, and a straw is placed inside the cage to quantify straw carrying (*see* Subheading 2). Baseline measurements of digging and straw carrying are obtained across the next 2 days (days 1–2) prior to initiating hormone treatment. Thereafter, digging and straw carrying are again measured systematically for the remainder of the experiment, according to the experimenter's specific objectives (*see* **Notes**, below). From days 3 to 21, rabbits receive 5 μg EB/day (0.2 mL from a solution of 25 μg EB/mL vegetable oil) by subcutaneous injection. From days 4 to 16, rabbits also receive 10 mg P/day (subcutaneous injection of 0.5 mL, from a solution of 20 mg P/mL vegetable oil), administered concurrently with EB. Alternatively, an independent solution of EB+P can be prepared (10 μg EB/mL + 20 mg P/mL vegetable oil); this solution is administered as a single subcutaneous injection of 0.5 mL. Control treatment would consist of concurrent injections of equivalent volumes of vegetable oil.

Hormonal Treatment 2 The same procedure is followed as described for Hormone treatment 1, except from days 11 to 21 Ovx females should also receive Br (1 or 3 mg/kg/day) + sodium carbonate buffer vehicle or Br + oPRL (3 mg/kg/day).

Hormonal Treatment 3 This protocol is identical to Hormonal Treatment 1, except that Ovx rabbits receive T or DHTP, instead of EB. Thus, from days 3 to 24, rabbits receive T (1 or 5 mg/day), DHTP (5 mg or 10 mg/day), or vegetable oil (0.5 mL/day). Rabbits that are assigned to T and DHTP groups also received P across days 2–15 (10 mg/day) [2].

Notes (1) All steroids are given s.c. and are dissolved in vegetable oil (25 μg/mL for EB; 20 mg/mL for P). (2) Br is dissolved in water: propylenglicol: methanol (6:3:1) solution. (3) oPRL is dissolved in sodium carbonate buffer 0.1 M, pH 7.6. (4) In order to avoid multiple injections, it is best to prepare a single solution of EB +P to use during the period when these two hormones are administered concurrently. (5) Here, we have described procedures as they were carried out in published studies, for which digging, straw carrying, and hairpulling were assessed on days 1 and 2 (to obtain pretreatment measurements), and on days 5, 9, 14, 17, 10, 19, 20, 21, and 22 [9] or days 4, 6, 11, 16, 18, 19, 20, 21, and 22 [2]. However, the experimenter may want to modify the schedule of test sessions depending on the specific experimental objectives, keeping in mind that digging steadily increases across the 13 days of EB+P administration, while straw carrying and hair-pulling are comparatively discretely expressed behaviors, being displayed across 3–4 days after P withdrawal. (6) Although food intake would not be considered a component of MB per se, it has been reported that this behavior decreases just before parturition, coinciding with hormonal changes that occur during this time (decrease in plasma P, increase in plasma PRL). Since this decline in food intake might serve as a useful model for physiological mechanisms that control appetite, it is recommended to assess prepartum changes in food intake in addition to maternal behaviors.

4 Protocol 3. Induction of Nest Building by Combined Central and Systemic Administration of Steroid Hormones

Studies in rodents indicate that the medial preoptic area and the adjacent bed nucleus of the stria terminalis (MPOA/BNST) are important for the neural circuit involved in the control of MB [25]. In the rabbit, both regions are rich in E receptor alpha [26], and the medial preoptic area also contains P receptors [27]. Central and peripheral administration of E and P stimulate MB in rodents, as well as in the rabbit [28]. The objective of the present protocol is to induce maternal nest building by the combined administration of intracerebral and systemic hormone administration. In this way, one can identify brain regions that are necessary and sufficient for hormone responsiveness.

In Ovx rabbits that had previously received intracerebral implants of EB directly into the MPOA/BNST, digging and straw carrying were induced after systemic administration and withdrawal of P [29]. Similarly, digging was induced by administration and withdrawal of P in Ovx does that had received EB implants into the nucleus accumbens. Specifically, this experimental manipulation induced an increase in digging during systemic P administration that peaked on day 18 (one day after the last P administration), then sharply decreased until day 23, the last day of recording. This response was observed in 72% of females implanted in the MPOA/BNST and 67% implanted in the nucleus accumbens. With regard to straw carrying, there was a slight increase on day 16, but after the last systemic injection of P on day 17, this behavior increased sharply on day 21 and subsequently decreased until day 25, the last day of recording. Straw carrying was induced in 53% of females implanted in the MPOA/BNST. No hairpulling was observed in these experiments.

4.1 Materials

(1) Multiparous adult female rabbits (3.0–3.5 kg body weight). (2) Individual and wire mesh cages and wooden or transparent acrylic box for the construction of the nest. (3) Ketamine and rompun for ovariectomy. (4) Cardboard and dry straw (*see* Subheading 2.1). (5) EB and P (Sigma St Louis, MO, USA). (6) Vegetable oil. (7) Penicillin. (8) Sodium pentobarbital to perfuse animals in order to verify the site of implant. (9) 0.9% saline. (10) Rabbit stereotaxic apparatus. (11) Stainless steel gauge 22 guide cannulae. (12) Inner stainless steel gauge 28 cannulae.

4.2 Procedure

Females are ovariectomized under anesthesia of ketamine (25 mg/kg) and rompun (8 mg/kg), and after surgery should be injected with 200,000 units of penicillin. One month later, they are anesthetized again and implanted bilaterally with stainless steel gauge 22 guide cannulae in the MPOA/BNST and nucleus accumbens, using a rabbit stereotaxic apparatus, using the following coordinates: anterior = 2, lateral = 1.5, dorsoventral = −2, with aid of a rabbit brain atlas [29, 30]. At the time of implantation of the guide cannulae, inner cannulae (gauge 28) filled with melted crystals of EB are introduced into them. After intracerebral implantation (experimental day 0), rabbits are injected s.c. with 200,000 units of penicillin and transferred briefly to a warm room for recovery from anesthesia. Across experimental days 3–17, EB-implanted females are injected s.c. with 10/mg day progesterone in 0.5 mL vegetable oil. The quantification of digging, straw carrying, and hairpulling could be performed on days 1, 2, 6, 9, 13, 16, 19, 21, 23, and 25, as described in Subheading 3. At the end of the experiment, animals are deeply anesthetized with pentobarbital and perfused transcardially (0.9% saline solution and 10% formaldehyde). The brain is removed from the skull, fixed, embedded in paraffin, and stained with cresyl violet to verify the implantation site.

Notes (1) For the implantation of cannulas with steroid hormones, the steroid crystals must be slightly melted with low heat and the melted crystals drawn into the cannula by capillary action. (2) For details of the measures of the cages and boxes *see* Subheading 2.1.

5 Protocol 4. Maternal Behavior at Parturition

The time at which parturition occurs in the rabbit, as in many species, is variable, but in contrast to many other species occurs very quickly (around 7–15 min), which makes it difficult to observe [31]. Although it is known that pregnancy lasts between 30 and

32 days [31, 32], a recent study of New Zealand white rabbits housed in laboratory conditions has shown that two-thirds of the pregnant rabbits studied gave birth on gestation day 31, regardless of litter size. From these mothers, 35.5% gave birth in the dark phase, and 64.5% in the light phase [3]. When labor begins, the pregnant rabbit often twitches her flanks or ventrum, adopts a crouching posture with her head down between her hind legs, and starts licking and ingesting the slight flow of blood fluid from the vulva. Once the first pup emerges, she assists with her muzzle, sniffs and licks the newborn's body and eats the placenta and amniotic membranes, as well as grooms her anogenital area and shows abdominal contraction [13, 31]. A mother rabbit delivers around 6–12 pups, most of them born alive but some stillbirths may also occur [12, 20]. Most pups are delivered without placenta and amniotic sac and show great mobility, gasping, and searching for a nipple and beginning suckle often while labor is still in progress. After the delivery of the last pup, the mother continues consuming the remaining membranes and placentas and licking the pups, while some reach a nipple and begin to suckle. After about 10 min, the doe leaves the maternal nest and does not return to the nest to nurse again until 24 h later [19, 33–35]. Like many species, the mother rabbit needs contact with her litter during parturition for the establishment and maintenance of MB across the postpartum period (*see* Subheading 6). However, little is known about the mechanisms of parturition and how they could influence the establishment of MB; these processes in the rabbit provide an important model for biological mechanisms by which events of parturition participate in the establishment of MB.

5.1 Materials

(1) Sexually active male rabbits. (2) Primiparous or multiparous pregnant rabbit does. (3) Maternal wire mesh cages. (4) Controlled light-dark cycle for animal housing room and food and water ad libitum (*See* Subheading 8.2.3). (5) Glass or acrylic nest box (30 cm long × 40 cm wide × 15 cm height). (6) Glass topped stand with a mirror below the cage at 45-degree angle from the floor. (7) Red light. (8) Dim white light. (9) Stopwatch. (10) Closed circuit video recording system. (11) Paper sheet to write up the latency, duration, and occurrence of behaviors; alternatively, a Behavioral Recording Software could be used (*see* Subheading 8.2.3).

5.2 Procedure

(1) Transfer a pregnant rabbit (day 29) into a maternal wire cage lacking a floor that is placed onto a transparent glass surface, which constitutes the top of the observation apparatus. This apparatus has a mirror facing upward at a 4-degree angle, mounted under the top surface of the apparatus. The mirror allows for easy viewing of the underside of the rabbit as she sits on the glass floor. (2) From gestational day 30 to parturition, the mother is continuously recorded (use a red light during the dark phase) with a video camera

or using a closed-circuit video. (3) Register behavior by hand using a datasheet and pen and paper, or with the assistance of Behavioral Recording Software. The following parameters and behaviors are considered: (a) length in days of pregnancy, (b) latency to delivering first pup, (c) duration of parturition, (d) delivery time per pup, (e) total number of live and stillborn pups, (f) number of pups born in anterior or posterior position, (g) frequency and time of vaginal retention (pups in the vulva for more than 5 s before delivery), (h) number of pups with- and without placenta and amniotic membranes, number of placentas ingested by the mother, (i) frequency and time spent licking the pups, (j) number of mothers nursing, (k) latency for the mother leave the nest and litter, after delivery of the last pup/placenta, and (l) percentage of mothers that nurse and the number of pups attached to the nipples [31].

Notes For registering behaviors using a Behavioral Recording Software, it is necessary to configure single keystrokes that code for each behavior described above; for example, "n" for nursing, "p" eat placenta, "l" for pup body licking, and "b" birth of a pup. (for details, refer to the software users' manual).

6 Protocol 5. Postpartum Maternal Behavior

In contrast to most mammals, postpartum maternal behavior of rabbits is unique because its expression is restricted to a daily (most typically at night) nursing bout that lasts 3–5 min and occurs with a circadian periodicity across at least the first 2 weeks postpartum [3, 4, 7, 34, 35].

After birth, the mother leaves the nest and the litter, covers the entrance of the burrow (if under natural or seminatural conditions), and returns approximately 24 h later to nurse her litter again. This process is repeated every day across the entire lactation. As has been demonstrated for rats and other species, the female rabbit requires two main types of factors for the establishment and maintenance of MB during the postpartum period [16]: hormonal and nonhormonal. Currently, it is known that certain hormonal changes that occur at the end of pregnancy (*see* above) and that regulate nest-building and are also involved in the development and maintenance of maternal responsiveness toward the offspring. Specifically, when plasma levels of PRL are experimentally decreased by systemic administration of Br across pregnancy days 26 to parturition (but not when administered across postpartum days 1–7), the frequencies of entrances into the nest box and of adoption of nursing posture during the first weeks of lactation are decreased; this effect is prevented if Br is administered simultaneously with intracerebroventricular infusion of PRL [36, 37].

With regard to nonhormonal factors, it had been reported that most primiparous, but not multiparous, mothers that were deprived of their offspring at parturition did not display nursing during the first 5 postpartum days [38]. Likewise, preventing the perception of sensory and social stimuli from the pups by anesthetizing the mother during daily nursing bouts across days 1–7 postpartum (but not across days 11–17 postpartum), disrupted the expression of MB in 66% of them [38]. Moreover, the litter size also impacts on the maintenance of nursing across early lactation. Thus, at lactation day 15, most of the mothers rearing six to eight pups displayed nursing, in contrast to those having less than four pups per litter [3]. These data support the proposal that PRL at the end of pregnancy, but not after parturition, is involved in the induction of and maintenance of MB, while sensory and social stimuli from the offspring at parturition and during early lactation are essential for the establishment and maintenance of MB during the postpartum period. Here, we describe methodologies used to assess MB during lactation, which could be used to investigate mechanisms by which hormones, sensory and social stimuli derived from the litter participate in the establishment and maintenance of maternal responsiveness in rabbits.

6.1 Materials

(1) Sexually experienced male rabbits. (2) Primiparous or multiparous adult female rabbits (3.0–3.5 kg body weight). (3) Individual and maternal wire mesh cages and wooden or transparent acrylic box for the construction of the nest. (4) Compressed cardboard cut to exactly the size of the floor of the wooden or acrylic box. (5) A sufficient supply of dry straw. (6) Balance in grams to weigh the cardboard, the straw, and litters. (7) Red light lamp or electric hot pad. (8) Container for holding the pups while weighing them. (9) Stopwatch. (10) Paper sheet to register the latency, duration, and occurrence of behaviors.

6.2 Procedure

Three weeks after mating with a sexually active male, (1) place the pregnant doe into a wire maternal cage that contains a wooden or transparent acrylic nest box that has a round or rectangular opening on one side. If possible, it is recommended to record all components of nest building to have a more comprehensive view of the maternal responsiveness of the experimental females (*see* Subheading 2). (2) Check each morning and evening for the presence of pups, to establish the approximate time of parturition. (3) 5–8 h after parturition, remove the pups from the nest box, and cull the litters to four to eight pups each (place the excess pups with a lactating adoptive mother). (4) House each litter in a medium-sized acrylic box and place them in a room near the vivarium under a red-light lamp or on top of on an electric pad with warm

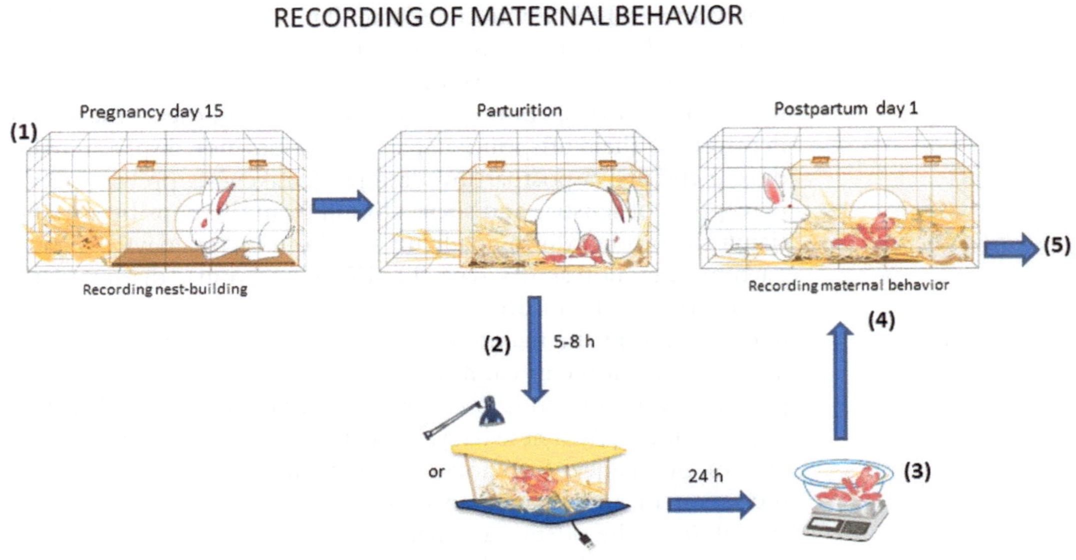

Fig. 3 Schematic representation of the recording of maternal behavior

heat (*see* Fig. 3, **step 2**). (5) 24 h later (postpartum day 1; PPD1) weigh the pups on a balance (**step 3**). (6) Place the litter into the nest box and record the following parameters and behaviors (**step 4**): (a) latency to enter the nest box, (b) latency to initiate nursing, (c) time spending suckling, (d) incidence of nursing, (e) number of false entrances into the nest box (defined as when the female enters the nest box and later exits without having nursed the pups), (f) sniffing pups, (g) occurrence of nest-building, and (h) milk output (*see* below). The test ends when the mother nurses the pups and then hops out of the nest box, or after 60 min if the mother fails to nurse the pups. An optional strategy is to do a spot-check every 30 min for at least 3 h. (7) Remove the pups from the nest box (Fig. 3, **step 5**) and weigh them (repeat **step 3**, Fig. 3) to calculate milk output: the difference in weight of the pups before and after nursing is a quantitative measure of milk output. (8) After weighing them, return the pups to their original holding container under warm heat (*see* Fig. 3, **step 2**). (9) Repeat the test 24 h later and on the following postpartum days, or according to the investigator's experimental design [7, 36–39].

7 Protocol 6. Postpartum Maternal Behavior: Free Access to the Nest and Continuous Video Recording

The two important aspects of MB in the rabbit are the circadian periodicity and the short duration of nursing. Thus, to determine the moment at which a mother rabbit spontaneously enters the nest box to nurse her offspring, several authors have utilized an experimental design that allowed the mother rabbit continuous free access to the nest box, while the duration of the light: dark cycle was varied. These studies confirmed that mother rabbits nurse spontaneously once a day (most often during the dark phase), and with a circadian periodicity [3, 34]. However, it was found that both processes depend primarily on the quantity and quality of the suckling stimulus. Thus, mothers that were rearing small litters (less than four pups), but not those that had six to eight pups, showed disturbances in the once-a-day nursing periodicity, increasing the number of daily entrances into the nest box [3]. Blocking the perception of suckling stimuli by anesthetizing the mother, decreasing the number of suckled pups, covering the nipples, or surgically removing the nipples, increased the duration of the nursing bout [4, 10, 40]. Here, we describe the method used by González-Mariscal and collaborators [3] to assess the above aspects of MB in rabbits.

7.1 Materials

(1) Sexually experienced male rabbits. (2) Pregnant New Zealand white adult rabbit does. (3) Individual and maternal wire mesh cages, and transparent acrylic box with an opening square on one side (*see* Subheading 8.2.3). (4) Controlled light-dark condition and food and water and libitum (*See* Subheading 8.2.3). (5) Video recording system connected to a computer. (6) Straw and compressed cardboard. (7) Statistical package for performing Raleigh analysis.

7.2 Procedure

(1) Introduce an acrylic nest box inside the rabbit's wire mesh maternal cage on day 15 of pregnancy. The nest box should contain a sheet of compressed cardboard covering its floor, and 100 g of straw for the construction of the nest should be placed inside the cage but outside of the nest box (*See* Subheading 2). (2) In front of the wire mesh cage, install a video camera, connected to a computer, to allow for continuous recording of activity across pregnancy day 29 to postpartum day 15. (3) On the day of parturition, the newborn pups should be removed from the nest (*See* Subheading 6) and the litter size should be adjusted to 1, 2, 4, or 6–8 pups (thereby forming four experimental groups that will allow for experimental comparisons between litters of different sizes). (4) Return the pups to the nest box and record the following parameters: (a) day and time of parturition, (b) latency to the first nursing after parturition, (c) number of times and time of the day when the mother entered the nest box, (d) occurrence of nursing, (e) duration of the nursing bout, and (f) number and duration of nonmaternal behaviors such as eating, drinking, grooming, or rearing. All these behaviors should be observed and registered across 5 min before and 5 min after nursing. (5) Play back each video during the morning following its recording (i.e., the video

recorded on pregnancy day 29 is played back on day 30), and the rabbit mother's behavior is registered on a data sheet. (6) Analyze the data using a Raleigh analysis in order to determine the vector that best describes the time of nursing [41].

8 Protocol 7. Induction of Maternal Behavior by Anosmia

In the rat, studies have shown that the olfactory bulb exerts a tonic inhibition on the expression of MB, and experimentally induced anosmia reduces the latency to show maternal behavior in virgin rats [42, 43]. Indeed, naive female rats exposed to foster pups for the first time display avoidance and rejection toward them. However, after 6–8 days of exposure to the pups, they begin to show an attraction toward them, retrieving them, and adopting a nursing posture. This process is called sensitization [16, 44]. The long latency (several days) to display maternal responsiveness is notably reduced by inducing anosmia through lesioning either the main or accessory olfactory bulb, or both [42, 43, 45]. However, in contrast to rats, nonpregnant, nonlactating rabbits do not show any indication of sensitization, even after several days of pup exposure. In order to assess the role of the main and accessory olfactory systems on the induction of maternal responsiveness in the female rabbit [46, 47] lesioned the AOB or the main olfactory epithelium (by infusion of $ZnSO_4$) of virgin or Ovx female rabbits. They found that lesioning the AOB induced maternal responsiveness (as indicated by the female rabbit entering the nest box and adopting a nursing posture over the pups), in one-third of the lesioned females within the first 3–13 consecutive days of pup exposure [46]. Similar effects were observed in 40–70% of virgin female rabbits that had been made transiently anosmic [47]. Interestingly, this same procedure did not have an effect on Ovx rabbits. In this protocol, we describe both procedures.

8.1 Lesion of the Accessory Olfactory Bulb

8.1.1 Materials

(1) Adult virgin female rabbits (3.0–3.5 kg body weight), donor rabbit mothers for providing pups for the sensitization procedure (at least three mothers for each experimental rabbit), as well as sexually active males for mating. (2) Individual and maternal wire mesh cages as well as wooden or acrylic box (for details of cage measurements and boxes *see* Subheading 8.2.3). (3) Medium polypropylene cage (44–50 × 26–28 × 16–20 cm) containing straw for housing the pups each day when they are removed from the maternal cage. (4) Ketamine and rompun. (5) Rabbit stereotaxic apparatus. (6) Dentist's drill. (7) Pasteur pipette. (8) Suction pump. (9) Penicillin. (10) Silk thread (00 gauge). (11) Sodium pentobarbital for sedating the animal prior to perfusion. (12) 0.9% saline solution. (13) Straw and compressed cardboard.

8.1.2 Procedure

Three weeks after ovariectomy, the AOB of females is surgically removed bilaterally. Females are placed on a rabbit stereotaxic apparatus (under general anesthesia), and an incision is made to the skin that covers the upper part of the nose (*see* Fig. 4). The portion of the skull that overlies the olfactory bulb is drilled and removed. The AOB can be identified as a flat, ovoid, and light-colored structure that rests on top of the main olfactory bulb. The AOB is removed bilaterally by applying a gentle suction using a pasteur pipette connected to a suction pump. After removing the AOB, close the surgical incision by suturing with silk thread (00 gauge). Transfer the rabbit to the maternal wire mesh cage with a nest box, and place straw inside the cage but outside of the nest box. After 7–10 days of the surgery, begin MB tests.

The steps to follow the test are displayed in Fig. 4. Briefly, (1) on the first day of testing (day 1) remove four pups of 1–4 days old from a donor mother (*see* Subheading 8.2.3 for details of mating and parturition, and Fig. 4, **step 1**). (2) Place them inside the nest box of the experimental doe (**step 2**). (3) Record MB for 1 h on a sheet (**step 3**) the following behavioral parameters: (a) entrances into the nest box, (b) sniffing or licking the pups, (c) aggression/cannibalism, (d) hovering/crouching posture over

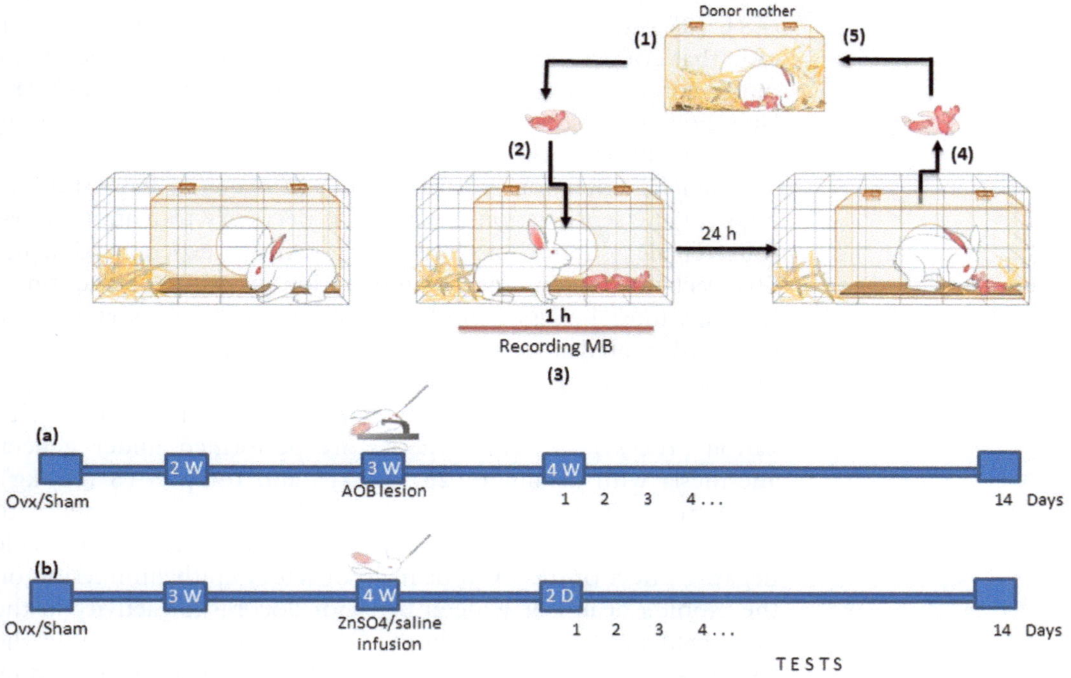

Fig. 4 Schematic representation of the protocol to recording maternal responsiveness in virgin with AOB lesion (**a**) or ZnS04 infusion (**b**) (*see* text for details). The MB tests must be repeated daily for 14 days. Olfactory tests of preference in the ZnS04 infusion experiment can be performed on days 3, 6, and 8 after injury. Ovx = ovariectomy, W = week, D = day, ZnS04 = zinc sulfate, AOB = accessory olfactory bulb

the litter, and (e) the time spent inside the nest box. (4) Allow the pups to remain inside the nest box during the next 24 h, at which time they should be removed (**step 4**) and returned to their biological mother and replaced by four newly fed pups from a donor mother (**step 5**). (5) Begin the next daily test session (repeat **steps 2–5**). (6) Repeat this procedure daily for at least the next 13 days (Fig. 4, [46]).

8.2 Lesion of the Main Olfactory Epithelium

8.2.1 Materials

(1) Adult virgin female rabbits (3.0–3.5 kg body weight) donor mother rabbits to provide pups for the sensitization procedure and sexually active males to mate them. (2) 5%$ZnSO_4$. (3) Saline solution. (4) Silk thread (0.0 gauge). (5) Polyethylene tubing (Clay Adams PE60; O.D. 1.22 mm; I.D. 0.76 mm). (6) Insulin syringe. (7) Rabbit pups of 1–4 days old from donor mothers (at least three foster mothers for each experimental female rabbit). (8) Straw and compressed cardboard.

8.2.2 Procedure

(1) Virgin female rabbits are ovariectomized and transferred to individual wire mesh cages. (2) Three weeks later, anosmia is induced by intranasal infusions of $ZnSO_4$. This procedure consists of gently holding the rabbit in a supine position and infusing into each nostril 1.0–1.5 mL of $ZnSO_4$ or saline solution. The rabbit is held in the supine position for 2 min after the infusion [48]. (3) After this, return the rabbit to its maternal wire mesh cage that contains an acrylic or wooden nest box, and place straw outside of the nest box. (4) Recordings of MB are initiated 1–2 days later. Follow the steps shown in Fig. 4 and described in Subheading 8.1 for recording MB [47].

General notes for Protocol 7: (1) In order to have stimulus pups for the sensitization process, it is necessary to mate multiparous females (donor mothers) with sexually experienced males 30 days before beginning the sensitization procedure. Since stimulus pups must be less than 4 days of age, it is necessary to have previously mated at least three donor female rabbits for each experimental subject (mating of donor females are done 4 days apart, so that pups of less than 4 days old are available for the entire sensitization process). (2) All surgeries are performed under general anesthesia with ketamine (25 mg/kg) and rompun (8 mg/kg). (3) At the end of every surgery, the rabbits must be injected with 200,000 units of penicillin. (4) In order to detect any possible negative effect of the AOB or main olfactory epithelium lesion on the rabbit's health or general behavior, locomotor activity in the open field and food intake should be assessed before and after the lesion [47, 49]. (5) In order to verify that anosmia is provoked by the lesion, it is suggested to carry out odor perception tests before and after the lesion [47–49]. (6) Two minutes after $ZnSO_4$ infusion, the snout of the rabbit should be held at a downward angle for some moments to allow the remaining solution to drain from the nostrils.

8.2.3 General Notes

(1) The chinchilla and New Zealand White rabbit have been the most frequently used breeds for the study of MB, but in theory any rabbit breed could be used, according to the specific objectives and practical concerns of the experimenter. It should be stressed, however, that breeds may differ with respect to gestation time, litter size, quality of nest construction, and other fundamental measures of MB. Therefore, if other breeds are used, the experimenter should carefully obtain normative data on the specific behavior(s) to be studied before initiating experiments. (2) All subjects used in these protocols must be maintained in controlled light/dark conditions (14 h light/10 h dark), natural temperature (15–25 °C), and given rabbit chow pellets and water ad libitum. (3) The basic criterion that defines a maternal rabbit is the display of pup nursing inside the nest box. (4) If the experimenter wishes to test the effect of a pharmacological (or other) challenge on MB, it is best to use female rabbits that have had at least one previous successful pregnancy and parturition, since there is often a significant improvement in nest quality with previous maternal experience. (5) In order to induce pregnancy, the female rabbit is mated with a sexually experienced male and is allowed to receive at least two ejaculations. (6) For all protocols that use foster pups, these pups must be housed and maintained in warm conditions, in order to assure that they are active at the time of the experiment and thereby able to provide optimal cues (e.g., suckling) for the appropriate display of MB by the mother rabbit. (7) Details on animals, housing conditions, wire mesh cages, wooden or acrylic nest box, antibiotics, water, food, and anesthetics are described in Subheading 2. (8) There are several Behavioral Recording Software that can be used for analyzing rabbit MB; examples include the following: (a) The Observer, Noldus Information Technology, Scolari, Sage Publication Software, (b) StopWatch, or (c) Behavioral Observation Research Interactive Software (BORIS), a free, versatile event-logging software [50].

9 Summary and Future Directions

The maternal rabbit expresses a number of unique behaviors that can be used to study neurobiological questions that have relevance beyond the study of maternal behavior. For example, maternal nest building has been studied in the rabbit more than in any other species and comprises three different components that are triggered by distinct hormonal cues. Such precise hormonal regulation of behavior can be used as a means to elucidate molecular mechanisms by which E and P interact at the cellular level to modulate prepartum behavior. Straw carrying behavior has been proposed as a model for elucidating the neurobiology that underlies compulsion in obsessive-compulsive disorder [51, 52]; indeed, the brain regions that are activated during straw carrying largely overlap with those that have been implicated in obsessive-compulsive symptoms [53–55].

In many species, the interaction between the mother and the offspring around the time of parturition is essential for establishing the maternal bond [56]. Since the mother rabbit expresses MB in a discrete (3–4 min duration) and predictable (once-a-day periodicity) manner, it is exceptionally accessible for experimental manipulations aimed at elucidating the sensory and neuroendocrine underpinnings of the formation and maintenance of MB. Thus, it has been shown that pup contact at parturition is necessary for the establishment of MB (expression of nursing behavior 1 day after parturition) in primiparous (but not multiparous) rabbits [38]. Moreover, MPOA and lateral septum activity are associated with maternal state, while magnocellular nuclei (supraoptic and paraventricular) activity is increased in response to nursing [28].

One particularly interesting behavior that has hardly been studied in the rabbit is placentophagia [13]. At parturition, all mother rabbits readily consume the placenta (or, in an experimental context, fresh chicken liver). In stark contrast, placenta and liver are not consumed prior to parturition, and placentophagia wanes across the first 5 postpartum days. This behavior could be used as a model for studying physiological mechanisms that control appetite and the neurobiological and neuroendocrine mechanisms that determine the salience of food-associated stimuli.

Finally, unlike rodents, once-a-day nursing in the rabbit offers unique opportunities to study how sensory and neuroendocrine factors interact for the establishment and maintenance of this circadian behavior, unusual among mammals. The extended time frame before, and then after, daily nursing has revealed that there is a specific suckling "threshold" for the maintenance of the circadian behavior [41]. Some of the behavioral and neurobiological consequences of this event have been explored revealing a remarkable array of neural structures and neuroendocrine cells (oxytocinergic and dopaminergic), synchronized to the single daily nursing bout at the central level [57–62]. However, very little is known about how peripheral hormones (e.g., glucocorticoids) influence the interaction between the sensory stimulus and circadian rhythms during lactation. At the core of the spontaneous return of the mother for nursing is the mesolimbic dopaminergic system that seems to "drive" mothers to return to the nest [61]. Thus, the rabbit model offers an extraordinary opportunity to study how a sensory stimulus contributes to the establishment of neural and neuroendocrine circuitry underlying the establishment of the maternal brain in the rabbit, a "circadian brain," which has scarcely been explored.

Thus, future studies of rabbit maternal behavior—maternal nest building, formation of maternal behavior, circadian rhythmicity of nursing, and others—should reveal neurobiological and neuroendocrine mechanisms that not only have relevance for understanding maternal behavior but also provide important insight into fundamental neurobiological questions.

Acknowledgments

The authors thank Angel Azael Melo Baza for the design and elaboration of the figures and to Laura García for her assistance.

References

1. Venge O (1963) The influence of nursing behavior and milk production on early growth in rabbits. Anim Behav 11:500–506
2. González-Mariscal G, Jiménez P, Beyer C, Rosenblatt JS (2003) Androgens stimulate specific aspects of maternal nest-building and reduce food intake in rabbits. Horm Behav 43:312–317
3. González-Mariscal G, Lemus AC, Vega Gonzalez A, Aguilar Roblero R (2013) Litter size determines circadian periodicity of nursing in rabbits. Chronobiol Int 30:711–718
4. González-Mariscal G, Caba M, Martínez-Gómez M, Bautista A, Hudson R (2016) Mothers and offspring: the rabbit as a model system in the study of mammalian maternal behavior and sibling interactions. Horm Behav 77:30–41
5. Ross S, Denenberg VH, Sawin PB, Meyer P (1955) Changes in nest building behaviour in multiparous rabbits. Br J Anim Behav 4:69–74
6. Denenberg VH, Huff RL, Ross S, Sawin PB, Zarrow MX (1963) Maternal behaviour in the rabbit: the quantification of nest-building. Anim Behav 11:494–499
7. González-Mariscal G, Díaz Sánchez V, Melo AI, Beyer C, Rosenblatt JS (1994) Maternal behavior in New Zealand white rabbits: quantification of somatic events, motor patterns, and steroid plasma levels. Physiol Behav 55: 1081–1089
8. Negatu Z, McNitt JI (2002) Hormone profiles and nest-building behavior during the periparturient period in rabbit does. Anim Reprod Sci 72:125–135
9. González-Mariscal G, Melo AI, Jiménez P, Beyer C, Rosenblatt JS (1996) Estradiol, progesterone, and prolactin regulate maternal nest-building in rabbits. J Neuroendocrinol 8: 901–907
10. González-Mariscal G (2007) Mother rabbits and their offspring: timing is everything. Dev Psychobiol 49:71–76
11. Anderson CO, Zarrow MX, Fuller GB, Denenberg VH (1971) Pituitary involvement in maternal nest-building in the rabbit. Horm Behav 2:183–189
12. Cross BA (1958) On the mechanism of labour in the rabbit. J Endocrinol 16:261–276
13. Melo AI, González-Mariscal G (2003) Placentophagia in rabbits: incidence across the reproductive cycle. Dev Psychobiol 43:37–43
14. Zarrow MX, Faroow A, Denenberg VH, Sawin PB, Ross S (1963) Maternal behaviour in the rabbit: endocrine control of maternal nest-building. J Reprod Fertil 6:375–383
15. Zarrow MX, Denenberg VH, Anderson CO (1965) Rabbit: frequency of suckling in the pup. Science 150:1835–1836
16. Rosenblatt JS (1967) Nonhormonal basis of maternal behavior in the rat. Science 156: 1512–1514
17. Benedek I, Altbäcker V, Molnár T (2021) Stress reactivity near birth affects nest building timing and offspring number and survival in the European rabbit (Oryctolagus cuniculus). PLoS One 16:e0246258
18. Seltmann MW, Rangassamy M, Zapka M, Hoffman KL, Rödel HG (2017) Timing of maternal nest building and perinatal offspring survival in a group-living small mammal. Behav Ecol Sociobiol 71:64–77
19. Zarrow MX, Denenberg VH, Kalberer WD (1965) Strain differences in the endocrine basis of maternal nest-building in the rabbit. J Reprod Fertil 10:397–401
20. Sawin PB, Crary DD (1953) Genetic and physiological background of reproduction in the rabbit: II. Some racial differences in the pattern of maternal behavior. Behaviour 6:128–146

21. Zarrow MX, Sawin PB, Ross S, Denenberg VH, Crary D, Wilson ED, Farook A (1961) Maternal behaviour in the rabbit: evidence for an endocrine basis of maternal nest building and additional data on maternal-nest building in the Dutch-belted race. J Reprod Fertil 2: 152–162
22. Sawin PB, Denenberg VH, Ross S, Hafter E, Zarrow MX (1960) Maternal behavior in the rabbit: hair loosening during gestation. Am J Physiol 198:1099
23. Sawin PB, Curran RH (1952) Genetic and physiological background of reproduction in the rabbit. I. The problem and its biological significance. J Exp Zool 120:165
24. Olson ME, Bruce J (1986) Ovariectomy, ovariohysterectomy and orchidectomy in rodents and rabbits. Can Vet J 27:523–527
25. Olazábal DE, Pereira M, Agrati D, Ferreira A et al (2013) Flexibility and adaptation of the neural substrate that supports maternal behavior in mammals. Neurosci Biobehav Rev 37: 1875–1892
26. Caba M, Beyer C, González-Mariscal G, Morrell JI (2003) Immunocytochemical detection of estrogen receptor-alpha in the female rabbit forebrain: topography and regulation by estradiol. Neuroendocrinology 77:208–222
27. Caba M, Rovirosa MJ, Beyer C, González-Mariscal G (2003) Immunocytochemical detection of progesterone receptor in the female rabbit forebrain: distribution and regulation by oestradiol and progesterone. J Neuroendocrinol 15:855–864
28. González-Mariscal G, Kinsley CH (2009) From indifference to ardor: the onset, maintenance, and meaning of the maternal brain. In: Pfaff DW, Arnold AP, Etgen AM, Fahrbach SE, Rubin RT (eds) Hormones, brain and behavior. Elsevier Academic Press, pp 109–136
29. González-Mariscal G, Chirino R, Rosenblatt JS, Beyer C (2005) Forebrain implants of estradiol stimulate maternal nest-building in ovariectomized rabbits. Horm Behav 47:272–279
30. Sawyer CH, Evertt JW, Green JD (1954) The rabbit diencephalon in stereotaxic coordinates. J Comp Neurol 101:801–824
31. Hudson R, Cruz Y, Lucio A, Ninomiya J, Martínez-Gómez M (1999) Temporal and behavioral patterning of parturition in rabbits and rats. Physiol Behav 66:599–604
32. Shanklin DR (1966) Oxytocin induction in pregnant rabbits, with special reference to the stillbirth rate. Am J Obstet Gynecol 94:242–248
33. Hudson R, Distel H (1989) The temporal pattern of suckling in rabbit pups: a model of circadian synchrony between mother and young. In: Reppert SM (ed) Development of circadian rhythmicity and photoperiodism in mammals. Perinatology Press, Boston, pp 83–102
34. Jilge B (1993) The ontogeny of circadian rhythms in the rabbit. J Biol Rhythms 8(3): 247–260; Jilge B (1995) Ontogeny of the rabbit's circadian rhythms without an external zeitgeber. Physiol Behav 58:131–140
35. Lincoln DW (1974) Suckling: a time-constant in the nursing behaviour of the rabbit. Physiol Behav 13:711–714
36. González-Mariscal G, Melo AI, Parlow AF, Beyer C, Rosenblatt JS (2000) Pharmacological evidence that prolactin acts from late gestation to promote maternal behaviour in rabbits. J Neuroendocrinol 12:983–992
37. González-Mariscal G, Chirino R, Flores Alonso JC, Rosenblatt JS, Beyer C (2004) Intracerebroventricular injections of prolactin counteract the antagonistic effect of bromocriptine on rabbit maternal behaviour. J Neuroendocrinol 16:949–955
38. González-Mariscal G, Melo AI, Chirino R, Jiménez P, Beyer C, Rosenblatt JS (1998) Importance of mother/young contact at parturition and across lactation for the expression of maternal behavior in rabbits. Dev Psychobiol 32:101–111
39. González-Mariscal G, Gallegos JA (2014) The maintenance and termination of maternal behavior in rabbits: involvement of suckling and progesterone. Physiol Behav 124:72–76
40. Findlay AL, Tallal PA (1971) Effect of reduced suckling stimulation on the duration of nursing in the rabbit. J Comp Physiol Psychol 76:236–241
41. González-Mariscal G, Lemus AC, Vega-González A, Aguilar-Roblero R (2013) Litter size determines circadian periodicity of nursing in rabbits. Chronobiol Int 30:7011–7718
42. Fleming AS, Rosenblatt JS (1974) Olfactory regulation of maternal behavior in rats. II. Effects of peripherally induced anosmia and lesions of the lateral olfactory tract in pup-induced virgins. J Comp Physiol Psychol 86:233–246
43. Mayer AD, Rosenblatt JS (1977) Effects of intranasal zinc sulfate on open field and maternal behavior in female rats. Physiol Behav 18: 101–109
44. Stolzenberg DS, Champagne FA (2016) Hormonal and non-hormonal bases of maternal behavior: the role of experience and epigenetic mechanisms. Horm Behav 77:204–210

45. Fleming A, Vaccarino F, Tambosso L, Chee P (1979) Vomeronasal and olfactory system modulation of maternal behavior in the rat. Science 203:372–374
46. González-Mariscal G, Chirino R, Beyer C, Rosenblatt JS (2004) Removal of the accessory olfactory bulbs promotes maternal behavior in virgin rabbits. Behav Brain Res 152:89–95
47. Chirino R, Beyer C, González-Mariscal G (2007) Lesion of the main olfactory epithelium facilitates maternal behavior in virgin rabbits. Behav Brain Res 180:127–132
48. Kurien BT, Scofield RH (1999) Mouse urine collection using clear plastic wrap. Lab Anim 33:83–86
49. Mulvaney BD, Heist HE (1971) Regeneration of rabbit olfactory epithelium. Am J Anat 131: 241–251
50. Friard O, Gamba M (2016) BORIS: a free, versatile open-source event-logging software for video/audio coding and live observations. Methods Ecol Evol 7:1325–1330
51. Hoffman KL, Rueda Morales RI (2009) Toward an understanding of the neurobiology of "just right" perceptions: nest building in the female rabbit as a possible model for compulsive behavior and the perception of task completion. Behav Brain Res 204:182–19153
52. Hoffman KL, Rueda Morales RI (2012) D1 and D2 dopamine receptor antagonists decrease behavioral bout duration, without altering the bout's repeated behavioral components, in a naturalistic model of repetitive and compulsive behavior. Behav Brain Res 230:1–10
53. Cano Ramírez H, Hoffman KL (2017) Activation of the orbitofrontal and anterior cingulate cortices during the expression of a naturalistic compulsive-like behavior in the rabbit. Behav Brain Res 320:67–74
54. Cano Ramírez H, Hoffman KL (2018) Activation of cortical and striatal regions during the expression of a naturalistic compulsive-like behavior in the rabbit. Behav Brain Res 351: 168–177
55. Cano-Ramírez H, Pérez-Martínez LP, Hoffman KL (2020) Dataset on c-Fos expression within components of corticostriatal thalamocortical circuits during the expression of a compulsive-like behavior in the female rabbit: brain-behavior relationships. Data Brief 34: 106696
56. Walter MH, Abele H, Plappert CF (2021) The role of oxytocin and the effect of stress during childbirth: neurobiological basics and implications for mother and child. Front Endocrinol 12:742236
57. Aguilar-Roblero R, González-Mariscal G (2020) Behavioral, neuroendocrine and physiological indicators of the circadian biology of male and female rabbits. Eur J Neurosci 51: 429–453
58. Caba M, González-Mariscal G (2009) The rabbit pup, a natural model of nursing-anticipatory activity. Eur J Neurosci 30:1697–1706
59. Meza E, Waliszewski SM, Caba M (2011) Circadian nursing induces PER1 protein in neuroendocrine tyrosine hydroxylase neurons in the rabbit doe. J Neuroendocrinol 23:472–480
60. Meza E, Aguirre J, Waliszewski S, Caba M (2015) Suckling induces a daily rhythm in the preoptic area and lateral septum but not in the bed nucleus of the stria terminalis in lactating rabbit does. Eur J Neurosci 41:196–204
61. Aguirre J, Meza E, Caba M (2017) Dopaminergic activation anticipates daily nursing in the rabbit. Eur J Neurosci 45:1396–1409
62. Caba M, González-Mariscal G, Meza E (2018) Circadian rhythms and clock genes in reproduction: insights from behavior and the female rabbit's brain. Front Endocrinol 9:106

Chapter 4

Basic Protocols to Study Parental Behavior in Rats

Angel I. Melo, Mario Caba, Francisco Castelán, and Margarita Martínez-Gómez

Abstract

Maternal care is an affiliative, motivated, and complex social behavior that has been conserved through evolution to ensure the survival of the offspring and promote their growth, development, and maturation. In the rodent, maternal behavior is primed during pregnancy by hormonal changes and subsequent neurochemical responses and is activated and maintained after parturition by sensorial and social stimuli from the pups. But taking care of the infant—or parental behavior—can also be induced by nonhormonal factors in any biologically related or unrelated individual. Although these factors act on the "maternal circuit," other neural systems are involved in the fine-tuning of this behavior, which allows the mother to be sensitive to the pups' cues, interpret their needs, and provide them with fast, and contingent care. The peripartum period is a critical time for the mother–infant dyad. On the one hand, the maternal brain responds actively to the mother–infant interaction, modifying its neural structure and functionality (neuronal plasticity). On the other hand, the pup's brain is altered by this interaction through epigenetic processes; such epigenetic alterations persist across the lifespan of the individual and may even be transmitted to the next generation (transgenerational transmission). Animal models are essential for understanding the complexity and flexibility of neurobehavioral processes that underlie parental responsiveness. In this chapter, our aim is to describe an array of basic behavioral protocols that are used to study parental behavior in the laboratory rat, the species most widely used to study neurobiological, neuroendocrine, cognitive, emotional, and behavioral bases of parental behavior.

Key words Parental behavior, Maternal behavior, Methods, Maternal aggression, Recording maternal behavior

1 Introduction

In mammals, the terms parental behavior or parenting refer to the care given to an immature organism by any individual of its social group, to ensure its survival during the critical preweaning period. More specifically, the biological father's care of the infant is referred to as *paternal behavior*, while infant care given by any other individual (unrelated nonpregnant female, unrelated male, or sibling) is called *alloparental care.* The term *maternal behavior* (MB) refers specifically to care given to the offspring by its mother [1–3].

Raúl G. Paredes et al. (eds.), *Animal Models of Reproductive Behavior*, Neuromethods, vol. 200,
https://doi.org/10.1007/978-1-0716-3234-5_4,

The mother is the most important protagonist of this biological process because she is the offspring's first social contact and provides nutrition and social interaction, which contribute to their development, growth, and socialization. *Maternal behavior* in mammals can be considered a complex social, affiliative, and goal-directed behavior highly conserved among species. It consists of the expression of multiple behaviors at the end of pregnancy and during and after parturition that provide nutrients, warmth, and protection, as well as sensory (tactile, odor, and auditive) and social stimuli. Moreover, the mother's milk provides hormones and immunological and bioactive components that are necessary for the development and maturation of metabolic, affective, cognitive, neuroendocrine, immune, reproductive, social, and behavioral systems [1–5]. MB is primed by the presence of high circulating levels of hormones of pregnancy: mainly estrogen, progesterone, and prolactin. Later, it is triggered by a change in the ratio of these hormones at the end of pregnancy. Specifically, across pregnancy, a series of neurobiological (neurogenesis and synaptic density), neuroendocrine (estrogen, oxytocin, prolactin, progesterone, and corticosterone), and neurochemical (dopamine and serotonin) changes occur in order to prepare the mother to be sensitive to the offspring's signals and to interpret their needs and intentions, thereby providing them with appropriate, rapid, and contingent care [2, 6–9]. After birth, the establishment and maintenance of MB is under control of the interchange of sensory and social cues between the mother and the offspring, which permits behavioral synchrony across lactation until the decline of MB [1, 10, 11]. This sensory and social intercommunication within the dyad allows for flexibility in the expression of MB, which adapts across lactation to the size of the litter and the pups' age, in addition to showing differences between rat strains [10].

Interestingly, parental responsiveness can be induced in intact nonpregnant and nonlactating females, in juveniles, in ovariectomized or hypophysectomized females, and even in male rats by continuous and repeated exposure to pups, a process called *sensitization* [12]. This process is an example of nonhormonal regulation of parental behavior. The study of parental behavior is not only a means to understand the endocrinological, psychobiological, and neurobiological mechanisms involved in the control of MB itself, but also contributes to our understanding of neurobehavioral mechanisms that underlie social behaviors, motivated behaviors, and cognitive processes, among others.

Thus, the study of mother–offspring interactions provides important information related to (1) mammalian social life (socialization); (2) genetic, cellular, and physiological substrates of MB; (3) transgenerational transmission of behavior (epigenetically); (4) brain plasticity during the perinatal period; (5) neural activity induced by infant cues in the female or male brain associated with the activation of the "parental circuit"; (6) executive functions such as empathy, theory of mind and social cognition during the mother–infant interaction; (7) critical periods of development; (8) ontogenetic processes across life; and (9) rewarding properties of pup-associated sensory cues and conditioned and unconditioned place-preference [2, 4, 9, 13–16].

Due to the complexity and dynamism of parental behavior, it is necessary to apply methodologies that allow for the quantitative and qualitative assessment of the biological basis of the onset, establishment, maintenance, and decline of behavioral components of MB. The level and investment of maternal care varies among species: as short as a few minutes per day (European rabbit), continuous (rodents), permanent (kangaroos), or for an extended period of time (humans and nonhuman primates) [2, 3, 14, 17]. In this chapter, we describe in detail six protocols for the study of MB in the laboratory rat, since this species has been the most widely studied nonhuman species in this regard. Studies of rat MB have provided important information on the behavioral, neuroendocrine, and neurobiological bases of MB [18–20]. The first protocol describes the most useful method to assess MB: continuous recording of MB without disturbing the nest, or after a brief mother–infant separation. The second protocol details the procedure for periodic (spot-check) recording of the lactating rat without disturbing the nest or the litter. Protocol 3 describes the methodology for assessing nest-building behavior, and the fourth protocol describes the process of sensitization, where a nonpregnant individual (male, female, or juvenile) displays MB after a prolonged period of exposure to foreign pups. Protocol 5 describes the methodology for inducing sensitization by hormonal priming. In Protocol 6, we describe the main method used for assessing maternal aggression, a behavior that is important for protecting offspring from intruders. Finally, given that maternal motivation is a fundamental component of MB, in the final section, we provide an overview of this aspect of MB, along with general descriptions of experimental strategies used to assess and quantify it.

1.1 Description of Parental Behaviors in Rats

A few days before parturition, and during the first 2 weeks thereafter, the female rat builds a maternal nest to have a safe place for the delivery and for her offspring [1]. During parturition, the mother licks the pup's body and anogenital areas to remove the amniotic liquid and membranes, consume the placentas and umbilical cord (*placentophagia),* clean the pup, and stimulate respiration [21]. Additionally, the mother appears to facilitate delivery of the pups and/or placentas retained in the vaginal channel by pulling them with her mouth [21, 22]. Once the offspring are born, the dam continuously licks each pup, *sniffs* them, and groups them into the nest (*grouping*). Sometimes she rebuilds the nest and permits the pups to lie under her ventrum while they search for the nipples, all the while adopting the posture of *hover over the pups*. The sensory stimulation of the pups to her ventrum activates the posture of *crouching* or *arched-back nursing* [23, 24] and she starts nursing pups. Finally, as the quantity and quality of pup stimuli increases, she adopts a well-defined, intense, quiescent, and upright nursing posture called *kyphosis* [24] or *high-arched-back nursing* [23–26]. Alternatively, the mother can display *passive nursing* (lying on her side or back while nursing some pups). Additionally, to protect the offspring, the mother expresses high levels of maternal aggression during the first half of lactation; thereafter such aggression gradually declines and disappears at weaning, around 21–28 days after parturition [2, 11, 27]. MB has been classified relative to the time of parturition (prepartum and postpartum MB), according to whom the behavior is addressed (directed to the offspring or some other individual), according to whether its expression involves locomotor activity or not (active behaviors and quiescence/passive behaviors) or if it is related to nursing (pronurturant and nurturant behaviors) [11, 18, 25, 28]. It must be considered that mother–offspring social interactions are a useful bioassay for understanding the impact of early experiences and the intergenerational transmission of behavioral traits. Thus, when mothering goes wrong due to any of a variety of factors, such as early adverse experiences of the mother or experimental manipulations used to induce an animal model of a mental disorder, this disturbed MB will in turn negatively affect the offspring's growth and development. For example, we [29, 30] and others [31–33] have observed in such cases that disturbed MB includes *atypical retrieval* (when the mother grasps and carries a pup by parts of its body other than its dorsal skin, such as by the head, mouth, leg, or ventrum), *atypical nest-bulding* (when the mother repeatedly rearranges nest and rebuilds the nest even while the pups are inside it), *re-retrieving pups* (the mother grasps and carries a pup seemingly without direction from the nest to some other location), and *cannibalism/infanticide*. Moreover, during the test sessions, the mother can also display nonmaternal behaviors such as eating, drinking water, rearing, climbing, digging, sniffing the cage, air, or bedding, self-grooming, and locomotor activity. If the frequency of such behaviors is high during test sessions, it could suggest an imbalance in the expression of nonmaternal and MB, or disrupted expression of MB.

1.2 Components of Parental Behaviors in Rats

Retrieval or Retrieving

In certain circumstances (such as due to an experimental manipulation), pups might be scattered outside of the maternal nest. In such cases, the typical response of the mother is to display retrieving, or pup retrieval. This behavior consists of the mother grasping the pup by its dorsal skin with her mouth and transporting it back to the nest (Fig. 1a).

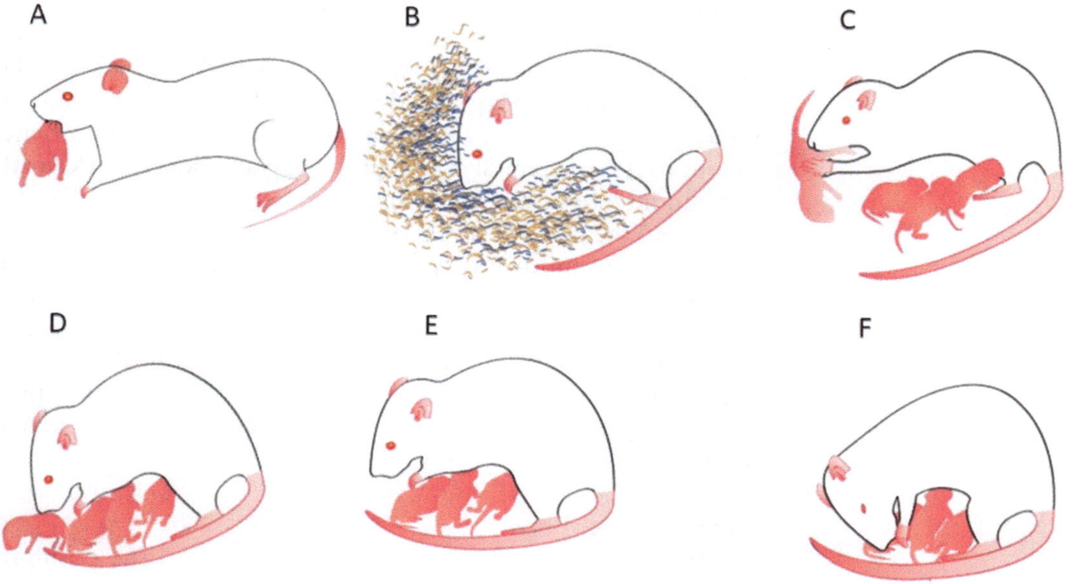

Fig. 1 Illustrations of the main components of maternal behavior in the rat. (**a**) Retrieving. (**b**) Nest building. (**c**) Anogenital licking. (**d**) Body licking and hovering over the pups. (**e**) Crouching. (**f**) Kyphosis

Mouthing/Grouping

Mouthing, or grouping, refers to using the mouth to reposition a pup across a short distance, such as within or around the nest [34]. The expression of retrieval and grouping involves motivational processes (maternal motivation) as well as cognitive functions. The mother is driven to move toward pups by her perception of the primary sensory signals emitted by them, gathering them, and transporting them to the nest, presumably relying on spatial memory processes in order to return them to the correct nest site [12, 35–37].

Nest-Building

Nest construction is the first prepartum, non-pup directed, and active maternal behavior; it is expressed as a complex sequence of behavioral components at the end of pregnancy as a result of specific hormonal changes (see above). It involves the mother carrying nest material, such as wood sawdust or chips with her mouth, while pushing and accommodating it with her paws or snout into a corner of the maternal cage (Fig. 1b). Despite the importance of this behavior, there are surprisingly few studies focused on determining its causal factors, their relationship with the quality of maternal care, and the survival of the progeny [38, 39]. The nest is generally rated according to the height of its walls [10] (*see* Subheading 3.3 and Fig. 4).

Pup Body Licking

The mother licks the pup's body from the head to the trunk and the ventrum (*body licking*). This behavior can be identified by the characteristic long licking movements of the mother's head. In addition, the dam grasps the pup with both forepaws and orients its snout to the anogenital region of the pup and licks this area with short and intense licking movements (*anogenital licking*). During this process, the pup reflexively becomes immobile, its legs splayed, and usually the mother consumes the expelled urine to replace water and electrolytes lost during nursing (Fig. 1c, d). This behavior has an important role for the development of the offspring. For example, the mother rats lick the anogenital region of male offspring more frequently than females; this stimulus promotes processes of sexual differentiation that later impact on adult sexual behavior [40–43].

Hover over the Pups

Once the mother retrieves all the pups to the nest, she positions herself over some or all the pups without arching her back (see above) and permits them to stay there, either resting or searching for the nipples. She may start licking the pups, perform self-grooming, and/or rearrange the nest material (Fig. 1c, d).

Nursing Behavior

Once one or two pups begin suckling while they are under the ventrum, the dam adopts a nursing posture in which she slightly arches her back and licks the pups frequently (*crouching posture*, Fig. 1e). Eventually, she begins to display a quiescent posture characterized by intense arching of the back, with legs splayed (*kyphosis posture*, Fig. 1f); during kyphosis, the dam ceases all other behaviors, including pup licking. Kyphosis is considered to be a passive MB, and milk letdown and pup nursing occur only during this behavior. Sometimes the mother lies on her side or back while nursing pups (passive nursing). Frequently, the mother alternates between different postures while she is nursing [44]. Interestingly, nonpregnant, nonlactating females do not display kyphosis; this is because this behavior is primarily triggered by sensory stimulation of the mother's ventrum associated with pup suckling [18, 19, 23, 24] (Fig. 1f).

2 General Considerations

The choice of an adequate method to study MB depends on the specific research question. Some general considerations for the study of MB are described below:

1. Time of light/dark cycle to record MB. Most of the studies that use Continuous Observation tests are performed during the first hours after lights on (12 h;12 h, lights on at 08:00 h), since rats are nocturnal animals, exploring, and foraging during the night and more likely to be close to the pups during the early morning. Mother rats also tend to display more attention to the pups during the mornings of early lactation [45, 46].
2. Age of pups. Most of the studies that apply sensitization, spot-check, or retrieval tests use young pups (1 to 6–8 days old) because nonpregnant, nonlactating, or males respond better to them and retrieve them faster than they do older pups [10, 47, 48]. With regard to lactating females, there is a correlation between time of lactation and age of pups and level of maternal responsiveness. Thus, maternal responsiveness is higher during early lactation when the pups are young, compared to during late lactation when the pups are older. Specifically, during early lactation, the latency to retrieve pups is short and the mother spends more time close to the pups and nursing them. Moreover, maternal aggressiveness is higher when the pups are young [36]. At late lactation (after postpartum days 10–12), maternal responsiveness begins to decline and the mother rat begins to display avoidance behaviors toward the pups, such as rejections, darting-hopping, shaking, or lying on their ventrum to avoid suckling by the pups. This decline of maternal behavior is similar in sensitized naïve females exposed to old pups (up to 10 postnatal days), although they exhibit an earlier decline [10, 11, 36].
3. Size of maternal cages. During the first 2 weeks of lactation, the mother spends more time close to the pups when they are housed in medium cages compared to when they are housed in small cages [48]. Interestingly, the behavior of virgin female rats is different than that of lactating rats: maternal responsiveness is faster and better in a small cage compared to a medium one [49].
4. For the studies that use continuous recording and a behavioral observation software, it is suggested that a subset of tests (around 10%) should be scored by two observers to avoid biases.
5. In the case of the undisturbed mother-pup separation tests that are repeated across several days of lactation, it is recommended that wood chip bedding changes be done twice per week, beginning at the second week of lactation.
6. Record within the experimental notebook the information on each experimental subject (identification number, age, reproductive state, experimental treatment, etc.). This will allow video recordings (which are identified by the subject's

identification number, with no additional information) to be analyzed by observers blinded to experimental treatment.

7. Choosing the specific test for assessing maternal responsiveness depends on the investigator's specific research question: different protocols of data collection emphasize different aspects of MB and are appropriate in different circumstances. For example, the spot-check protocol allows for the relatively rapid analysis of extended periods (many hours) of behavioral testing; however, it is not appropriate for analyzing brief behavioral events such as licking or mouthing (which can be easily missed by spot-checking) or for accurately quantifying duration of a particular behavior. Continuous Observation is more appropriate for brief behavioral events, precise measures of duration, assessing the quality of maternal behavior, and analyzing behavioral sequences.

8. There are important differences in maternal behavior between rat strains. Thus, Wistar, Long Evans, and Sprague-Dawley rats differ importantly with regards to the expression of maternal care. Long Evans rats spend more time being close to pups and nursing them, compared to the other two strains [50] and they also display more pup licking compared to Fisher-344 rats [51]. Moreover, there are important differences in fear and anxiety behaviors that could impact the expression of MB: Long Evans rats show greater emotionality than Wistar and Sprague-Dawleys rats, and it is known that this characteristic can affect the expression of maternal care [50]. In addition, Roman low-avoidance rats spend more time close to pups, but the Roman high-avoidance rats display more side-nursing behavior, than their counterparts [52]. These behavioral differences could be since Roman low avoidance displays more anxiety behavior in conditions of stress, compared to Roman high avoidance [53].

9. All experiments of MB should use the same litter size and the same pup sex ratio (mother rats display more licking directed at male pups compared to female pups) for each experimental subject. A litter size of 6–10 pups promotes good maternal care and positively affects the pups' development (e.g., pups reared in litter of 6–9 pups have increased myelin thickness and greater compound action potential of the sensory sural nerve, compared to pups from litters of 3 or 12 pups) [54]. Additionally, pups reared in litters of up to 12 pups receive less maternal licking compared to pups reared in litters of 4 pups or less, further indicating that litter size alters the expression of maternal behaviors [18, 28, 54, 55] that ultimately affect the pups' phenotype [55]. These recommendations also apply for the sensitization test.

10. Time of the day to perform the MB tests. Although most rats response to the challenger of be deprived of pups for around 3–5 min and reexposed later on, it is better to perform the tests during the first hours after the lights on.
11. All protocols must be reviewed and approved by an Institutional Animal Care and Use Committee.
12. Rats must be housed in Animal Housing Facilities under standard conditions: ambient temperature of 22 °C, humidity of 40–50%, 12:12 light: dark photoperiod, and rat food pellets and water provided ad libitum.

3 Protocols

3.1 Protocol 1: Continuous Observation of MB

This method is the most widely used one, since it provides a detailed analysis of MB, including behavioral latencies, frequencies, and duration of each component of MB. This procedure challenges the mother to display active components of MB, and if the period of observation is long (up to 30 min), it is possible to record passive components as well [56, 57]. MB can be recorded under undisturbed conditions, or after different periods of mother–offspring separation (3–5 min to 4 h). The time of sessions can vary from 10, 15, 30, and 60 min to 4 h, and for several sessions per day. The *undisturbed* or *without mother–offspring separation test* consists in continuously recording the expression of different components of MB during the test session, without disturbing the mother or pups, using behavioral analysis software (see below). The *Continuous Observation with a mother–litter separation*, which is the most frequently used method and the one that we describe here, consists in first removing the pups from the nest, weighing them (if the separation lasts more than 5 min, they should be placed on a warm electric pad), and then returning them to the maternal cage in order to begin the test (see below and Fig. 2 for details). If the investigator wishes to obtain information about milk production, they must first swab the anogenital region of the pups with a wet brush upon removing them from the nest, to stimulate urination and defecation. The pups are then weighed before returning them to the maternal cage to begin the test. After a period of 25–30 min (or the time required for the mother to nurse the pups), the pups are again removed from the nest and weighed as a litter. The difference in the weight before and after nursing is a quantitative measure of milk output. Continuous observer methods typically use a software for behavioral analysis (see below); alternatively, behaviors can be scored by hand using paper data sheets [20].

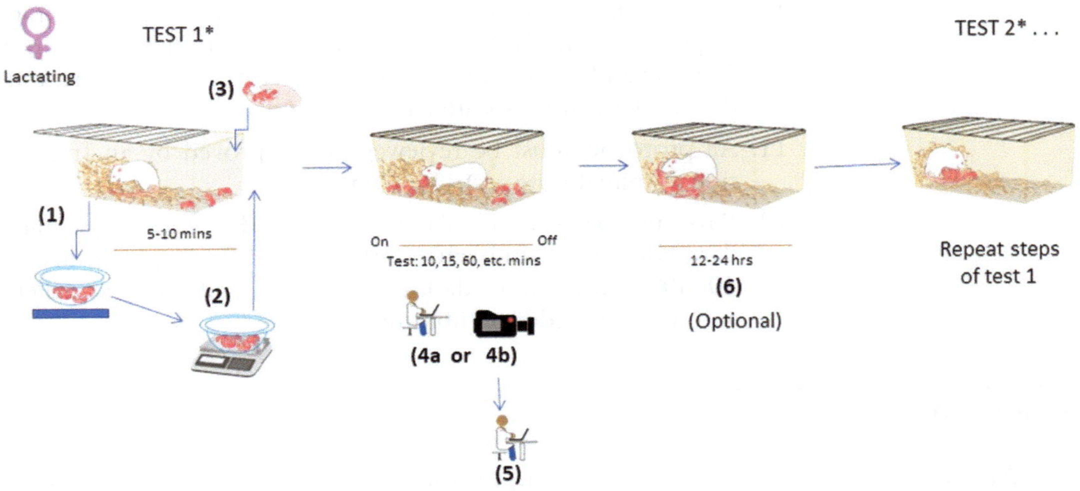

Fig. 2 Schematic representation of the recording of rat maternal behavior using the continuous maternal observation test. *PPD = postpartum day

Materials
(1) Lactating rats. (2) Litter of 6–10 pups with equal numbers of male and females. (3) Medium polypropylene maternal cage (44–50 × 26–28 × 16–20 cm). (4) Wood chips (enough to cover the floor of the cage with a 1–1.5 cm high). (5) Small bowl. (6) Warm electric pad or incubator. (7) Small balance. (8) Behavioral recording Software (e.g., *The Observer*, Noldus Information Technology, Scolari, Sage Publication Software or software *Stop-Watch*, ODLog behavioral analysis software, Macropod Inc., USA). (9) Video camera. (10) Laptop computer dedicated exclusively to registering behavior. (11) Paper data sheets divided into columns (by time period) and rows (by behavior) [20], and (12) stopwatch.

Procedure (Fig. 2)
(1) Separate the pups from the nest, leaving the mother in the maternal cage, and place the pups into a small bowl warmed on an electric pad or inside an incubator for 5–10 min (**step 1**) or longer. (2) Meanwhile, turn the laptop on, open the behavioral observational software, and write up the name of the file using a code beginning with the number of the subject, followed by the reproductive condition, the number of test or postpartum day (e.g., PPD3) and prepare the software to be ready to start the recording. (3) Weigh the pups (**step 2**). If information on milk output per nursing bout is needed; the pups must first be swabbed with a wet brush in the anogenital region to stimulate urination and defecation, and then dried with a paper towel before weighing them (not shown in Fig. 2). (4) Return the pups to the maternal cage and scatter them in the side of the cage opposite to where the nest is

located, or opposite to the location of the mother rat if a nest is not present (**step 3**). (5) If the observer wishes to record MB directly (in vivo), they should sit in front of the maternal cage, using a small table on which is placed the laptop, and start the software to begin recording behavior (**step 4a**). The observer makes single keystrokes that code for each behavior of interest that is displayed by the mother rat; for example, "r" for retrieving, "h" for hover over the pups, "b" for body licking, and "a" for anogenital region (for details on the configuration of the software, refer to the software users' manual). (6) If the observer prefers to film the session for later analysis (this is the recommended method in **step 4b**), one or two video cameras must be placed in front of and laterally to the maternal cage. Another option is placing mirrors at an angle on each side of the cage so that the sides can be viewed from above and using just one video camera mounted above the cage (zenital camera). Just before beginning the test session, both cameras should be started. Later, recordings can be played back and analyzed by one or more observers blinded to the experimental treatment, using the behavior analysis software (**step 5**). (7) If the investigator wishes to quantify behavior across an extended period, a second option is to do spot-checking across the next 12–24 h (**step 6**). (8) Allow the pups to remain with the mother until the next scheduled test, when this entire procedure can be repeated.

3.2 Protocol 2: Spot-Check Observation

This method does not require sophisticated equipment, as it makes use of a time-sampling procedure to record MB by pen and paper. Spot-check observation is useful for recording multiple observations of many rats at the same time, and across an extended time (days), such as across lactation, and in an undisturbed situation [18–20, 23, 41, 58]. In addition, this method makes it possible to collect information on the mother's MB (mainly nonactive components), nonmaternal behavior, or absence of behavior, over a long period of time (for 5 s every 2–5 min for 2–5 h/day or longer). The observer must record the presence (with a cross or a check mark or a number, e.g., "1") or the absence (with no mark or "0") of each of the maternal behaviors in a data sheet having columns that correspond to fixed observation times, and rows corresponding to individual behaviors. However, there are some drawbacks that the investigator must consider before deciding to use this method: (1) The behavior of the mothers can be altered by the entrance and exit of the observer to the observation room. (2) It is not possible to register relatively brief behaviors such as carrying the young or, in general, nest construction.

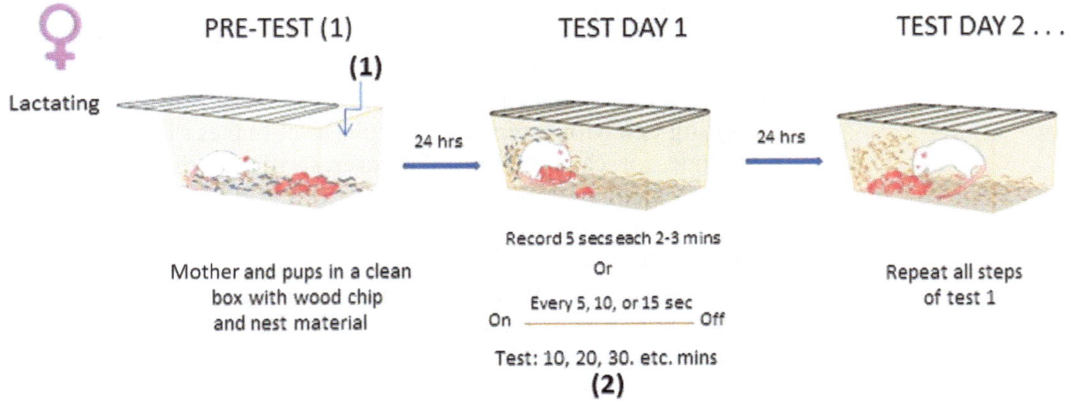

Fig. 3 Schematic representation of the recording of rat maternal behavior by the spot-check method

Materials

(1) Lactating rats. (2) Litter of 6–10 pups with equal numbers of males and females. (3) Medium polypropylene maternal cage (44–50 × 26–28 × 16–20 cm). (4) Wood chips (enough to cover the floor of the cage with a 1–1.5 cm high). (5) Paper data sheets divided into columns (time) and rows (behavior) (*see* ref. 19 for data sheets examples). (6) stopwatch.

Procedure (Fig. 3)

(1) Place the experimental rat and her litter in a clean maternal cage with wood chips and nest material (if the observer wishes to record nest-building; *see* Subheading 3.3) (**step 1**). (2) 24 h later start the first test. Taking care not to disturb the mother rat, the observer must get close to the maternal cage, start the stopwatch, and register on the data sheet the behaviors described above that the mother displays during the following 5 s of observation. (3) Every 2–3 min thereafter, this procedure is repeated, until the end of the test session (10, 20, 30, etc. minutes; Test 1). An alternative way to record MB using this method consists of registering the behavior of the mother each 5, 10, or 15 s until the end of the test session. (4) 24 h later repeat the steps of Test 1.

3.3 Protocol 3: Nest-Building Test

Although nest-building behavior might be observed during the Continuous Observation (Protocol (1) or spot-check (Protocol (2), the present protocol will allow for a more detailed assessment of this behavior. Basically, a mother rat or a nonpregnant, non-lactating female rat under specific treatment (hormones, drugs, etc.) is exposed to nest material, which is scattered around the cage. Later 4–24 h, the observer must score the quality of the nest [10, 39]. Depending on the research question, this procedure could be done once or repeated daily. For example, to see the ontogeny of nest-building during pregnancy, or to assess the capacity of the hormones of pregnancy (estrogen, progesterone, and prolactin) to induce the expression of nest-building, the test should be started at midpregnancy and repeated every 2–3 days until parturition or applied during or after hormonal treatment (*see* Subheading 3.5), respectively.

Materials

(1) Lactating or nonpregnant, nonlactating rats. (2) Medium cages (44–50 × 26–28 × 16–20 cm) of polypropylene with bedding of wood chips. (3) 8–10 pups in the case of lactating rats (in the case of nonlactating rats, it is also possible to use pups from "donor" mothers). (4) Nest material: pieces of paper towel, cotton pads (most often used for mice), or strings of yarn that can facilitate the quantification of the nest (unpublished data). (5) If the researcher wishes to assess the effect of hormones of pregnancy in ovariectomized female rats, the following materials will be needed: Estradiol benzoate, progesterone, prolactin (*see* Subheading 3.5), and insulin syringes of 1 ml.

Procedure (Fig. 4)

(1) Re-house the experimental rat into a clean polypropylene cage that has been covered with wood chips (**step 1**). (2) 24 or 48 h later, scatter nest material onto the floor of the cage (**step 2**). (3) 4–24 h later, score the quality of the nest (**step 3**). (4) Destroy the nest and remove the nest material, and add new nest material scattered around the floor of the cage (**step 4** as in **step 1**) and repeat **step 3** (4–24 h later). These procedures can be repeated daily, according to the objectives of the experiment.

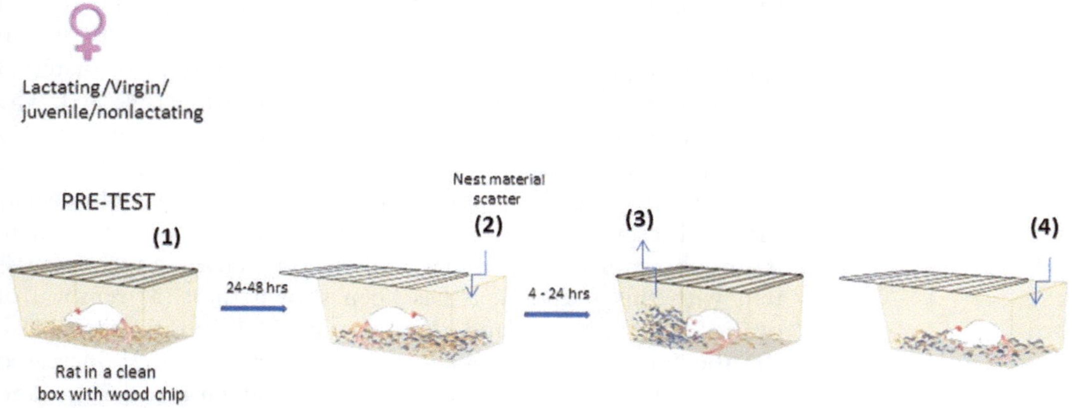

Fig. 4 Schematic representation of the recording of nest-building behavior in the maternal rat

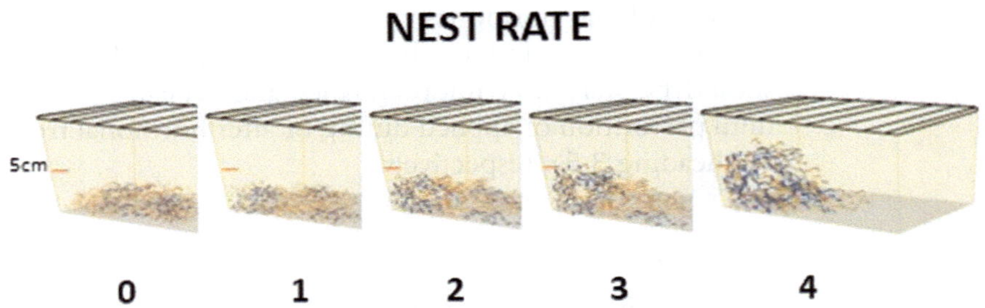

Fig. 5 Schematic representation of how rat maternal nest quality is scored

The nest quality can be scored as follows (Fig. 5): Score 0 = no nest; 1 = Poor nest, some wood chips are put together into one corner of the cage and some strips of nest material (e.g., paper towel) are incorporated into an incipient and flat nest; 2 = Fair or partial nest, more wood chips are added to the still flat nest and most of the pieces of nest material have been incorporated into it; 3 = Good nest, the nest has low walls (<5 cm) but it has a form of "nest" and all strips of nest material are at the nest site and mixed in with the wood chips; 4 = Excellent nest, the walls of the nest are high (>5 cm) and all the strips of nest material are used [10, 36].

3.4 Protocol 4: Sensitization Process

The objective of this protocol is to induce parental responsiveness in a nonpregnant rat toward donor pups. This protocol can be applied to nonpregnant intact female, nonlactating intact female, juvenile female, ovariectomized, or hypophysectomized females, juvenile males, intact or gonadectomized male rats [12, 37, 59]. When exposed to unfamiliar pups for the first time, the typical response of a female or male rat is fear of them and rejection (neophobia). However, extended, and repeated exposure toward them decreases fear and avoidance responses and increases an approach response that eventually triggers the display of maternal responsiveness; this process is termed sensitization [34, 36]. The latency to display MB depends on the condition of the rats: it is shorter in juvenile rats (female and males) and it is longer in males [12, 59, 60]. The observation that nonpregnant rats could display maternal responsiveness led Rosenblatt [12] to propose that there is a nonhormonal basis of maternal responsiveness in rats. Consistent with this proposal, it has been shown that: (a) removal of ovaries or pituitary gland does not prevent sensitization [12], (b) the forebrain of juvenile and adult virgin express increased c-fos and fosB expression after sensitization [61], and (c) sensitization does not alter the estrous cycle in intact female rats. Additionally, sensitization induced maternal behavior in transgenic mice that lack the aromatase gene (thereby preventing the synthesis of estradiol) [62].

Materials

(1) Nonpregnant rats (nonlactating female, juvenile female, juvenile male, or adult male). (2) Litter of 10–12 pups with equal numbers of male and females, obtained from donor mothers (the number of female donors should be the double to the number of experimental rats that will be sensitized). (3) Medium polypropylene maternal cage (44–50 × 26–28 × 16–20 cm). (4) Wood chips for bedding (enough to cover the floor of the cage with a 1–1.5 cm high). (5) If necessary, nest material (e.g., pieces of paper towel). (6) Small bowl. (7) Behavioral recording software (*see* Subheading 3.1 for details on the software). (8) Video camera and tripod. (9) Laptop computer dedicated exclusively to registering behavior.

Procedure (Fig. 6)

(1) Place a rat from each experimental group into a clean polypropylene box previously filled with bedding material, and, if required, 8–10 pieces of paper towel as nest-building material (**step 1**). (2) 24–48 h later (Test Day 1), five to six pups (ideally, 1 day old) taken from a lactating donor are scattered in the side of the cage opposite to the nest, or opposite to where the rat sleeps (**step 2**). Begin observing and live recording the behavior of the rat (MB or nonmaternal behavior following Protocol 1) directly into the laptop using computer-based event recording (See type of software most used in Protocol 1) (**step 3a**); alternatively, videotape the 10-, 15-, 30-, or 60-min test (**step 3b**) for later playback and analysis using software for behavior analysis (**step 4**). Foster pups are left for 12 h or overnight with the experimental rat (*see* above). (3) 24 h

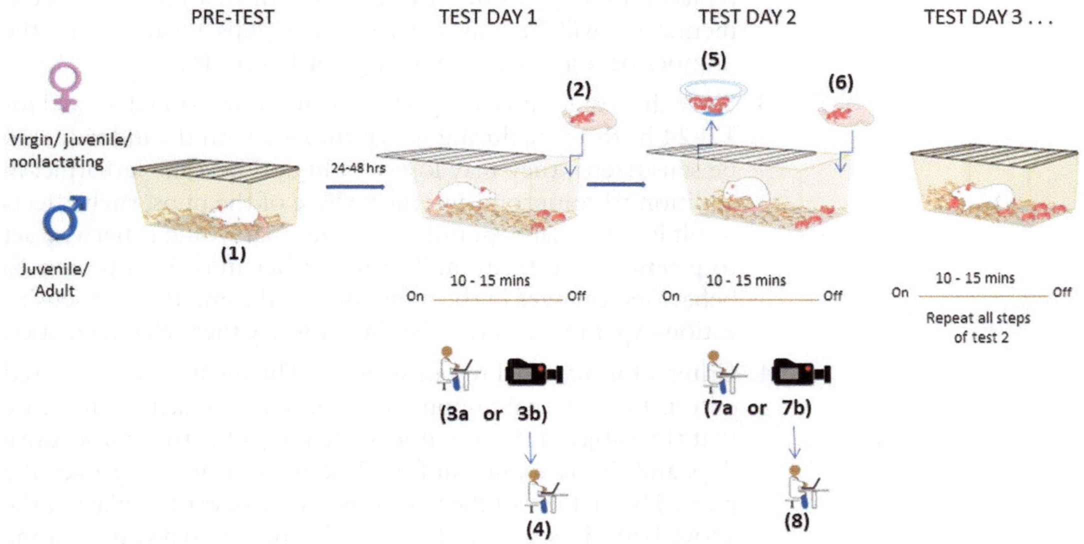

Fig. 6 Schematic representation of the sensitization protocol and recording of maternal behavior in sensitized maternal rats

later foster pups are removed from the cage and placed into a small bowl (**step 5**) and returned to the donor mother. (4) Take another five or six newly fed pups from the same donor mother and scatter them in the side of the cage opposite to the nest site (**step 6**). Start observation and live recording the behavior of the rat (**step 7a**) or videotape the test for later analysis (**step 7b**). Record the behavior of the rat (**step 8**) as was done the previous day (**step 4**). (5) Repeat **steps 5–8** several times each day until the experimental subject displays maternal responsiveness (see the criteria described below), or until having completed 14–21 consecutive test days if the subject does not show MB.

Specific Considerations for Protocols 4 and 5

1. It is suggested to start the tests during the first hours of the light phase.
2. In order to always have pups of 1 to 5–6 days old for the tests, 5–6 days after the first cohort of donor females were mated, a second cohort of donor females should be mated. In addition, since adult males need typically require more than 10 days of pup exposure to become sensitized, the investigator would need mate a third cohort of donor females. Thus, once the foster pups from the first cohort are older than 5 or 6 days, they should be replaced by pups from the mothers of the second cohort. For example, for Test Day 1 1-day-old pups should be used for Test Day 2, 2-day-old pups, etc., until Test Day 5 or 6, when the pups should be replaced with pups from the second donor. In turn, these pups are used for 5 or 6 days and then replaced by pups from a third donor. In this way, the experimental rat will be exposed to young pups regardless of the number of tests that the investigator has to do.
3. Since the foster pups used in this protocol are food deprived for 12–24 h (the time during which they are with the individual to be sensitized), they may lose weight and become lethargic; in addition to compromising the welfare of the pups, such effects result in low-quality pup-derived cues that would otherwise act to promote sensitization. The researcher must be attentive to behavioral changes in the pups and avoid using them in sensitization experiments if they begin to display these characteristics.
4. Criteria for maternal responsiveness: The most frequently used criteria for maternal responsiveness induced by sensitization are that the subject retrieves at least two pups for two consecutive days and displays pup sniffing, licking, and hovering over the pups. The first day of the two consecutive days of displaying the above behaviors is considered the latency (in days) to become maternal or paternal [18].

5. If the experimental rat injures or cannibalizes a pup, the test should be terminated. Additionally, male rats take more time to become parental than female rats; juvenile (female or male) rats take less time to become parental than adult rats. There are important inter-strain differences in the latency to become maternal or paternal.

3.5 Protocol 5: Alternative Sensitization Protocol to Induce MB by Hormones

A new mother responds promptly, contingently, and appropriately to her pups once they are delivered. However, although a nonpregnant female can display MB after sensitization (see above), this process has a slow onset (around 8–10 days), and the quality of MB is inferior compared to female rats that delivered pups naturally. The key difference between these two situations resides in the presence or absence of the hormones of pregnancy. As mentioned above, an abrupt decline in circulating progesterone occurs close to parturition, along with increasing levels of estradiol and important increases in prolactin, placental lactogens, and oxytocin, as well prolactin and oxytocin after parturition [2, 37, 63, 64]. Therefore, to induce maternal behavior in nonmaternal female rats with a short onset latency, it is necessary to mimic the patterns of these hormones that occur during natural pregnancy and parturition. Such hormones can be administered systemically by injection or by implants of silastic tubing filled with these hormones; both methods allow for extended systemic exposure to estradiol and progesterone. To mimic the prepartum decline in progesterone, progesterone administration is simply stopped, or the silastic tubing filled with progesterone is removed [2, 18, 64]. The most technically sophisticated method to induce a rapid maternal responsiveness in a nonmaternal rat is by implanting estradiol in medial preoptic area (key neural area involved in the expression of MB [65], or by infusing oxytocin into the cerebrospinal fluid of virgin or estrogen-primed ovariectomized rats [20, 66]. Another method to rapidly and efficiently induce maternal responsiveness consists in removing the uterus (hysterectomy) near the end of pregnancy (day 17), which induces a decline of progesterone and an increase of estrogens [63].

We briefly describe the method to induce MB by implanting silastic capsules filled with 17-beta-estradiol (E2) and progesterone (P) [18, 64, 67].

Materials

(1) Nonlactating intact female rats. (2) E2 (crystalline, Sigma Chemical Co. St. Louis, Mo). (3) P (crystalline, Sigma Chemical Co. St. Louis, Mo). (4) Silastic Tubing (0.058 in. ID × 0.0077 in. OD, Down-Corning, Michigan). (5) Silastic Medical Adhesive type A, Factor II Inc. Dow Corning. (6) Medium polypropylene maternal cage (44–50 × 26–28 × 16–20 cm). (7) Drugs and medicines for surgery: xylazine (Rompun v/v, Miles Laboratories) and ketamine (Ketazet, Bristol Laboratories) and gentamycin sulfate

(antibiotic). (8) Wood chips for bedding (enough to cover the floor of the cage with a 1–1.5 cm high). (9) Nest material (if necessary for objectives of experiment; e.g., pieces of paper towel *see* Subheading 3.3). (10) Material and instruments for bilateral ovariectomy and for preparing and implanting the silastic capsules (surgery kit) [68]. (11) All of the materials used in the protocol for sensitization (*see* Subheading 3.4).

Procedure

(1) Perform the bilateral ovariectomy (for details of surgery *see* ref. 57). (2) Cut silastic tubing of 20 mm and 30 mm lengths and filled with estradiol or progesterone, respectively, and seal with silastic medical adhesive. (3) Wash the steroid-filled sylastic capsules with ethanol and incubate them for 48 h in PBS (pH 7.0) before implantation. (4) Under anesthesia (halothane or a mix of ketamine plus xylazine), implant the E2 capsule subcutaneously 7–8 days after ovariectomy (Experimental day 0): make a small incision in the skin of the rat's neck with a scalpel, insert the capsule, and close by the incision with a suture. (5) On day 3, implant 3 30 mm P-filled capsules subcutaneously. (6) On the morning of day 13, remove the P-filled capsules that had been implanted. (7) On day 14, behavioral tests of sensitization are performed (*see* Subheading 3.4).

3.6 Protocol 6: Maternal Aggression Test

Maternal aggression is an agonistic, motivated, effective, and complex behavior that falls within the repertoire of maternal behaviors. Its main goal is to protect the offspring from harm from potential intruders. Maternal aggression increases at the end of pregnancy, reaches its highest levels during the first 2 weeks postpartum, and declines thereafter, reaching premating levels by the time weaning occurs [69]. It is modulated by multiple factors, including hormones of pregnancy, stimuli from the pups (differing according to age of litter), and from the intruder (differing according to intruder age and sex) [70]. The study of maternal aggression allows for determining the state of aggressiveness of the mother and for the assessment of the role of sensory and neuroendocrine factors that are involved in the regulation of maternal aggression itself, as well as other aspects of maternal care [69, 71]. Moreover, studies of maternal aggression that apply pharmacological manipulations provide an opportunity to investigate pharmacological strategies to treat pathological aggression in humans. In addition, this test provides an interesting opportunity to study two social behaviors that are mutually exclusive: maternal aggression and sexual behavior (female rat has a postpartum estrus). Moreover, maternal behavior can even be simultaneously assessed. Simultaneous testing of maternal behavior, along with maternal aggression and sexual behavior with a sexually active male, is possible only if testing is performed 12 h after parturition (for details of the variants of this test *see* ref. 72). The maternal aggression test most frequently utilizes a resident-intruder paradigm, like that used for studying territorial/offensive aggression in males [71]. The best time to do the test is during the first 2 weeks after parturition, during the light period (although in male rats this test is typically done during the first part of the dark period), and in the presence of the pups [69, 73, 74]. The test consists of introducing a male or female conspecific, having a similar weight to that of the maternal female rat, into the maternal cage while the pups remain inside it, and recording the behaviors of the mother during the following 15–30 min. The maternal aggression test can be performed on a single day during early lactation [75] or across several days [56]. In order to have a more comprehensive view of the maternal state of the mother, it is useful to record maternal aggression and maternal behavior of the same animals on alternate days. Although the behavioral components of maternal aggression and offensive/territorial aggression in males are similar, the maternal rat expresses both offensive and defensive behaviors. Offensive behaviors include a rapid approach to intruders, sniffing the intruder's genital area, adopting a posture of aggression, and displaying lateral threats, lateral attacks, boxing, bites (to the head, neck, and body), kicks, and pins. Defensive behaviors by the mother include displaying an upright posture in front of the intruder (frontal attack) and trying to induce the intruder to adopt incomplete and complete submission postures (a lower latency to express such behaviors, and a longer duration of their display indicates higher aggression for details and definitions of behaviors) [69, 71, 73, 75–78].

Materials

(1) Lactating female rats with their litters (8–10 pups per litter, same number of males and females). (2) Male or female intruder rats (equal to the number of experimental mothers) of similar weight and age of the mothers. (3) Medium polypropylene maternal cage (44–50 × 26–28 × 16–20 cm). (4) Wood chips for bedding. (5) Small table. (6) Behavioral recording Software (*see* Subheading 3.1). (7) Two video cameras and tripods. (8) Laptop computer for exclusive use to register behaviors. (9) Test room and Vivarium with all facilities.

Procedure (Fig. 7)

(1) One hour before the test the experimental mother rat must be transported to the test room (**step 1**) and food and water may be removed from the cage, to avoid distractions for the rats (optional). (2) 10–15 min before test, the cage with the mother and the litter is moved to a small table for habituation (**step 2**) and one video camera is placed in front of the cage and the other lateral to it

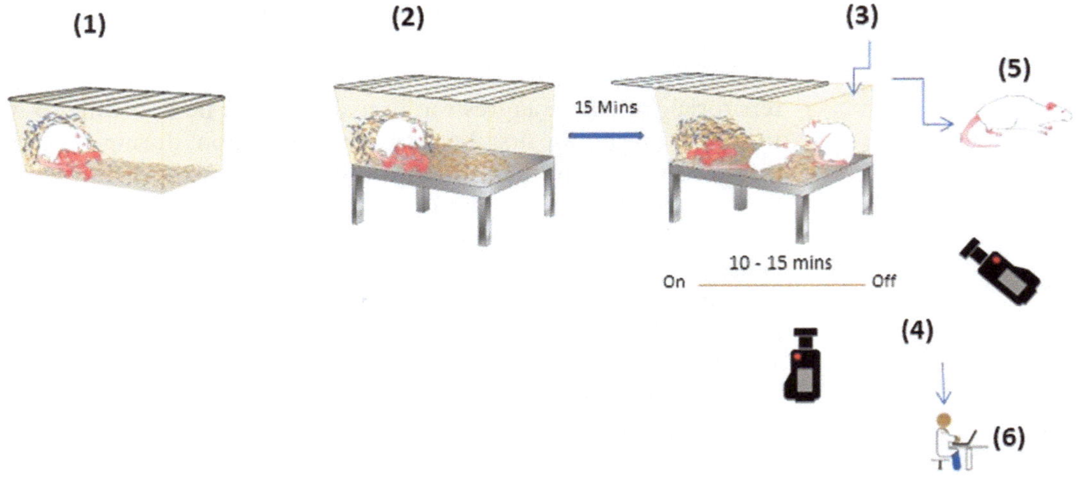

Fig. 7 Schematic representation of the record of maternal aggression in the rat

(**step 3**). As mentioned in the previous protocol, another option is to install mirrors laterally to the cage, so that the side view is reflected upward; this way, a single camera mounted above the cage will capture both top and side views. (3) Video record for a few seconds on a small piece of paper that has the identification number of the experimental animal written on it (do not include the experimental condition, so that later the video can be analyzed by an observer blinded to experimental treatment). (4) Introduce the intruder (female or male) into the maternal cage, in the side of the cage opposite to the maternal nest and litter and video record for the following 10–15 min (**step 4**). It is suggested to leave the young in the nest, as this is the natural situation and one that provides a broader view of the mother's aggression. As far as we know, it is unlikely that the pups will be harmed. For example, we have done this procedure using artificially reared mothers (an experimental manipulation that increases the aggressiveness of the rat) and we never have seen any injuries to the pups [75]. (5) Remove the intruder rat from the maternal cage at the end of the test (**step 5**) and designate the intruder rat for another use (never use the same intruder with the same resident for more than one test). (6) An observer blinded to experimental treatment should playback the video recording in slow motion, using frame-by-frame advancement. A laptop computer with behavioral software (*see* Subheading 3.1) is used to register the behaviors of the resident and intruder rats described above. Latency, frequency, and duration of each behavior are quantified. If the objectives of the experiment require repeated testing across 2 or more days, simply repeat all the steps above described.

3.7 Protocol 7: Maternal Motivation

Olazábal et al. [79] proposed the concept of a "maternal motivational state" that persists, without satiation, across pregnancy and lactation. This motivational state promotes the display of specific components of maternal behavior, including nest building, licking/grooming of offspring, retrieval/transport, crouching postures, nursing, guarding/maternal aggression, and call vocalization. The mother's behavior alternates between these active behaviors (e.g., retrieval and maternal aggression) and passive behaviors (e.g., crouching and nursing), depending on the presence of particular triggering stimuli. For example, in the rodent, pup retrieval behavior is triggered by pups scattered outside the nest. This active maternal behavior persists as long as the triggering stimulus (scattered pups) is present. When all pups have been retrieved, the mother's behavior then becomes directed toward a different motivated behavioral component of maternal behavior (e.g., crouching over the pups grouped in the nest). Thus, a persistent maternal motivational state promotes the display of a set of species-specific maternal behaviors that have the overarching goal of caring for the offspring. Successfully performing any one of these behavioral repertoires is not associated with the satiation of the maternal motivational state; rather, successful performance removes or neutralizes the triggering stimulus, thereby allowing the mother to switch to a different maternal or nonmaternal behavior. Considering this conceptualization of maternal motivation, these investigators propose that the maternal motivational state (induced by either internal hormonal milieu or by extended exposure to newborn pups) can be assessed by observing active as well as passive motivated components of maternal behavior.

Nevertheless, studies of maternal motivation in rodents have focused mainly on active components of maternal behavior (which have traditionally been referred to as "appetitive" components of maternal behavior): pup retrieval, gaining access to pups, and the incentive salience of pup stimuli compared to other competing nonmaternal stimuli. Quantitative measures of maternal motivation include the latency to retrieve scattered pups, the amount of physical effort a mother will perform to have access to pups, and the preference for pup stimuli relative to other nonmaternal stimuli. Although an exhaustive description of each of these methodologies is beyond the scope of this chapter, we will briefly summarize these procedures and direct the reader to published studies that provide more detailed descriptions.

Pup retrieval. Certain parameters of pup retrieval behavior recorded during the general procedure for assessing maternal behavior (*see* Subheading 3.1) can be used as measures of maternal motivation. The latency to retrieve the first pup, the total number of pups retrieved, and the latency to retrieve the last pup all reflect the mother's motivational drive to complete the task of returning scattered pups to the nest: shorter latencies and more pups retrieved would reflect higher motivation [e.g., 80].

Effort spent to gain access to maternal stimuli. Motivation is often operationally represented by the amount of effort an animal will spend to obtain a certain objective or reward: higher motivation to achieve a goal is associated with more effort spent toward achieving it. In this context, a modified version of the pup retrieval test has been applied in which a barrier is placed between the pups and the mother rat. Thus, to gain access to the pups, the mother rat must climb over the barrier, which effectively increases the amount of effort that the rat must spend to retrieve the pups. The latency to retrieve the first and last pups, and the number of pups retrieved would be measures of motivation [81]. A similar protocol makes use of an operant chamber, in which the female rat is trained to press a bar one time to gain access to and retrieve one pup. The first study that used this procedure [82] tested the effects of lesions to the medial preoptic area, the amygdala, and the nucleus accumbens on the mother rat's display of bar pressing to gain access to a pup. In that particular study, the bar-press reinforcement schedule was one pup delivered for one bar press (a fixed ratio, or "FR-1" schedule). However, after training on a FR-1 schedule, an investigator could gradually increase the number of bar presses required (FR-2, FR-5, FR-10, etc.) until the mother rat no longer exerts the effort required to receive a pup. The "break point" could be used as a quantitative measure of maternal motivation, as it has been used in other experimental contexts [e.g., 83].

Preference test. The motivational value of maternal stimuli versus nonmaternal stimuli can be assessed using a simple preference test. In this test paradigm, the maternal rat can freely choose to be near maternal stimuli (e.g., pups) and nonmaternal stimuli (e.g., neutral objects or other competing stimuli). The preference test can be carried out in a three-chambered apparatus in which one chamber is empty and the other two contain maternal or nonmaternal stimuli [e.g., 84]; alternatively, the test can be done in an open field arena in which the stimuli are contained within small cages that allow the interchange of visual, auditory, and olfactory information, but not physical contact between the mother rat and the stimuli [e.g., 85].

Conditioned place preference. This procedure is carried out using a plexiglass three-chambered apparatus. Transparent walls divide the left, center, and right chambers, and each wall has an opening through which the animal can pass freely between chambers. The left and right chambers have distinct contextual cues, such as horizontal versus vertical stripes on the walls and different tactile characteristics of the floors. A series of conditioning sessions are carried out, in which maternal stimuli (e.g., newborn pup) and nonmaternal stimuli (e.g., a neutral object or access to cocaine) are placed into opposing left and right chambers. During these conditioning sessions, the rat learns the association between the unconditioned stimulus (pup and neutral object) and its spatial context (left or right chamber) and develops a "place preference" for the chamber corresponding to the stimulus with the highest incentive salience, or motivational value. Finally, during the test session, the rat is placed inside the apparatus, without the stimuli being present. The total time spent in each of the chambers (left, center, and right) across the 60 min test session is recorded, and the proportion of time spent in each chamber is calculated. The proportion of time spent in the pup-associated chamber serves as a quantitative measure of the female's motivation to seek pup stimuli, as well as a measure of the relative motivational value (incentive salience) of pup-associated cues relative to competing nonmaternal stimuli [e.g., 86, 87].

4 Summary

The study of maternal behavior has crossed its traditional borders due to its complexity, flexibility, versatility, and the importance of its intergenerational transmission. It is now possible to take a broader perspective of maternal behavior than that of affiliative social behavior and its implications for offspring development. Thus, the study of maternal behavior can be seen from neuroendocrine, neurobiological, affective, cognitive, genetic, chronobiological, psychiatric, and, of course, behavioral points of view. The study of maternal behavior can even provide insight into processes of neuronal plasticity that occur in the offspring and in the mother's brain itself. The authors of this work hope that it will be very useful for students and researchers who are starting in this field.

Acknowledgments

The authors thank Dr. Kurt L. Hoffman for editing this manuscript, Angel Azael Melo Baza for the design and elaboration of the figures, and Laura García for her technical assistance.

References

1. Rosenblatt JS, Lehrman DS (1963) Maternal behavior of the laboratory rat. Maternal behavior in mammals. John Wiley & Sons, New York, pp 8–57
2. González-Mariscal G, Melo AI (2013) Parental behavior. In: Pfaff D (ed) Neuroscience in the 21st century, vol 4, 1st edn. Springer-Verlag, New York, pp 2069–2100
3. Feldman R (2016) The neurobiology of mammalian parenting and the biosocial context of human caregiving. Horm Behav 77:3–17. https://doi.org/10.1016/j.yhbeh.2015.10.001
4. Fleming AS, O'Day DH, Kraemer GW (1999) Neurobiology of mother-infant interactions: experience and central nervous system plasticity across development and generations. Neurosci Biobehav Rev 23(5):673–685. https://doi.org/10.1016/s0149-7634(99)00011-1
5. Fleming AS, Corter C (1988) Factors influencing maternal responsiveness in humans: usefulness of an animal model. Psychoneuroendocrinology 13(1–2):189–212. https://doi.org/10.1016/0306-4530(88)90014-5
6. Siegel HI, Rosenblatt JS (1975) Hormonal basis of hysterectomy-induced maternal behavior during pregnancy in the rat. Horm Behav 6:211–222. https://doi.org/10.1016/0018-506X(75)90008-2
7. Numan M, Numan MJ, Pliakou N, Stolzenberg DS, Mullins OJ, Murphy JM et al (2005) The effects of D1 or D2 dopamine receptor antagonism in the medial preoptic area, ventral pallidum, or nucleus accumbens on the maternal retrieval response and other aspects of maternal behavior in rats. Behav Neurosci 119:1588–1604. https://doi.org/10.1037/0735-7044.119.6.1588
8. Bridges RS (2015) Neuroendocrine regulation of maternal behavior. Front Neuroendocrinol 36:178–196. https://doi.org/10.1016/j.yfrne.2014.11.007
9. González-Mariscal G, Caba M, Hoffman KL, Melo AI (2017) Parental behavior. In: Pfaff DW, Joëls M (eds) Hormones brain and behavior, vol 1, 3rd edn. Academic Press, Oxford, pp 83–116. Chapter 1.03
10. Reisbick S, Rosenblatt JS, Mayer AD (1975) Decline of maternal behavior in the virgin and lactating rat. J Comp Physiol Psychol 89(7):722–732. https://doi.org/10.1037/h0077059
11. Li M (2020) Psychological and neurobiological mechanisms underlying the decline of maternal behavior. Neurosci Biobehav Rev 116:164–181. https://doi.org/10.1016/j.neubiorev.2020.06.009
12. Rosenblatt JS (1967) Nonhormonal basis of maternal behavior in the rat. Science 156:1512–1514
13. Lonstein JS, Lévy F, Fleming AS (2015) Common and divergent psychobiological mechanisms underlying maternal behaviors in non-human and human mammals. Horm Behav 73:156–185. https://doi.org/10.1016/j.yhbeh.2015.06.011
14. Barrett J, Fleming AS (2011) Annual research review: all mothers are not created equal: neural and psychobiological perspectives on mothering and the importance of individual differences. J Child Psychol Psychiatry 52(4):368–397. https://doi.org/10.1111/j.1469-7610.2010.02306.x
15. Kuroda KO, Tachikawa K, Yoshida S, Tsuneoka Y, Numan M (2011) Neuromolecular basis of parental behavior in laboratory mice and rats: with special emphasis on technical issues of using mouse genetics. Prog Neuro-Psychopharmacol Biol Psychiatry 35(5):1205–1231. https://doi.org/10.1016/j.pnpbp.2011.02.008
16. Bridges RS (2020) The behavioral neuroendocrinology of maternal behavior: past accomplishments and future directions. Horm Behav 120:104662. https://doi.org/10.1016/j.yhbeh.2019.104662
17. González-Mariscal G, Caba M, Martínez-Gómez M, Bautista A, Hudson R (2016) Mothers and offspring: the rabbit as a model system in the study of mammalian maternal behavior and sibling interactions. Horm Behav 77:30–41. https://doi.org/10.1016/j.yhbeh.2015.05.011
18. Lonstein JS, Fleming AS (2001) Parental behaviors in rat and mice. Curr Protoc Neurosci 8.15. 226:1–15
19. Ferreira A, Agrati D, Uriarte N, Pereira M, Zuluaga MJ (2012) The rat as a model for studying maternal behavior. In: Cruz-Morales SE, Arriaga-Ramìrez JC (eds) Behavioral animal models. Research Signpost, Trivandrum, Kerala, pp 73–88
20. Capone F, Bonsignore LT, Cirulli F (2005) Methods in the analysis of maternal behavior in the rodent. Curr Protoc Toxicol Chapter 13:Unit13.9. https://doi.org/10.1002/0471140856.tx1309s26

21. Kristal MB (1980) Placentophagia: a biobehavioral enigma (or De gustibus non disputandumest). Neurosci Biobehav Rev 4:141–150
22. Hudson R, Cruz Y, Lucio RA, Ninomiya J, Martínez-Gómez M (1999) Temporal and behavioral patterning of parturition in rabbits and rats. Physiol Behav 66(4):599–604
23. Stern JM, Johnson SK (1989) Perioral somatosensory determinants of nursing behavior in Norway rats (Rattus norvegicus). J Comp Psychol 103(3):269–280. https://doi.org/10.1037/0735-7036.103.3.269
24. Stern JM, Johnson SK (1990) Ventral somatosensory determinants of nursing behavior in Norway rats. I. Effects of variations in the quality and quantity of pup stimuli. Physiol Behav 47(5):993–1011. https://doi.org/10.1016/0031-9384(90)90026-z
25. Stern JM, Lonstein JS (2001) Neural mediation of nursing and related maternal behaviors. Prog Brain Res 133:263–278. https://doi.org/10.1016/s0079-6123(01)33020-0
26. Numan M, Fleming AS, Lévy F (2006) Maternal behavior. In: Neill JD (ed) Knobil and Neill's physiology of reproduction, vol 2, 3rd edn. Elsevier, San Diego, pp 1921–1923. Chapter 35
27. Ferreira A, Hansen SJ (1986) Sensory control of maternal aggression in Rattus norvegivus. J Comp Psychol 100(2):173–177
28. Stern JM (1996) Somatosensation and maternal care in Norway rats. In: Slater PJ, Rosenblatt JS, Snowden CT (eds) Parental care: evolution, mechanisms, and adaptive significance: Adv Study Par Behav, vol 25. Academic Press, New York, pp 243–294
29. Ugarte A, Eguibar JR, Cortés Mdel C, León-Chávez BA, Melo AI (2011) Comparative analysis of maternal care in the high-yawning (HY) and low-yawning (LY) sublines from Sprague-Dawley rats. Dev Psychobiol 53(2):105–117. https://doi.org/10.1002/dev.20497
30. Sánchez-Olguin CP, Zamudio SR, Guzmán-Velázquez S, Márquez-Portillo M, Caba-Flores MD, Camacho-Abrego I, Flores G, Melo AI (2022) Neonatal ventral hippocampus lesion disrupts maternal behavior in rats: an animal model of schizophrenia. 64:e22283. https://doi.org/10.1002/dev.22283
31. Kimble DP, Rogers L, Hendrickson CW (1967) Hippocampal lesions disrupt maternal not sexual behavior in the albino rat. J Comp Physiol Psychol 63(3):401–407. https://doi.org/10.1037/h0024605
32. Terlecki LJ, Sainsbury RS (1978) Effects of fimbria lesions on maternal behavior in the rat. Physiol Behav 21(1):89–97. https://doi.org/10.1016/0031-9384(78)90281-0
33. Afonso VM, Sison M, Lovic V, Fleming AS (2007) Medial prefrontal cortex lesions in the female rat affect sexual and maternal behavior and their sequential organization. Behav Neurosci 121(3):515–526. https://doi.org/10.1037/0735-7044.121.3.515
34. Lonstein JS, Wagner CK, De Vries GJ (1999) Comparison of the "nursing" and other parental behaviors of nulliparous and lactating female rats. Horm Behav 36(3):242–251. https://doi.org/10.1006/hbeh.1999.1544
35. Beach FA, Jaynes J (1956) Studies of maternal retrieving in rats. III Sensory cues involved in the lactating female's response to her young. Behavior 10:104–124
36. Fleming AS, Rosenblatt JS (1974) Maternal behavior in the virgin and lactating rat. J Comp Physiol Psychol 86:957–972
37. Stolzenberg DS, Champagne FA (2016) Hormonal and non-hormonal bases of maternal behavior: the role of experience and epigenetic mechanisms. Horm Behav 77:204–210. https://doi.org/10.1016/j.yhbeh.2015.07.005
38. Seltmann MW, Rangassamy M, Zapka M, Hoffman KL, Rodel HG (2017) Timing of maternal nest building and perinatal offspring in a group-living small mammals. Behav Ecol Sociobiol 71:64. https://doi.org/10.1007/s00265-017-2296-2
39. Schwabe K, Boldt L, Bleich A, van Dijk RM, Helgers SOA, Häger C, Nowakowska M, Riedesel AK, Schönhoff K, Struve B, Wittek J, Potschka H (2020) Nest-building performance in rats: impact of vendor experience and sex. Lab Anim 54(1):17–25. https://doi.org/10.1177/0023677219862004
40. Fleming AS, Kraemer GW, Gonzalez A, Lovic V, Rees S, Melo AI (2002) Mothers beget mothers: the transmission of behavior and its neurobiology across generations. Pharmacol Biochem Behav 73:61–75. https://doi.org/10.1016/s0091-3057(02)00793-1
41. Caldji C, Tannenbaum B, Sharma S, Francis D, Plotsky P, Meaney MJ (1998) Maternal care during infancy regulates the development of neural systems mediating the expression offearfulness in the rat. Proc Natl Acad Sci U S A 95:5335–5340. https://doi.org/10.1073/pnas.95.9.5335
42. Lucio RA, Aguilar-Pérez C, Lomanowska AM, Rodríguez-Piedracruz V, Flores-Jiménez M, Hoffman KL, Melo AI (2022) Tactile stimulation prevents disruptions in male rat copulatory behavior induced by artificial rearing. Int J

Impot Res 27. https://doi.org/10.1038/s41443-022-00530-2

43. Moore CL (1984) Maternal contributions to the development of masculine sexual behavior in laboratory rats. Dev Psychobiol 17(4): 347–356. https://doi.org/10.1002/dev.420170403

44. Stern JM, Dix L, Bellomo C, Thramann C (1992) Ventral trunk somatosensory determinants of nursing in Norway rats: II Role of nipple and surrounding sensations. Psychobiology 20:71–80. https://doi.org/10.3758/BF03327165

45. Grota LJ, Ader R (1969) Contiuous recording of maternal behavior in rattus norvegicus. Anim Behav 17:722–729. https://doi.org/10.1016/S0003-3472(69)80019-9

46. Grota LJ, Ader R (1974) Behavior of lactating rats in a dual-chambered maternity cage. Horm Behav 5(4):275–282. https://doi.org/10.1016/0018-506x(74)90014-2

47. Stern JM, Mackinnon DA (1978) Sensory regulation of maternal behavior in rats: effects of pup age. Dev Psychobiol 11(6):579–586. https://doi.org/10.1002/dev.420110607

48. Stern JM, Lonstein JS (1996) Nursing behavior in rats is affected in a small nest box and with hyperthermic pups. Dev Psychobiol 29: 101–122. https://doi.org/10.1002/(SICI)1098-2302(199603)29:2<101:AID-DEV2>3.0.CO;2-W

49. Terkel J, Rosenblatt JS (1971) Aspects of nonhormonal maternal behavior in the rat. Horm Behav 2:161–171. https://doi.org/10.1016/0018-506X(71)90031-6

50. McIver AH, Jeffrey WE (1967) Strain differences in maternal behavior in rats. Behaviour 28(1):210–216. https://doi.org/10.1163/156853967x00244

51. Moore CL, Wong L, Daum MC, Leclair OU (1997) Mother-infant interactions in two strains of rats: implications for dissociating mechanism and function of a maternal pattern. Dev Psychobiol 30(4):301–312

52. Driscoll P, Fümm H, Bättig K (1979) Maternal behavior in two rat lines selected for differences in the acquisition of two-way avoidance. Experientia 35(6):786–788. https://doi.org/10.1007/BF01968248

53. Ferré P, Fernández-Teruel A, Escorihuela RM, Driscoll P, Corda MG, Giorgi O, Tobeña A (1995) Behavior of the Roman/Verh high- and low-avoidance rat lines in anxiety tests: relationship with defecation and self-grooming. Physiol Behav 58(6):1209–1213. https://doi.org/10.1016/0031-9384(95)02068-3

54. Martínez-Álvarez V, Segura-Alegría B, Rodríguez-Torres EE, Porras-Graciela M, Aguirre-Benítez E, González del Pliego M, Hudson R, Quiroz-González S, Melo AI, Jiménez Estrada I (2022) Mother and sibling interactions during the preweaning period influence myelination and impulse propagation of the sensory sural nerve in the adult rat. Dev Psychobiol 64(7):e22316. https://doi.org/10.1002/dev.22316

55. Enes-Marques S, Giusti-Paiva A (2018) Litter size reduction accentuates maternal care and alters behavioral and physiological phenotypes in rat adult offspring. J Physiol Sci 68(6): 789–798. https://doi.org/10.1007/s12576-018-0594-8

56. Pereira M, Ferreira A (2006) Demanding pups improve maternal behavioral impairments in sensitized and haloperidol-treated lactating female rats. Behav Brain Res 175:139

57. Afonso VM, King SJ, Novakov M, Burton CL, Fleming AS (2011) Accumbal dopamine function in postpartum rats that were raised without their mothers. Horm Behav 60(5): 632–643. https://doi.org/10.1016/j.yhbeh.2011.08.016

58. Champagne FA, Francis DD, Mar A, Meaney MJ (2003) Variations in maternal care in the rat as a mediating influence for the effects of environment on development. Physiol Behav 79: 359–371. https://doi.org/10.1016/s0031-9384(03)00149-5

59. Mayer AD, Rosenblatt JS (1979) Ontogeny of maternal behavior in the laboratory rat: early origins in 18- to 27-day-old young. Dev Psychobiol 12:407–424. https://doi.org/10.1002/dev.420120502

60. Mayer AD, Freeman NC, Rosenblatt JS (1979) Ontogeny of maternal behavior in the laboratory rat: factors underlying changes in responsiveness from 30 to 90 days. Dev Psychobiol 12(5):425–439. https://doi.org/10.1002/dev.420120503

61. Kalinichev M, Rosenblatt JS, Nakabeppu Y, Morrell JI (2000) Induction of c-fos-like and fosB-like immunoreactivity reveals forebrain neuronal populations involved differentially in pup-mediated maternal behavior in juvenile and adult rats. J Comp Neurol 416(1):45–78. https://doi.org/10.1002/(sici)1096-9861(20000103)416:1<45:aid-cne5>3.0.co;2-k

62. Stolzenberg DS, Rissman EF (2011) Oestrogen-independent, experience-induced maternal behaviour in female mice. J Neuroendocrinol 23(4):345–354. https://doi.org/10.1111/j.1365-2826.2011.02112.x

63. Bridges RS, Rosenblatt JS, Feder HH (1978) Serum progesterone levels and maternal

behavior after pregnancy termination: behavioral stimulation after progesterone withdrawal and inhibition by progesterone maintenance. Endocrinology 102:258. https://doi.org/10.1210/endo-102-1-258

64. Bridges RS (1984) A quantitative analysis of the roles of dosage, sequence, and duration of estradiol and progesterone exposure in the regulation of maternal behavior in the rat. Endocrinology 114(3):930–940. https://doi.org/10.1210/endo-114-3-930
65. Numan M, Rosenblatt JS, Komisaruk BR (1977) Medial preoptic area and onset of maternal behavior in the rat. J Comp Physiol Psychol 91(1):146–164. https://doi.org/10.1037/h0077304
66. Pedersen CA, Prange AJ Jr (1979) Induction of maternal behavior in virgin rats after intracerebroventricular administration of oxytocin. Proc Natl Acad Sci U S A 76(12):6661–6665. https://doi.org/10.1073/pnas.76.12.6661
67. Novakov M, Fleming AS (2005) The effects of early rearing environment on the hormonal induction of maternal behavior in virgin rats. Horm Behav 48(5):528–536. https://doi.org/10.1016/j.yhbeh.2005.04.011
68. Sophocleous A, Idris AI (2019) Ovariectomy/orchiectomy in rodents. Methods Mol Biol 1914:261–267. https://doi.org/10.1007/978-1-4939-8997-3_13
69. Bosch OJ (2013) Maternal aggression in rodents: brain oxytocin and vasopressin mediate pup defence. Philos Trans R Soc Lond Ser B Biol Sci 368(1631):20130085. https://doi.org/10.1098/rstb.2013.0085
70. Lonstein JS, Gammie SC (2002) Sensory, hormonal, and neural control of maternal aggression in laboratory rodents. Neurosci Biobehav Rev 26(8):869–888. https://doi.org/10.1016/s0149-7634(02)00087-8
71. de Almeida RM, Ferreira A, Agrati D (2014) Sensory, hormonal, and neural basis of maternal aggression in rodents. Curr Top Behav Neurosci 17:111–130. https://doi.org/10.1007/7854_2014_312
72. Agrati D, Fernández-Guasti A, Ferreño M, Ferreira A (2011) Coexpression of sexual behavior and maternal aggression: the ambivalence of sexually active mother rats toward male intruders. Behav Neurosci 125(3):446–451. https://doi.org/10.1037/a0023085
73. Erskine MS, Barfield RJ, Goldman BD (1978) Intraspecific fighting during late pregnancy and lactation in rats and effects of litter removal. Behav Biol 23(2):206–218. https://doi.org/10.1016/s0091-6773(78)91814-x
74. Ferreira A, Hansen S (1986) Sensory control of maternal aggression in Rattus norvegicus. J Comp Psychol 2:173–177. https://doi.org/10.1037/0735-7036.100.2.173
75. Melo AI, Hernández-Curiel M, Hoffman KL (2009) Maternal and peer contact during the postnatal period participate in the normal development of maternal aggression maternal behavior and the behavioral response to novelty. Behav Brain Res 201(1):14–21. https://doi.org/10.1016/j.bbr.2009.01.023
76. Quadros MI, Takahashi A, Klaus A, Miczek A (2010) Serotonin and aggression. In: Múller CP, Jacobs BJ (eds) Handbook of behavioral neuroscience, vol 21. Elsevier, pp 687–713. https://doi.org/10.1016/S1569-7339(10)70106-5. Chapter 4.10
77. Olivier B, Mos J, van Oorschot R (1985) Maternal aggression in rats: effects of chlordiazepoxide and fluprazine. Psychopharmacology 86(1–2):68–76. https://doi.org/10.1007/BF00431686. PMID: 3927368
78. Kolunie JM, Stern JM, Barfield RJ (1994) Maternal aggression in rats: effects of visual or auditory deprivation of the mother and dyadic pattern of ultrasonic vocalizations. Behav Neural Biol 62(1):41–49. https://doi.org/10.1016/s0163-1047(05)80057-3
79. Olazábal DE, Pereira M, Agrati D, Ferreira A, Fleming AS, González-Mariscal G, Lévy F, Lucion AB, Morrell JI, Numan M, Uriarte N (2013) New theoretical and experimental approaches on maternal motivation in mammals. Neurosci Biobehav Rev 37(8):1860–1874. https://doi.org/10.1016/j.neubiorev.04.003
80. Bayerl DS, Klampfl SM, Bosch OJ (2014) Central V1b receptor antagonism in lactating rats: impairment of maternal care but not of maternal aggression. J Neuroendocrinol 26(12):918–926. https://doi.org/10.1111/jne.12226
81. Salais-López H, Abellán-Álvaro M, Bellés M, Lanuza E, Agustin-Pavon C, Martínez-García F (2021) Maternal motivation: exploring the roles of prolactin and pup stimuli. Neuroendocrinology 111(9):805–830. https://doi.org/10.1159/000510038
82. Lee A, Clancy S, Fleming AS (2000) Mother rats bar-press for pups: effects of lesions of the mpoa and limbic sites on maternal behavior and operant responding for pup-reinforcement. Behav Brain Res 108(2):215–231. https://doi.org/10.1016/s0166-4328(99)00170-9
83. Hecht GS, Spear NE, Spear LP (1999) Changes in progressive ratio responding for intravenous cocaine throughout the

reproductive process in female rats. Dev Psychobiol 35(2):136–145. https://doi.org/10.1002/(SICI)1098-2302(199909)35:2%3C136::AIDDEV6%3E3.0.CO;2-K

84. Ferreño M, Pose S, Agrati D, Zuluaga MJ, Ferreira A, Uriarte N (2018) Incentive value of newborn pups relative to juveniles for mother rats raising overlapping litters. Behav Process 157:333–336. https://doi.org/10.1016/j.beproc.2018.07.016

85. Gao J, Nie L, Li Y, Li M (2020) Serotonin 5-HT_{2A} and 5-HT_{2C} receptors regulate rat maternal behavior through distinct behavioral and neural mechanisms. Neuropharmacology 162:107848. https://doi.org/10.1016/j.neuropharm.2019.107848

86. Seip KM, Morrell JI (2008) Exposure to pups influences the strength of maternal motivation in virgin female rats. Physiol Behav 95(4):599–608. https://doi.org/10.1016/j.physbeh.2008.09.003

87. Wansaw MP, Pereira M, Morrell JI (2008) Characterization of maternal motivation in the lactating rat: contrasts between early and late postpartum responses. Horm Behav 54(2):294–301. https://doi.org/10.1016/j.yhbeh.2008.03.005

Chapter 5

The Bruce Effect: Complementary Roles of Olfactory Memory and Male-Sourced Estradiol

Denys deCatanzaro

Abstract

Uterine implantation of fertilized ova is a highly sensitive transition from minimal maternal investment to increasing devotion of resources to pregnancy. It involves a finely tuned interplay in the uterus between progesterone (P4) and the powerful estrogen, 17β-estradiol (E2). P4 generally supports implantation, whereas minute elevations of E2 terminate it. Diverse maternal stressors can cause implantation to fail, usually by raising the female's endogenous E2:P4 ratio. This is viewed as an ancient mammalian adaptation that forestalls adverse outcomes when circumstances are not propitious to maternal and fetal/offspring health.

In 1959, it was discovered that exposing inseminated female mice to novel males (those other than the sire) caused implantation failure (the Bruce effect). Novel males of a genetic strain distinct from the sire's strain induced the strongest effects. This led to the notion that the inseminated female imprints on the sire's odor and reacts differently to novel males' odors (the olfactory memory hypothesis). Females experiencing the Bruce effect showed degeneration of the ovarian corpora lutea, which produce P4 and are normally sustained by pituitary prolactin pulses. The effect was shown to be mitigated by rendering the female anosmic or by giving her exogenous prolactin or P4. This led to decades of research examining constituents of male urine that signal individual differences, receptors in the olfactory system that transduce chemical messages to neural signals, and neural pathways that bring this information to the hypothalamus, which controls pituitary prolactin. Unfortunately, much of this research was conducted without reference to the phenomenon of stress-induced implantation failures, and some of it was confounded by human handling and other stress-inducing procedures.

Later, it was discovered that male urine and seminal emissions contain substantial quantities of bioactive E2, especially when males have not recently mated and are near females. Manipulations that diminish male urinary E2 and those that reduce female reactivity to E2 can prevent the Bruce effect. When tritium-labeled E2 (^{3}H-E2) is administered to males, untreated females housed briefly with these males show substantial radioactivity in the uterus and other tissues where estrogen receptors are abundant. This steroid transfer occurs without mediation by the brain, as E2 is a small, polar, and lipophilic molecule that is readily absorbed percutaneously, nasally, and vaginally directly into circulation. Moreover, when novel males are housed directly with females and allowed to mate, relatively high levels of E2 are deposited directly into the female's reproductive tract.

Raúl G. Paredes et al. (eds.), *Animal Models of Reproductive Behavior*, Neuromethods, vol. 200,
https://doi.org/10.1007/978-1-0716-3234-5_5,

The olfactory memory and the male-sourced-E2 hypotheses are not mutually exclusive alternatives. In fact, they are complementary, as a high E2:P4 ratio impedes implantation via several known uterine mechanisms. Just as there are multiple and often redundant mechanisms that subserve other fundamental adaptations (e.g., hunger, thirst, mating behavior, and circadian rhythm), the Bruce effect is multicausal.

Key words Bruce effect, Implantation, Estradiol, Progesterone, Olfactory memory, Steroid transfer

1 Introduction

While studying other phenomena, Hilda Bruce [1] incidentally observed that recently inseminated female mice that were housed with novel ("strange") adult males often failed to show intrauterine ovo-implantation and returned to estrus 4–5 days after the original mating. This "pregnancy block" was not induced if the female returned to the original sire and was greatest if the novel male was of a different genetic strain from the sire [2]. Females were most vulnerable within 48 h of coitus and the reaction rapidly disappeared after the start of blastocyst implantation [3]. The effect was shown to be androgen dependent, as castrated males showed a reduced capacity to induce the effect [4] and androgen treatment of ovariectomized females enabled them to induce it [5]. Typically, not all novel-male-exposed females lost pregnancy. Bruce and Parrott [6] observed that pregnancy was retained in the face of novel males when females were previously rendered anosmic by removing their olfactory bulbs. Female nasal exposure to novel male urine alone may be sufficient to induce the effect, although this required stringent conditions. Parkes and Bruce [7] observed that newly mated females tended to lose pregnancy when housed each for up to 3 days in soiled cages vacated by five novel males.

Increasingly, the Bruce effect has been found in diverse mammals, both wild and domestic, albeit with some variation in form. These species include various rodents [8–12], ungulates [13–15], and at least one primate species [16]. This review will primarily focus on laboratory mice, where the data on mechanisms are richest due to large samples of animals examined in many controlled laboratory procedures. It will discuss evidence supporting the role of female olfactory memory in the Bruce effect, which is largely the work of neuroscientists examining olfactory receptors of male urinary proteins and peptides, pathways in the brain, and transduction of neural to endocrine signals that bear upon implantation. It will then discuss work proving that bioactive estrogens are found in male urine and seminal emissions. These excretions are targeted behaviorally toward females and absorbed nasally, percutaneously, and vaginally into the females' circulation. These estrogens arrive in the uterus where estrogens are known to disrupt implantation. I will then conclude by showing how the two lines or work are complementary. But before examining the mechanisms of the Bruce effect, it is important to review hormonal mechanisms that control the success or failure of implantation.

2 Female Adaptations Around Implantation and Stress

In the days following insemination and conception, a female's investment in pregnancy is minimal, and there is a substantial possibility that pregnancy will end. Female mammals long ago evolved endogenous mechanisms that forestall adverse outcomes in the health and survival of mother and potential offspring. Implantation is fragile and vulnerable to various stressors, that is, circumstances that challenge the female's coping mechanisms, typically activating her sympathetic nervous system and the hypothalamic–pituitary–adrenocortical axis. Very often the mechanisms of the Bruce effect have been studied in isolation, without mention of the strong resemblance of this effect to stress-induced implantation failure.

2.1 Diverse Stressors Can Prevent Implantation

Fertilization occurs in the fallopian tubes, then the conceptus undergoes some cell divisions and migrates toward the uterus, where in order to thrive it must implant at the blastocyst phase in the uterine epithelium. Implantation begins a transition to progressively increasing devotion of the mother's physiological resources. A wealth of evidence from diverse mammals indicates that various stressors can terminate a pregnancy during the peri-implantation period [17]. Experiments with laboratory rodents demonstrate that implantation can be prevented by physical restraint [18], environmental and social changes [19], predator exposure [20], and human handling [21], among other stressors. In domestic herbivores, implantation failure can occur after exposure to stressors such as extreme temperature [22] and herding, restraint, and transportation [23, 24]. Note that postimplantation stressors can also produce spontaneous abortion (miscarriage), fetal resorption, and a variety of developmental disorders; these involve diverse mechanisms that are distinct from those of implantation failure and include actions of cortisol, reductions in P4, and maternal immune cells that attack the embryo or fetus [25–28].

2.2 Mechanisms of Implantation Success or Failure

Successful implantation involves coordinated interplay in the uterus of progesterone (P4) and the most powerful estrogen, 17-β-estradiol (E2), with finely tuned dynamics that depend upon timing and relative concentrations of these steroids. Generally, a high E2:P4 ratio undermines implantation [20, 29]. Although E2 is critical in the preparation of the uterus for implantation, excessive E2 can disrupt processes that are necessary for successful implantation. Elevated E2 can affect the rate of passage of fertilized ova down the fallopian tube, such that blastocysts arrive prematurely in an unprepared uterus [30], and it can also directly jeopardize the survival of the embryo [31]. In a healthy pregnancy, the uterine lumen, the fluid-filled space inside the uterus, closes in around blastocysts prior to implantation, facilitating their contact with the uterine wall [32]. P4 drives this process by causing the efflux of fluid to the epithelial cells and uterine glands, whereas E2 undermines this process by inducing fluid secretion into the lumen [33–36]. Also, epithelial cells in the peri-implantation uterus secrete a number of adhesion proteins, including cadherins and integrins, that essentially glue the blastocysts to the epithelium and reinforce the closure of the uterine lumen [37–41]. These proteins are promoted by P4 and downregulated by E2 [39, 42–44].

Stress perturbs the E2:P4 ratio, and both E2 (or its androgen precursors) and P4 can come from the adrenal cortex as well as the ovaries. The adrenal glands increase their production of steroids and other hormones during stress, but exogenous doses of classic stress hormones like epinephrine (adrenaline) and glucocorticoids (cortisol, corticosterone) have relatively little impact on implantation [45, 46]. However, when given subcutaneously daily on days 1–5 after insemination, the adrenocortical androgens androstenedione and DHEA, respectively, disrupted implantation at 500 μg and 100 μg doses [46]. For testosterone, the minimal effective dose was 27 μg, whereas for estrone or estriol, it was 9 μg, and for E2 it was merely 37 ng [46, 47]. Ma et al. [48] found that, when appropriately timed, a single dose of 10 ng E2 closed the implantation window, rendering the uterus refractory to blastocyst implantation. However, the impacts of doses of E2 can be modulated by counteracting influences of P4 [20, 29].

In female rats, exposure to acute swimming stress was associated with increased E2 [49]. A similar effect was found in female mice exposed to a relatively severe stressor (predator exposure) but not when they were exposed to a milder stressor (isolation on an elevated platform) [50]. In inseminated laboratory rodents, various chronic stressors [20, 51, 52] or acute immune challenges [53] have been reported to increase E2. When inseminated female mice were subjected to peri-implantation restraint stress, concurrently injecting them with estrogen antibodies mitigated the loss of pregnancy [54]. P4 is also perturbed by stressors, but its dynamics depend on when the hormone is measured relative to the stress and the female's reproductive state. P4 is the metabolic precursor of glucocorticoids in the adrenal cortex, so a transitory rise in P4 is typically seen just after an acute stressor [50]. However, in recently inseminated mice, major stressors can suppress P4 [20, 53, 55]. Prolactin, whose actions in early pregnancy support P4 production by the ovarian corpora lutea, was also found to be greatly reduced by immune challenge or fasting stress [55]. In some circumstances, implantation failure can be prevented by administering exogenous P4 [21, 56, 57], exogenous P4 plus a very low dose of E2 [20], or a P4 analog [58] in conjunction with the stressor.

3 The Role of Female Olfactory Memory in the Bruce Effect

Early research became focused on "the olfactory memory hypothesis," the idea that during mating, a female forms a strong memory trace of the odor of the sire during the pregnancy. She then may react to distinctive male odors through a chain of neural and endocrine processes that suppress her P4 levels, leading to implantation failure and a return to estrus. The most commonly cited mechanism through which this would occur is summarized in Fig. 1, with more detail given below.

3.1 Urinary Markers of Individual Differences

It has long been known that novel male mice of different genetic strains differ in their ability to induce the Bruce effect [59], and that female mice are more susceptible to the Bruce effect when the genetic strain of the novel male differs from that of the sire [7, 60]. Many studies also suggest that mice can discriminate among the odors of different genetic strains of mice, if not among individuals [61–64].

Male mouse urine contains substantial quantities of major urinary proteins (MUPs), which are androgen dependent, less common in females, and thought to communicate maleness to conspecifics [65]. MUPs are highly polymorphic across individual male mice [66], which makes them strong candidates as stimuli for female discrimination among individuals [67, 68]. Evidence suggests that one MUP ("darcin") with the mass of 18,893 Da, acts as a pheromone that stimulates female memory and sexual attraction to an individual male's odor [69]. While they likely contribute, evidence indicates that MUPs are less critical in the Bruce effect than low-molecular-weight (LMW) male urinary constituents. Marchlewska-Koj [70] salted proteins out of male-urine, finding that these proteins alone could end a pregnancy when delivered to the inseminated female's nasal area. She later [71] compared the effects of a control solution, a male urinary fraction containing high-molecular-weight (HMW) proteins, a fraction containing LMW proteins, and a fraction containing urinary peptides and other constituents of low-molecular weight. Only the last of these fractions terminated pregnancy in inseminated females. She concluded that the active pheromone was either a low-molecular-weight peptide or, notably, a non-peptide substance appended to such peptides. Later, Peele et al. [72] found that exposing inseminated females to HMW proteins (MUPs) from male urine did not readily block pregnancy, unless they were combined with low-molecular-weight (LMW) urinary constituents taken from either familiar or unfamiliar males. LMW constituents from familiar or unfamiliar males on their own blocked pregnancy in 50–62% of subjects, compared to 100% of subjects when exposure was to

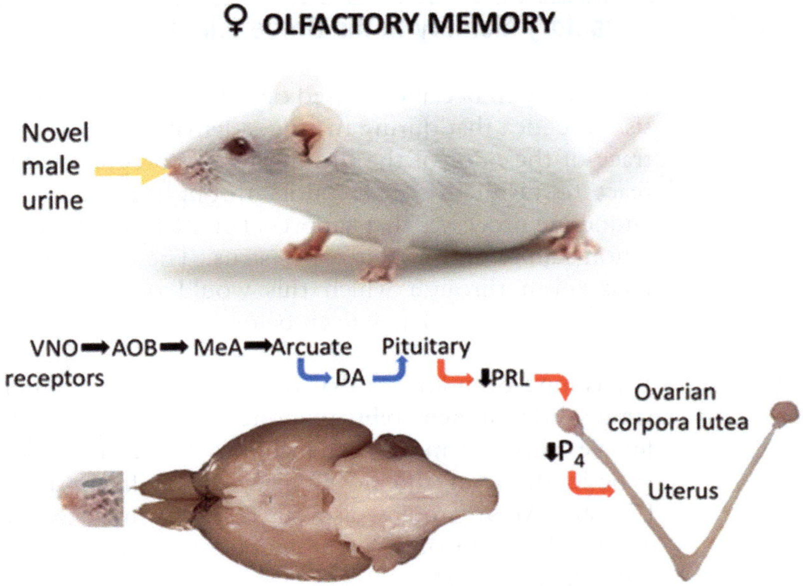

Fig. 1 Basic chain of mechanisms from female detection of novel males' odors to failure of implantation. The female is believed to form a memory trace of the sire during insemination. When subsequently confronted with a novel male, constituents of his urine will stimulate sensory neurons in her vomeronasal organ (VNO), which activates an excitatory pathway to the accessory olfactory bulb (AOB), then the medial amygdala (MeA), then the arcuate nucleus of the hypothalamus. This increases tuberoinfundibular dopamine (DA), which suppresses prolactin (PRL) pulses from the anterior pituitary that normally sustain ovarian corpora lutea that produce progesterone (P4). P4 is essential for the implantation of ova during the blastocyst phase

combined HMW and LMW constituents from novel males [72, 73].

Some studies suggest that males' differential excretion of small peptides related to the major histocompatibility (MHC) complex might account for inseminated females' discrimination among different genetic strains of mice. Leinders-Zufall et al. [74] examined two peptides, one characteristic of C57BL/6 strain mice and another of BALB/c strain mice, and found that certain vomeronasal organ cells of female C57BL/6 mice are exceptionally sensitive to these peptides in vitro. In vivo experimentation in that laboratory showed that supplementing urine from same-strain males with peptides characteristic of novel-strain males could enhance its capacity to disrupt pregnancy. Further work [74, 75] showed that nasal application of urine from an unfamiliar strain resulted in a high level of pregnancy failure in inseminated females, whereas failure rates were low following the application of the familiar strain's urine. The authors noted, however, that pregnancy disruptions did not occur in earlier pilot studies when these peptides were administered nasally in water as opposed to male urine, which again suggests that a urinary constituent unknown to these researchers was critical for the induction of pregnancy loss.

In Subheading 4, I will review evidence that E2 is the critical constituent of male urine that is necessary for pregnancy loss, potentially interacting with MUPs and these MHC peptides. At 272.4 Da, E2 certainly qualifies as an LMW urinary constituent, and it is lipophilic and readily absorbed directly into circulation after dermal, nasal, or vaginal exposure.

3.2 The Vomeronasal Organ

Another major line of research pursuing Bruce's discovery focused on the actions of male urinary substances on the olfactory system of the inseminated females and their transduction through neural mechanisms to processes that could bear upon intrauterine ovo-implantation. Work particularly became focused on the female's vomeronasal organ (VNO), part of the accessory olfactory system found at the base of the nasal septum. The VNO is a vasomotor pump that aspirates substances toward nasal receptors as an autonomic response to objects and odors that catch the animal's attention [76–78]. Bellringer et al. [79] and Lloyd-Thomas and Keverne [80] lesioned the VNO of female mice via bilateral incisions in the palate, burning away the organ with a diathermic tool and then filling the gap with polycyanacrylate cement. They subsequently observed that the Bruce effect was absent in such females and concluded that the effect must require specific odor receptors in the VNO. There is indeed a diversity of potential receptors in the VNO [81, 82]. Nevertheless, it is very difficult to exclude the possibility that collateral damage from VNO lesions could render inseminated females unreactive to novel males, regardless of the actions of male urinary peptides on this organ. Surely, destroying the VNO prevents the animal from sucking up whatever agents induce the effect, be they peptides, steroids, or something else. Moreover, VNO lesioning is likely to damage blood circulation. As evidence reviewed in Subheading 4 indicates, absorption by the nasal vasculature of E2 from male urine results in the direct transfer of this powerful steroid into circulation, without neural mediation.

3.3 Pathways in the Brain

The VNO ablation data led to many studies of the inseminated females' olfactory, neural, and endocrine responses to novel-male odors [63, 67]. Neuroscientists interested in the Bruce effect focused on pathways in the brain that convey signals from the VNO to the hypothalamus and pituitary. Brennan [63] reviewed studies suggesting that female VNO receptors responding to novel male odors activated an excitatory pathway via the accessory olfactory bulb (AOB) and corticomedial amygdala to the arcuate nucleus of the hypothalamus. In response, the release of dopamine by tuberoinfundibular neurons suppresses anterior pituitary prolactin pulses that are essential for ovarian corpora luteal production of P4. One influential study demonstrating that pathway was that of Luo et al. [83], who implanted microelectrodes in the accessory olfactory bulb of mice, which allowed recording of activity from single neurons. The mice were then exposed to conspecifics that differed by sex, genetic makeup, and hormonal state. The neuronal firing was modulated by physical contact with male and female mice, with individual neurons reacting selectively to specific combinations of sex and genetic strain. They inferred that mice and other mammals encode social and reproductive information received from a sensory activity from the VNO.

Another influential study was that of Li et al. [84], who placed stimulating electrodes in the AOB of anesthetized female mice at least 1 week prior to mating them. The stage of the estrous cycle of each mouse was monitored twice daily via vaginal smears. In the early days after mating, in the absence of any males, electrical stimulation of the AOB for 4 h reduced the number of females showing implanted embryos, mimicking the Bruce effect, but only when such stimulation was coincident with known times of prolactin surges.

Neural circuitry involved in female olfactory memory is undoubtedly more complex than that shown in Fig. 1. A review by Baum and Bakker [85] cited evidence that a second parallel circuit involving the main olfactory system detects volatile pheromones from conspecifics. While volatile substances could be a part of the overall constituents of urine that communicate individual identity of male mice to females, most evidence suggests that the major male urinary constituents underlying the Bruce effect are not volatile.

3.4 Transduction of Neural to Endocrine Signals

Bruce and Parkes [86] posited that novel males' odors caused blastocyst implantation to fail due to suppression of P4 levels at the uterus, mediated by neurochemical suppression of pituitary prolactin pulses that support the development and maintenance of ovarian corpora lutea. When they injected female mice with prolactin during the initial days of pregnancy, implantation failure was largely prevented in the presence of novel males. However, the injection of P4 was much less effective. Work from another laboratory [87–90] showed that implantation failure in mice during exposure to males or their urine could be prevented by exogenous prolactin or P4, depending on the timing of the injections. They also found signs of degeneration of the corpora lutea associated with pregnancy loss. Milligan [91] showed that male-induced implantation failure in the vole correlated with degeneration of the corpora lutea, consistent with prolactin suppression. Rosser et al. [92] found that exposure to males' soiled bedding disrupted implantation in mice when it was coincident with natural prolactin surges, but not when it occurred at other preimplantation times. Arcuate tuberoinfundibular dopamine suppresses pituitary prolactin, and it has been hypothesized that this may link the female's neural responses to male odors to hormonal actions that suppress implantation. Giving inseminated females the dopamine agonist, bromocriptine, mimics the influence of novel males on implantation [79, 92]. Giving them pimozide, which blocks dopaminergic transmission, prevents the Bruce effect [93]. Work by Wersinger et al. [94] examined how knockout (KO) females for oxytocin and vasopressin reacted in the Bruce effect. The oxytocin receptor (Oxt) KO females lost pregnancy regardless of whether the sire or a novel male was introduced 24 h after insemination. Two different vasopressin receptor knockouts were examined; Avpr1b KO females retained their pregnancies in the presence of a novel male, but the Avpr1a KO females showed a normal Bruce effect. They concluded that both Oxt and Avprlb receptors are critical for the Bruce effect but in different manners.

3.5 Concerns about Stressful Procedures in Some Olfactory Memory Research

Often the work by neuroscientists has not been integrated with growing knowledge from reproductive biology of uterine and steroid hormone dynamics that tightly regulate the success and failure of blastocyst development and implantation. Much older work needs to be scrutinized carefully because of invasive procedures, such as daily vaginal smears, injections, handling during nasal application of urine or its constituents, and neural interventions. Sometimes the relevant methods are not fully described. This is unfortunate given evidence that blastocyst implantation can be disrupted by human handling [27] and physical restraint [18], while stressors can elevate E2 [49, 50] and lower P4 [20, 55]. One early example was the study of Parkes and Bruce [7], who reported pregnancy loss in females housed in soiled cages vacated by five novel males, but that the effect was not observed unless "the stimulus was intensified by housing the animals in tall glass jars, with reduced ventilation, on cloth bedding highly retentive of animal smells, and renewing the soiled container twice daily for the 3-day-exposure period" (p. 303). Invasive procedures such as castration, hormone replacement, and various neural manipulations certainly have a role in research on mechanisms, but should be avoided during the window of implantation after insemination. As mentioned above, surgical damage to some of the female's most vital olfactory organs such as the VNO inevitably involves collateral damage to blood circulation. Studies involving chronically implanted microelectrodes [83, 84] had appropriate control procedures to prevent confounds, but such stressors will certainly alter the background endocrine state, activating pituitary–adrenocortical axis hormones and the sympathetic nervous system, and potentially also perturbing ovarian hormones. Data show that even minor stressors will perturb P4 simply because it is a precursor in the synthesis of cortisol and corticosterone [50]. As discussed in Subheading 2, major chronic stressors can elevate E2 and reduce P4 concentrations.

4 The Role of Male-Excreted Estradiol in the Bruce Effect

As reviewed in Subheading 2, female mammals have longstanding adaptations to terminate their own pregnancies when circumstances are not favorable for gestation, mediated peri-implantation by an increased E2:P4 ratio. I will now review evidence that non-sire (novel) males have evolved to excrete E2 and direct it to inseminated females such that they will return to estrus and be potentially re-inseminated. The mechanisms through which male-sourced E2 impedes implantation are summarized in Fig. 2, and the data that provide support will now be discussed.

4.1 General Methods

My laboratory began to study the Bruce effect after years of working on stress-induced implantation failure. In mice, the two effects closely resemble each other, as in both cases implantation fails probabilistically (not all females are affected) and in an all-or-none fashion (females either have no pups or normally sized litters) (cf. 20, 95). My students and I have strived to minimize human handling of animals, avoid invasive procedures during the window of implantation, and measure pregnancy outcomes either by allowing females to bear their litters or by sacrificing animals after that window and counting implantation sites.

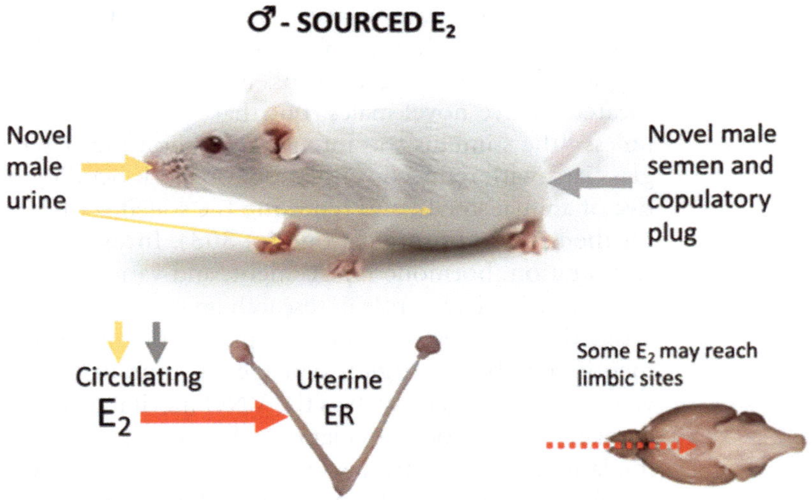

Fig. 2 Mechanisms via which male-sourced estradiol (E2) disrupts implantation. Male urine reliably contains substantial quantities of bioactive E2, which enters directly into the female's circulation after nasal and/or dermal exposure. This E2 is found especially in organs with substantial concentrations of estrogen receptors (ER), particularly the uterus. Male semen and copulatory plugs contain even more substantial concentrations of E2 than those in urine. E2 received intravaginally also transfers into circulation, but semen and the copulatory plug can directly interact with uterine ER. Minute elevations of E2 above optimal concentrations undermine uterine receptivity to blastocysts through several mechanisms. Comparatively little E2 reaches the brain after nasal exposure and mating, and almost none does so after percutaneous absorption

When novel males are directly housed with previously inseminated females, they tend to replace the pregnancy [2, 96]; when they are housed across a wire-mesh grid from females that prevents mating, the pregnancy is simply lost [96]. Unless specified otherwise, the data in this section are from indirect exposure, where males are housed above the females and separated by wire mesh, which exposes females to males' excretions and allows behavioral interaction but not mating. Of course, direct exposure is the most naturalistic method. The Bruce effect is not just another example of stress-induced pregnancy failure as it involves special dynamics related to novel males' excretions (as well as female olfactory memory).

4.2 Bioactive E2 Is Reliably Found in Male Excretions

E2 has been present in every one of thousands of urine samples from males measured in my lab, and urinary E2 concentrations in male and female mice are typically in overlapping ranges. However, adult males, especially dominant ones, tend to produce larger quantities of urine than females [97], and as discussed below, males' urination patterns and urinary E2 concentrations are socially more dynamic. Testosterone, the metabolic precursor of E2, is also reliably found in male mouse urine [95]. Female urine also contains P4 [98], which may counteract potential impacts of E2 when females are exposed to each other.

As mentioned, minute doses of exogenous E2 undermine implantation [47, 48], mimicking the Bruce effect [95]. This effect occurs when E2 is applied subcutaneously [46] or nasally [47]. As noted above, Ma et al. [48] found that the implantation window in mice could be closed by a single dose of 10 ng E2. The bioavailability of nasally applied E2 in rats has been shown to exceed 50% of an equivalent i.v. dose [99]. High levels of bioactive, unconjugated E2, and other estrogens occur naturally in male excretions, such as urine in mice [100, 101]. Although estrogen conjugates are common in larger, more complex mammals such as humans, they are sparse in female mouse urine and not detectable in the urine of male mice [100]. Interestingly, axillary perspiration of young men [102, 103] and seminal emissions of diverse mammals [104] contain very substantial concentrations of unconjugated E2.

4.3 Differences Between Sires and Novel Males in Behavior and Urinary Steroids

Wherever males and females are allowed to interact, behavior contributes to the Bruce effect. Female behavior plays a critical role. In one study [105], inseminated females that avoided both unfamiliar and familiar male scents during critical periods of susceptibility to the Bruce effect maintained their pregnancy, but those that spent more time with the unfamiliar male scent did not.

In male mice, urination is often a social response. Whereas female mice urinate in large puddles as do juvenile males, adult males in social conditions urinate in small droplets, dispersing them around their cages [106–108]. Sires are generally quite passive when they are housed nearby or directly with females that they have recently inseminated [109, 110], but novel males become agitated and may direct urine at females, attempting to mount and intromit if not separated by wire mesh [96]. Urinary E2 concentrations in novel-male urine begin to rise within 3 days of habitation with females in adjacent compartments separated by a wire-mesh grid [95, 101] (*see* Fig. 3). Quite possibly, this increase in E2 is due to the growing presence of seminal emissions, as males' compartments can become smeared with milky-white deposits [104, 111]. Novel males housed near females also progressively show polydipsia and polyurea, increasing their water consumption and urinary output, directing their urine repeatedly across the grid toward females [101]. In contrast, the rise in urinary E2 concentrations does not happen in sires similarly housed across a grid from their inseminated mates. If the sire is concurrently present when the female is exposed to a novel male, the Bruce effect may not occur, at least in part due to the sire's aggression toward the intruder [110].

A large part of the differences between sires and novel males is the simple fact that the sires have recently mated, whereas the novel males have not. This was clear in experiments where a state of "sexual satiety" was induced in a subset of novel males while others had been sexually deprived [60]. When males were mated repeatedly with other females, then presented to recently inseminated females, they failed to produce a Bruce effect. This was likely due to some combination of reduced motivation to mate [112] and reduced urinary E2. Although E2 was not measured in this study, it is known that its precursor, testosterone, decreases in concentration following ejaculation [113, 114].

4.4 Manipulations of Male Urinary E2 and Female Reactivity to E2 Have Predictable Effects

Castration diminishes the ability of novel male mice to disrupt implantation, an effect that occurs gradually in the weeks following surgery in conjunction with declines in urinary testosterone and E2, with apparent asymptotes 6 weeks following surgery [115]. Without restoring testosterone levels, restoration of normal urinary E2 concentrations in castrated males via intramuscular E2 injections reinstates their ability to disrupt implantation [116]. Similarly, giving intact novel males a drug (anastrozole) that blocks aromatase, the enzyme that naturally converts testosterone to E2, reduces their urinary E2 concentrations and diminishes their capacity to disrupt implantation [117]. It is also possible to make inseminated females less reactive to novel males by making them resistant to E2; when female mice were injected with exogenous antibodies to E2 during exposure to novel males, most of them retained their pregnancies [118].

Notably, bilateral removal of the prominent androgen-dependent sex-accessory glands, the preputials, did not diminish novel males' capacity to disrupt implantation, both when males were separated from females by a grid and when they were directly

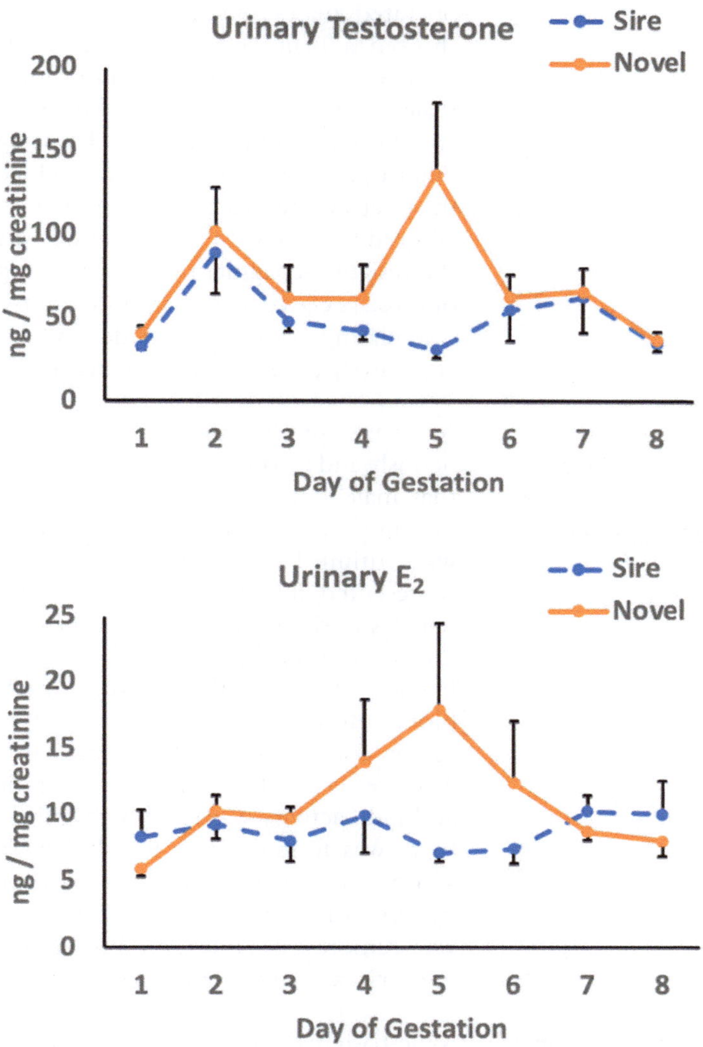

Fig. 3 Urinary testosterone and E2 concentrations in sires and novel males during the first 8 days of gestation. Each male was housed across a wire-mesh grid from a recently inseminated female. Each sire and novel male was alone with the female, with females randomly assigned to being with one or the other. Concentrations were adjusted for urinary creatinine following conventions to correct for differential hydration. Creatinine measures did not differ significantly between conditions. Error bars indicate S.E.M. There was a significant rise in urinary E2 concentrations by day 5 among the novel males compared to sires, which showed more constant measures across days. (From deCatanzaro et al. [95])

exposed and allowed to mate and fertilize females [96]. Later, we exposed inseminated females (separated by a grid) to (1) males with their vesicular-coagulating glands removed, (2) males with both preputials and vesicular-coagulating glands removed, or (3) sham-operated males, finding that females in all of these conditions showed a Bruce effect when compared to controls [119]. In a separate test, fertility was not impeded in preputialectomized males, but half of the males without vesicular-coagulating glands could not inseminate females. These data clearly rule out the possibility that emissions uniquely produced by the preputials contain the pheromone that disrupts implantation. As novel males were not allowed to mate with females in this study, emissions uniquely from the vesicular-coagulating gland were not addressed. However, both of these sex-accessory glands do emit E2, particularly the vesicular-coagulating gland as it provides a substantial repository of E2 in the copulatory plug (*see* Subheading 4.6).

4.5 E2 Transfers from Novel Males to Blood and Organs of Cohabiting Females

As a small and lipophilic molecule, E2 is readily absorbed percutaneously and nasally [120–124]. Experiments with mice have shown that male-sourced E2 transfers to the female's blood and tissues during simple cohabitation for a few days. Male mice were injected with tritium-labeled estradiol (^{3}H- E2) using a very low dose that represented just a fraction of their endogenous E2. Untreated females were then housed with these males, either directly or across a wire-mesh grid [122, 125]. In either housing arrangement, the females subsequently showed significant radioactivity in their blood, uterus, and other tissues, roughly corresponding in quantity to the concentration of estrogen receptors (ER) in these tissues. The uterus has the highest concentration of ER in the female body, and radioactivity was reliably found there. Relatively little radioactivity was found in the females' brain tissues. Species generality among mammals of male-to-female transfer of ^{3}H-E2 was tested by replicating these procedures in a species that is phylogenetically very remote from mice, big brown bats, and transfer was observed in every case where females cohabited with treated males [126].

4.6 E2 Is Abundant in Semen and Directly Reaches the Uterus During Mating

Male-to-female E2 transfer is especially rapid and pronounced during mating [104]. In rodents, domestic herbivores, and primates, high concentrations of E2 and other estrogens are naturally found in rete testes fluid, which is secreted by the testes into the epididymis, as well in ejaculated semen [127–132]. When male mice were injected with ^{3}H-E2 then each allowed to mate with a sexually receptive female, the females' blood serum and tissues showed substantial levels of radioactivity [104]. The transfer began when males mounted females, intensified with increasing numbers of intromissions, then spiked to very high levels in the uterus following ejaculation. Semen collected from the uterus showed very high concentrations of ^{3}H-E2, as did the copulatory plug that male mice insert into the vaginal canal just after insemination [104]. E2 from semen and the copulatory plug thus arrives directly in the reproductive tract, without passing first through circulation. Figure 4 provides excerpts of the data. In another

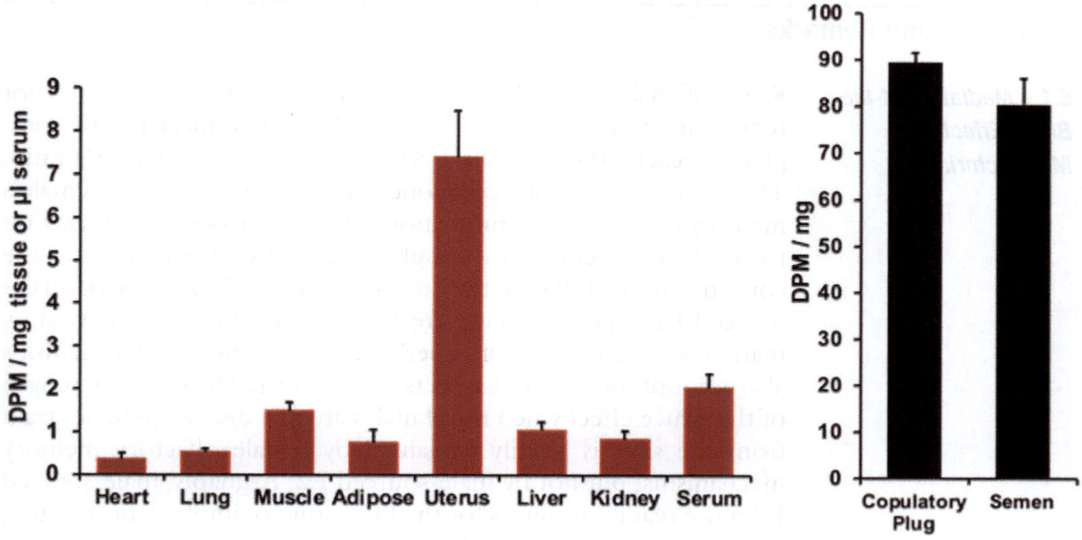

Fig. 4 Females absorb very substantial amounts of male-sourced E2 during mating. In order to ensure timed sexual receptivity, females were ovariectomized (OVX), then a few weeks later they were injected with E2 48 h and P4 6 h prior to mating. Each male had been injected with a small quantity of tritiated estradiol (^{3}H-E2) 24 and 72 h before mating. Each female's blood and tissues were obtained beginning approximately 5 min after insemination. The whole copulatory plug was extracted, and semen was scraped from the uterus above the plug. Tissues were prepared for scintillation counting, where radioactivity was measured in disintegrations per minute (DPM). A concentration of 1 DPM/mg tissue or 1 μL/DPM serum is equivalent to 1.38 pg E2 tissue or 1.38 pg/mL serum. Error bars indicate S.E.M. Two control conditions were run: (1) OVX females made receptive with replacement E2 and P4 that mated to ejaculation with males not given ^{3}H-E2, and (2) OVX females not given replacement E2 and P4 and therefore not sexually receptive, then each housed with a ^{3}H-E2-treated male, such that durations of exposure matched those of the experimental females. In both of these controls, radioactivity measures were <0.0 DPM for all tissues and serum. (From deCatanzaro and Pollock [104])

experiment (not shown), E2 was observed to linger in the females' reproductive tract and organs for at least 18 h, still being released from remnants of the copulatory plugs. That is noteworthy given the exceptional concentration of ER in the uterus and the various actions of minute elevations of E2 that terminate blastocyst implantation (discussed in Subheading 2). This spike in uterine E2 would block pregnancies sired by previous males and stimulate estrus and ovulation.

Although the major influences of male-sourced E2 are at the uterus where implantation is directly prevented by small elevations of E2, there could also be some impacts via the brain, particularly the hypothalamus. Pfaff [133] showed that minute doses of E2 induce female sexual receptivity via receptors at the ventromedial hypothalamus. It is also well known that actions of E2 at the hypothalamus can stimulate LH release from the anterior pituitary, which can stimulate ovulation.

5 Concluding Remarks

5.1 Mediation of the Bruce Effect Is Multifactorial

Roles of male-sourced E2 and female olfactory memory do not represent competing hypotheses. Rather, these mechanisms complement each other by collectively raising the female's E2:P4 ratio. The olfactory memory component relies largely on the female's memory of the sire, discrimination among male odors, and reaction to odors of novel males through a chain of events that lower the concentration of P4 at the uterus. Female absorption of male-sourced E2 is proven to elevate E2 at the uterus, as supported by many convergent lines of experimentation. The two mechanisms also account for different aspects of the effect. The greater strength of the Bruce effect when novel males are of a distinct genetic strain from the sires is readily explained by female olfactory memory mechanisms, but not by male-sourced E2. Arguably, male-sourced E2 more readily accounts for the full nature of implantation failure, especially when novel males and females interact naturally and mate. An additional factor is possible insofar as novel males stress previously inseminated females through their agitation, aggressive mounting, and direction of excretions [111]. This may mobilize the female's endogenous adrenocortical steroids, adding endogenous E2 to male-sourced E2. Chronic stress can also suppress P4, and possibly the female's perception that a male other than the sire is attempting to mate constitutes a stressor (*see* Subheading 5.3).

5.2 Direct Exposure to Novel Males Produces the Clearest Effects

The most naturalistic method for the study of the Bruce effect involves direct exposure of inseminated females to novel males, allowing them to interact freely and mate to insemination. In the quest to determine mechanisms, researchers have often used paradigms quite removed from direct exposure. In my lab, we have found that the Bruce effect is weakened probabilistically (fewer females show implantation failure) by any factors that remove the female from direct contact with the novel male. When females are separated from males by two grids spaced 0.5 cm apart, there are fewer implantation failures than when there is just one grid. When more novel males are housed in the vicinity of females, the effect is greater [96]. When novel males are above females and separated by a grid, the effect is clear, but when the locations are inverted, with females above males and in less contact with their behavior and excretions, most pregnancies are successful [96].

Male urine alone produces much weaker effects. This was clear in Bruce's original work. As mentioned, Parkes and Bruce [7] reported implantation failure in females housed in soiled cages vacated by five novel males but had to restrict their ventilation and renew the bedding frequently to produce an effect. Such manipulations surely stress females in manners unrelated to pheromones, via human handling, bacteria, physical restraint, and poor air quality. In another lab, urine alone was reported to be sufficient to disrupt implantation, either when dripped into females' bedding or painted on the female's nasal area [5, 87, 134]; however, questions can be raised about human handling in these procedures and indirect assessments of implantation through experimenter-interpreted vaginal-smear cell counts. Disruptions of implantation by nasal application of proteins and other substances extracted from male urine [70, 71] likely involved concentrates of the critical urinary constituents. Elsewhere [135], inseminated females had their nasal regions painted with water, urine from isolated novel males, or urine from males housed in proximity to females. Respectively 87.5%, 78.5%, and 57.1% produced litters, which is consistent with rises of male urinary E2 after a few days of female exposure (e.g., Fig. 3).

5.3 The State of Research on Male-Sourced E2 and Female Olfactory Memory

The evidence for a critical role in the Bruce effect of male-sourced E2 at the uterus is now clear and at a point of closure. E2 is the most potent natural disruptor of implantation through direct actions at the uterus. Exogenous E2 is consistently present in male mouse urine. The quantities of male urinary E2 approach values are sufficient to disrupt early pregnancy in nearby inseminated females, and testosterone in male urine reinforces E2's actions [95]. When housed for 3–4 days next to females through a wire grid, the concentration of E2 rises in male urine, which they actively directed at females [101]. Even greater quantities of unconjugated E2 are found in male semen and copulatory plugs, and they are delivered directly to the female's reproductive tract during mating [104].

The most conclusive aspect of the olfactory memory mechanism is that novel male exposure leads to dopaminergic suppression of prolactin pulses that sustain the corpora lutea, causing P4 suppression. Bruce first noticed the roles of prolactin, the corpora lutea, and P4, and the evidence has been reinforced by subsequent researchers. Females can discriminate among male odors to some degree, but the specific configurations of molecules that contribute to this are yet to be determined. One of the main proponents of the olfactory memory hypothesis [67] cautioned that despite findings that MHC peptides influence mate recognition, "the pregnancy blocking and individuality chemosignals remain to be identified" (p. 287). Both MUPs and MHC peptides may play roles [67, 68]. Perhaps it is mistaken to seek any single compound or even set of compounds as *the* specific pheromone(s), as the female's memory may instead involve some sort of gestalt, or configuration of multiple olfactory and possibly other stimuli. Perhaps the inseminated female's perception that a male other than the sire is attempting to mate simply constitutes a stressor, setting off the same mechanisms as do other stressors. Indeed, suppression of prolactin surges and P4 are also characteristic of other major stressors that disrupt implantation [20, 53, 55].

5.4 A Note on Evolutionary and Comparative Factors

There have been several reviews examining evolutionary advantages and disadvantages for males and females in the Bruce effect [136–140]. For novel males, the reproductive advantage is gained by terminating pregnancies sired by rival males and driving females into estrus then re-inseminating them. Obviously, original sires lose fitness through the Bruce effect, and when they are present they actively work to prevent the effect [110]. For females, the most commonly cited evolutionary contingencies involve pressures to remember the sire and to identify and react to novel males, terminating implantation and thereby forestalling the heavy costs of potential postpartum infanticide. Infanticide is known to be perpetrated by novel male mice encountering a litter they have not sired, and this phenomenon has been observed in diverse mammals [141]. There has also been some discussion that inbred females may gain hybrid vigor for their offspring by breeding with outbred males.

These possible explanations suggest mutually reinforcing factors that co-evolved in the two sexes. Ideally, empirical approaches to evolution will be more fruitful than speculation and modeling. An empirical approach requires a clear understanding of mechanisms and their conservation or variation across taxa. These mechanisms, both physiological and behavioral adaptations in males and females, are the evolved phenotypes. More comparative research on mechanisms would be very valuable. As pointed out elsewhere [140], one challenge comes from observations of pregnancy losses among some female primates when dominant males arrive in a group, without any obvious female contact with the males' excretions; however, these cases could be stress induced. In at least one large primate, humans, estrogens in urine are largely conjugated, but bioactive E2 can be present in very high concentrations in young men's axillary perspiration [102, 103].

I suggest an additional evolutionary explanation of the Bruce effect. Dynamics of prolactin, E2, and P4 and their roles in implantation are highly conserved among mammals. As detailed in Subheading 2, female mammals evolved long ago to raise their endogenous E2:P4 ratio in the peri-implantation period when stressed, thereby terminating implantation and forestalling various adverse outcomes. The Bruce effect essentially uses the same hormonal mechanisms in the female's reproductive tract. Male mammals must have subsequently evolved to exploit earlier female adaptations, simply by excreting females' most powerful steroid, E2. Males excreting substantial quantities of E2 in females' presence and during sexual activity would have had reproductive advantages relative to males not doing so. E2 is much more potent than testosterone at low doses [46, 142], and males could simply aromatize some of their testosterone to E2. Male-sourced E2 is not only adapted to manipulate peri-implantation females; it can also promote reproductive maturation (the Vandenbergh effect), sexual receptivity, ovulation, and initial preparation of the uterus for implantation [142].

Acknowledgments

This research was supported by grants from the Natural Sciences and Engineering Research Council throughout my career. Specific grant numbers are given in the cited papers. I am very grateful for the contributions of so many of my students, without which this research would never have progressed so far. Their names are first authors or co-authors of the papers cited from my lab. I also thank Jennifer deCatanzaro for her ongoing support and proofreading.

References

1. Bruce HM (1959) An exteroceptive block to pregnancy in the mouse. Nature 184:105
2. Bruce HM (1960) A block to pregnancy in mice caused by the proximity of strange males. J Reprod Fertil 1:96–103
3. Bruce HM (1961) Time relations in the pregnancy-block induced in mice by strange males. J Reprod Fertil 2:138–142
4. Bruce HM (1965) Effect of castration on the reproductive pheromones of male mice. J Reprod Fertil 10:141–143
5. Dominic CJ (1965) The origin of the pheromones causing pregnancy block in mice. J Reprod Fertil 10:469–472
6. Bruce HM, Parrott DMV (1960) Role of olfactory sense in pregnancy block by strange males. Science 131:1526
7. Parkes AS, Bruce HM (1962) Pregnancy block in female mice placed in boxes soiled by males. J Reprod Fertil 4:303–308
8. Bronson FH, Eleftheriou BE (1963) Influence of strange males on implantation in the deermouse. Gen Comp Endocrinol 3:515–518
9. Clulow FV, Langford PE (1971) Pregnancy-block in the meadow vole, *Microtus pennsylvanicus*. J Reprod Fertil 24:275–277
10. Heske EJ, Nelson RJ (1984) Pregnancy interruption in *Microtus ochrogaster*: laboratory artifact or field phenomenon. Biol Reprod 31:97–103
11. Marashi V, Rülicke T (2012) The Bruce effect in Norway rats. Biol Reprod 86:1–5
12. Rohrbach C (1982) Investigation of the Bruce effect in the Mongolian gerbil (*Meriones unguiculatus*). J Reprod Fertil 65:411–417
13. Al-Gubory KH (1998) Effects of the presence of rams during pregnancy on lambing performance in ewes. Anim Reprod Sci 52:205–211
14. Bartos L, Bartosova J, Pluhacek J, Sindelarova J (2011) Promiscuous behaviour disrupts pregnancy block in domestic horse mares. Behav Ecol Sociobiol 65:1567–1572
15. Berger J (1983) Induced abortion and social factors in wild horses. Nature 303:59–61
16. Roberts EK, Lu A, Bergman TJ, Beehner JC (2012) A Bruce effect in wild geladas. Science 335:1222–1225
17. deCatanzaro D, MacNiven E (1992) Psychogenic pregnancy disruptions in mammals. Neurosci Biobehav Rev 16:43–53
18. Wiebold JL, Stanfield PH, Becker WC, Hillers JK (1986) The effect of restraint stress in early pregnancy in mice. J Reprod Fertil 78:185–192
19. Bronson FH, Eleftheriou BE, Garick EI (1964) Effects of intra- and inter-specific social stimulation on implantation in deermice. J Reprod Fertil 8:23–27
20. Thorpe JB, Burgess PS, Sadkowski M, deCatanzaro D (2013) Estrogen-progesterone balance in the context of blastocyst implantation failure induced by predator stress. Psychoneuroendocrinology 38:3048–3056
21. Runner ML (1959) Embryocidal effect of handling pregnant mice and its prevention with progesterone. Anat Rec 133:330–331
22. Alliston CW, Ullberg LC (1961) Early pregnancy loss in sheep at ambient temperatures of 70° and 90°F as determined by embryo transfer. J Anim Sci 20:608–613
23. Doney JM, Smith WF, Gunn RG (1976) Effects of post-mating environmental stress

or administration of ACTH on early embryonic loss in sheep. J Agric Sci 87:133–136
24. Moberg GP (1976) Effects of environment and management stress on reproduction m the dairy cow. J Dairy Sci 59:1618–1624
25. Arck PC (2001) Stress and pregnancy loss: role of immune mediators, hormones, and neurotransmitters. Am J Reprod Immunol 46:117–123
26. Clark DA (2008) Immunological factors in pregnancy wastage: fact or fiction. Am J Reprod Immunol 59:277–300
27. Edwards H, Reburn CJ, Wynne-Edwards KE (1995) Daily patterns of pituitary prolactin secretion and their role in regulating maternal serum progesterone concentrations across pregnancy in the Djungarian hamster (*Phodopus campbelli*). Biol Reprod 52:814–823
28. Robertson SA (2010) Immune regulation of conception and embryo implantation – all about quality control? J Reprod Immunol 85:51–57
29. Gidley-Baird AA, O'Neill C, Sinosich MJ, Porter RN, Pike IL, Saunders DM (1986) Failure of implantation in human in vitro fertilization and embryo transplant patients: the effects of altered progesterone/estrogen ratios in humans and mice. Fertil Steril 45: 69–74
30. Ortiz ME, Villalon M, Croxatto HB (1979) Ovum transport and fertility following post-ovulatory treatment with estradiol in rats. Biol Reprod 21:1163–1167
31. Valbuena D, Martin J, de Pablo JL, Remohi J, Pellicer A, Simon C (2001) Increasing levels of estradiol are deleterious to embryonic implantation because they directly affect the embryo. Fertil Steril 76:962–968
32. Rajabi N, Thorpe JB, Foster WG, deCatanzaro D (2014) Novel male exposure reduces uterine e-cadherin, increases uterine luminal area, and diminishes progesterone levels while disrupting blastocyst implantation in inseminated mice. J Steroid Biochem Mol Biol 139: 107–113
33. Martin L, Finn CA, Carter J (1970) Effects of progesterone and oestradiol-17β on the luminal epithelium of the mouse uterus. J Reprod Fertil 21:461–469
34. Naftalin RJ, Pedley KC, Pocock VJ, Milligan SR (2002) Progesterone stimulation of fluid absorption by the rat uterine gland. Reproduction 123:633–638
35. Parr MB (1983) Relationship of uterine closure to ovarian hormones and endocytosis in the rat. J Reprod Fertil 68:185–188
36. Salleh N, Baines DL, Naftalin RJ, Milligan SR (2005) The hormonal control of uterine luminal fluid secretion and absorption. J Membr Biol 206:17–28
37. Aplin JD (1997) Adhesion molecules in implantation. Rev Reprod 2:84–93
38. Hyland RA, Shaw TJ, Png FY, Murphy CR (1998) Pan-cadherin concentrates apically in uterine epithelial cells during uterine closure in the rat. Acta Histochem 100:75–81
39. Jha RK, Titus S, Saxena D, Kumar PG, Laloraya M (2006) Profiling of e-cadherin, β-catenin and Ca2+ in embryo–uterine interactions at implantation. FEBS Lett 580: 5653–5660
40. Paria BC, Zhao X, Das SK, Dey SK, Yoshinaga K (1999) Zonula occludens-1 and e-cadherin are coordinately expressed in the mouse uterus with the initiation of implantation and decidualization. Dev Biol 208:488–501
41. Rahnama F, Thompson B, Steiner M, Shafiei F, Lobie PE, Mitchell MD (2009) Epigenetic regulation of e-cadherin controls endometrial receptivity. Endocrinology 150: 1466–1472
42. Potter SW, Gaza G, Morris JE (1996) Estradiol induces e-cadherin degradation in mouse uterine epithelium during the estrous cycle and early pregnancy. J Cell Physiol 169:1–14
43. Srinivasan KR, Blesson CS, Fatima I, Kitchlu S, Jain SK, Mehrotra PK, Dwivedi A (2009) Expression of αvβ3 integrin in rat endometrial epithelial cells and its functional role during implantation. Gen Comp Endocrinol 160:124–133
44. Widra EA, Weeraratna A, Stepp MA, Stillman RJ, Patierno SR (1997) Modulation of implantation- associated integrin expression but not uteroglobin by steroid hormones in an endometrial cell line. Mol Hum Reprod 3: 563–568
45. deCatanzaro D, Graham C (1992) Influence of exogenous epinephrine on two reproductive parameters in female mice: disruption of receptivity but not early pregnancy. Horm Behav 26:330–338
46. deCatanzaro D, MacNiven E, Ricciuti F (1991) Comparison of the adverse effects of adrenal and ovarian steroids upon early pregnancy in mice. Psychoneuroendocrinology 16:525–536
47. deCatanzaro D, Baptista MAS, Vella ES (2001) Administration of minute quantities of 17β-estradiol on the nasal area terminates early pregnancy in inseminated female mice. Pharmacol Biochem Behav 69:503–509

48. Ma W, Song H, Das SK, Paria BC, Dey SK (2003) Estrogen is a critical determinant that specifies the duration of the window of uterine receptivity for implantation. Proc Natl Acad Sci U S A 100:2963–2968
49. Shors TJ, Pickett J, Wood G, Paczynski M (1999) Acute stress persistently enhances estrogen levels in the female rat. Stress 3: 163–171
50. Thorpe JB, Gould KE, Borman ED, deCatanzaro D (2014) Circulating and urinary adrenal corticosterone, progesterone, and estradiol in response to acute stress in female mice (*Mus musculus*). Horm Metab Res 46: 211–218
51. MacNiven E, deCatanzaro D, Younglai EV (1992) Chronic stress increases estrogen and other steroids in inseminated rats. Physiol Behav 52:159–162
52. Misdrahi D, Pardon MC, Perez-Diaz F, Hanoun N, Cohen-Salmon C (2005) Prepartum chronic ultramild stress increases corticosterone and estradiol levels in gestating mice: implications for postpartum depressive disorders. Psychiatry Res 137:123–130
53. Agrawal V, Jaiswal MK, Jaiswal YK (2011) Lipopolysaccharide induces alterations in ovaries and serum level of progesterone and 17β-estradiol in the mouse. Fertil Steril 95: 1471–1474
54. deCatanzaro D, MacNiven E, Goodison T, Richardson D (1994) Estrogen antibodies reduce vulnerability to stress-induced failure of intrauterine implantation in inseminated mice. Physiol Behav 55:35–38
55. Parker VJ, Menzies JRW, Douglas AJ (2011) Differential changes in the hypothalamic-pituitary- adrenal axis and prolactin responses to stress in early pregnant mice. J Neuroendocrinol 23:1066–1078
56. MacNiven E, deCatanzaro D (1990) Reversal of stress-induced pregnancy blocks in mice by progesterone and metyrapone. Physiol Behav 47:443–448
57. Pratt NC, Lisk RD (1991) Role of progesterone in mediating stress-related litter deficits in the golden hamster (*Mesocricetus auratus*). J Reprod Fertil 92:139–146
58. Joachim R, Zenclussen AC, Polgar B, Douglas AJ, Fest S, Knackstedt M, Klapp BF, Arck PC (2003) The progesterone derivative dydrogesterone abrogates murine stress-triggered abortion by inducing a Th2 biased local immune response. Steroids 68:931–940
59. Marsden HM, Bronson FH (1965) Strange male block to pregnancy: its absence in inbred mouse strains. Nature 207:878
60. Spironello E, deCatanzaro D (1999) Sexual satiety diminishes the capacity of novel males to disrupt early pregnancy. J Comp Psychol 113:218–222
61. Bakker J (2003) Sexual differentiation of the neuroendocrine mechanisms regulating mate recognition in mammals. J Neuroendocrinol 15:615–621
62. Baum MJ, Keverne EB (2002) Sex differences in attraction thresholds for volatile odors from male and female mouse urine. Horm Behav 41:213–219
63. Brennan PA (2004) The nose knows who's who: chemosensory individuality and mate recognition in mice. Horm Behav 46: 231–240
64. Hurst JL (2009) Female recognition and assessment of males through scent. Behav Brain Res 200:295–303
65. Findlayson JS, Asofsky R, Potter M, Runner CC (1965) Major urinary protein complex of normal mice: origin. Science 149:981–982
66. Robertson DHL, Hurst JL, Bolgar MS, Gaskell SJ, Beynon RJ (1997) Molecular heterogeneity of urinary proteins in wild house mouse populations. Rapid Commun Mass Spectrom 11:786–790
67. Brennan PA (2009) Outstanding issues surrounding vomeronasal mechanisms of pregnancy block and individual recognition in mice. Behav Brain Res 200:287–294
68. Cheetham SA, Thom MD, Jury F, Ollier WER, Beynon RJ, Hurst JL (2007) The genetic basis of individual-recognition signals in the mouse. Curr Biol 17:1771–1777
69. Roberts SA, Simpson DM, Armstrong SD, Davidson AJ, Robertson DH, McLean L, Beynon RJ, Hurst JL (2010) Darcin: a male pheromone that stimulates female memory and sexual attraction to an individual male's odour. BMC Biol 8:75
70. Marchlewska-Koj A (1977) Pregnancy block elicited by urinary proteins of male mice. Biol Reprod 17:729–732
71. Marchlewska-Koj A (1981) Pregnancy block elicited by male urinary peptides in mice. J Reprod Fertil 61:221–224
72. Peele P, Salazar I, Mimmack M, Keverne EB, Brennan PA (2003) Low molecular weight constituents of male mouse urine mediate the pregnancy block effect and convey information about the identity of the mating male. Eur J Neurosci 18:622–628
73. Brennan PA, Peele P (2003) Towards an understanding of the pregnancy-blocking urinary chemosignals of mice. Biochem Soc Trans 31:152–155

74. Leinders-Zufall T, Brennan P, Widmayer P, Chandramani SP, Maul-Pavicic A, Jäger M, Li X-H, Breer H, Zufall F (2004) MHC class I peptides as chemosensory signals in the vomeronasal organ. Science 306:1022–1037
75. Kelliher KR, Spehr M, Li X-H, Zufall F, Leinders-Zufall T (2006) Pheromonal recognition memory induced by TRPC2-independent vomeronasal sensing. Eur J Neurosci 23:3385–3390
76. Meredith M (1994) Chronic recording of vomeronasal pump activation in awake behaving hamsters. Physiol Behav 56:345–354
77. Meredith M, O'Connell RJ (1979) Efferent control of stimulus access to the hamster vomeronasal organ. J Physiol 286:301–316
78. Meredith M, Marques DM, O'Connell RJ, Stern FL (1980) Vomeronasal pump: significance for male hamster sexual behavior. Science 207:1224–1226
79. Bellringer JF, Pratt HPM, Keverne EB (1980) Involvement of the vomeronasal organ and prolactin in pheromonal induction of delayed implantation in mice. J Reprod Fertil 59:223–228
80. Lloyd-Thomas A, Keverne EB (1982) Role of the brain and accessory olfactory system in the block to pregnancy in mice. Neuroscience 7:907–913
81. Herrada G, Dulac C (1997) A novel family of putative pheromone receptors in mammals with topographically organized and sexually dimorphic distribution. Cell 90:763–777
82. Rodriguez I, Del Punta K, Rothman A, Ishii T, Mombaerts P (2002) Multiple new and isolated families within the mouse superfamily of V1R vomeronasal receptors. Nat Neurosci 5:134–140
83. Luo M, Fee MS, Katz LC (2003) Encoding pheromonal signals in the accessory olfactory bulb of behaving mice. Science 299:1196–1201
84. Li C-S, Kaba H, Seto K (1994) Effective induction of pregnancy block by electrical stimulation of the mouse accessory olfactory bulb coincident with prolactin surges. Neurosci Lett 176:5–8
85. Baum MJ, Bakker J (2013) Roles of sex and gonadal steroids in mammalian pheromonal communication. Front Neuroendocrinol 34:268–284
86. Bruce HM, Parkes AS (1960) Hormonal factors in exteroceptive block to pregnancy in mice. J Endocrinol 20:xxix–xxx
87. Dominic CJ (1966) Observations of the reproductive pheromones of mice: II. Neuroendocrine factors involved in the olfactory block to pregnancy. J Reprod Fertil 11:415–421
88. Rajendren G, Dominic CJ (1987) The male-induced pregnancy block (the Bruce effect) in mice: re-evaluation of the ability of exogenous pregnancy in preventing implantation failure. Exp Clin Endocrinol 89:188–196
89. Rajendren G, Dominic CJ (1988) The male-induced implantation failure (the Bruce effect) in mice: investigations on luteal failure in pregnancy-blocked females. Exp Clin Endocrinol 92:13–19
90. Rajendren G, Dominic CJ (1993) The male-induced pregnancy failure (the Bruce effect) in mice: effect of exogenous progesterone on maintenance of pregnancy in male-exposed females. Exp Clin Endocrinol 101:356–359
91. Milligan SR (1976) Pregnancy blocking in the vole, Microtus agrestis. II. Ovarian, uterine and vaginal changes. J Reprod Fertil 46:97–100
92. Rosser AE, Remfry CJ, Keverne EB (1989) Restricted exposure of mice to primer pheromones coincident with prolactin surges blocks pregnancy by changing hypothalamic dopamine release. J Reprod Fertil 87:553–559
93. Marchlewska-Koj A, Jemiolo B (1978) Evidence for the involvement of dopaminergic neurons in the pregnancy block effect. Neuroendocrinology 26:186–192
94. Wersinger SR, Temple JL, Caldwell HK, Young WS 3rd (2008) Inactivation of the oxytocin and vasopressin (Avp) 1b receptor genes, but not the Avp 1a receptor genes, differentially impairs the Bruce effect in laboratory mice. Endocrinology 149:116–121
95. deCatanzaro D, Beaton EA, Khan A, Vella E (2006) Urinary oestradiol and testosterone levels from novel male mice approach values sufficient to disrupt pregnancy in nearby inseminated females. Reproduction 132:309–317
96. deCatanzaro D, Zacharias R, Muir C (1996) Disruption of early pregnancy by direct and indirect exposure to novel males in mice: comparison of influences of preputialectomized and intact males. J Reprod Fertil 106:269–274
97. Drickamer LC (1995) Rates of urine excretion by house mice (*Mus domesticus*): differences by age, sex, social status, and reproductive condition. J Chem Ecol 21:1481–1493
98. deCatanzaro D, Muir C, Beaton E, Jetha M, Nadella K (2003) Enzymeimmunoassay of oestradiol, testosterone and progesterone in urine samples from female mice before and

after insemination. Reproduction 126:407–414

99. Bawarshi-Nassar NN, Hussain AA, Crooks PA (1989) Nasal absorption and metabolism of progesterone and 17β-estradiol in the rat. Drug Metab Dispos 17:248–254
100. Muir C, Vella ES, Pisani N, deCatanzaro D (2001) Enzyme immunoassay of 17-β-estradiol, estrone conjugates, and testosterone in urinary and fecal samples from male and female mice. Horm Metab Res 33:653–658
101. deCatanzaro D, Khan A, Berger RG, Lewis E (2009) Exposure to developing females induces polyuria, polydipsia, and altered urinary levels of creatinine, 17β-estradiol, and testosterone in adult male mice (*Mus musculus*). Horm Behav 55:240–247
102. Muir CC, Treasurywala K, McAllister S, Sutherland J, Dukas L, Berger RG, Khan A, deCatanzaro D (2008) Enzyme immunoassay of testosterone, 17β-estradiol, and progesterone in perspiration and urine of preadolescents and young adults: exceptional levels in men's axillary perspiration. Horm Metab Res 40:819–826
103. Elliott B, Muir C, deCatanzaro D (2017) Sources of variance within and among young men in concentrations of 17β-estradiol and testosterone in axillary perspiration. Physiol Behav 173:23–29
104. deCatanzaro D, Pollock T (2016) Absorption and distribution of estradiol from male seminal emissions during mating. J Endocrinol 231:245–257
105. Becker SD, Hurst JL (2009) Female behaviour plays a critical role in controlling murine pregnancy block. Proc R Soc B 276:1723–1729
106. Arakawa H, Arakawa K, Blanchard DC, Blanchard RJ (2007) Scent marking behavior in male C57BL/6J mice: sexual and developmental determination. Behav Brain Res 182:73–79
107. Desjardins C, Maruniuk JA, Bronson FH (1973) Social rank in house mice: differentiation revealed by ultraviolet visualization of urinary marking patterns. Science 182:939–941
108. Maruniuk JA, Owen K, Bronson FH, Desjardins C (1974) Urinary marking in male house mice: responses to novel environmental and social stimuli. Physiol Behav 12:1035–1039
109. deCatanzaro D, Storey A (1989) Partial mediation of strange-male-induced pregnancy blocks by sexual activity in mice (*Mus musculus*). J Comp Psychol 13:381–388
110. deCatanzaro D, Murji T (2004) Inseminated female mice investigate rather than avoid novel males that disrupt pregnancy, but sires protect pregnancy. J Comp Psychol 118:251–257
111. deCatanzaro D, Muir C, Spironello E, Binger T, Thomas J (2000) Intense arousal of novel male mice in proximity to previously inseminated females: inactivation of males via chlorpromazine does not diminish the capacity to disrupt pregnancy. Psychobiology 28:110–114
112. McGill TE, Blight WC (1963) Effects of genotype on the recovery of sex drive in the male mouse. J Comp Physiol Psychol 56:887–888
113. Batty J (1978) Acute changes in plasma testosterone levels and their relation to measures of sexual behaviour in the male house mouse (*Mus musculus*). Anim Behav 26:349–357
114. Bliss EL, Frischat A, Samuels L (1972) Brain and testicular function. Life Sci 11:231–238
115. Spironello-Vella E, deCatanzaro D (2001) Novel male mice show gradual decline in the capacity to disrupt early pregnancy and in urinary excretion of testosterone and 17-β-estradiol during the weeks immediately following castration. Horm Metab Res 33:681–686
116. Thorpe JB, deCatanzaro D (2012) Oestradiol treatment restores the capacity of castrated males to induce both the Vandenbergh and the Bruce effects in mice (*Mus musculus*). Reproduction 143:123–132
117. Beaton EA, deCatanzaro D (2005) Novel males' capacity to disrupt early pregnancy in mice (*Mus musculus*) is attenuated via a chronic reduction of males' urinary 17-β-estradiol. Psychoneuroendocrinology 30:688–697
118. deCatanzaro D, Muir C, O'Brien J, Williams S (1995) Strange-male-induced pregnancy disruption in mice: reduction of vulnerability by 17β-estradiol antibodies. Physiol Behav 58:401–404
119. Zacharias R, deCatanzaro D, Muir C (2001) Novel male mice disrupt pregnancy despite removal of vesicular-coagulating and preputial glands. Physiol Behav 68:285–290
120. Anand Kumar TC, David GFX, Umberkomen B, Saini KB (1974) Uptake of radioactivity by body tissues and fluids in rhesus monkeys after intravenous injection or intranasal spray of tritium- labelled oestradiol and progesterone. Curr Sci India 43:435–439

121. Guzzo AC, Berger RG, deCatanzaro D (2010) Excretion and binding of tritium-labelled oestradiol in mice (*Mus musculus*): implications for the Bruce effect. Reproduction 139:255–263
122. Guzzo AC, Jheon J, Imtiaz F, deCatanzaro D (2012) Oestradiol transmission from males to females in the context of the Bruce and Vandenbergh effects in mice (*Mus musculus*). Reproduction 143:539–548
123. Hueber F, Besnard M, Schaefer H, Wepierre J (1994) Percutaneous absorption of estradiol and progesterone in normal and appendage-free skin of the hairless rat: lack of importance of nutritional blood flow. Skin Pharmacol 7: 245–256
124. Scheuplein RJ, Blank I, Brauner G, MacFarlane D (1969) Percutaneous absorption of steroids. J Invest Dermatol 52:63–70
125. Guzzo AC, Pollock T, deCatanzaro D (2013) Transfer of [3H]estradiol-17β and [3H]progesterone from conspecifics to cohabiting female mice. J Endocrinol 217:1–10
126. deCatanzaro D, Pollock T, Greville LJ, Faure PA (2014) Estradiol transfer from male big brown bats (*Eptesicus fuscus*) to the reproductive and brain tissues of cohabiting females, and its action as a pheromone. Gen Comp Endocrinol 208:126–133
127. Claus R, Hoang-Vu C, Ellendorff F, Meyer HD, Schopper D, Weiler U (1987) Seminal oestrogens in the boar: origin and functions in the sow. J Steroid Biochem 27:331–335
128. Eiler H, Graves CN (1977) Oestrogen content of semen and the effect of exogenous oestradiol-17β on the oestrogen and androgen concentration in semen and blood plasma of bulls. J Reprod Fertil 50:17–21
129. Free MJ, Jaffe RA (1979) Collection of rete testis fluid from rats without previous efferent duct ligation. Biol Reprod 20:269–278
130. Lemazurier E, Moslemi S, Sourdaine P, Desjardins I, Plainfosse B, Seralini G-E (2002) Free and conjugated estrogens and androgens in stallion semen. Gen Comp Endocrinol 125:272–282
131. Reiffsteck A, Dehennin L, Scholler R (1982) Estrogens in seminal plasma of human and animal species: identification and quantitative estimation by gas chromatography–mass spectrometry associated with stable isotope dilution. J Steroid Biochem 17:567–572
132. Waites GMH, Einer-Jensen N (1974) Collection and analysis of rete testis fluid from macaque monkeys. J Reprod Fertil 41:505–508
133. Pfaff DW (1980) Estrogens and brain function: neural analysis of a hormone-controlled mammalian reproductive behavior. Springer-Verlag, New York
134. Dominic CJ (1966) Observations of the reproductive pheromones of mice: I. Source. J Reprod Fertil 11:407–414
135. deCatanzaro D, Muir C, Sullivan C, Boissy A (1999) Pheromones and novel-male-induced pregnancy disruptions in mice: exposure to females is necessary for urine alone to induce an effect. Physiol Behav 66:153–157
136. Schwagmeyer PL (1979) The Bruce effect: an evaluation of male/female advantages. Am Nat 114:932–938
137. Labov JB (1981) Pregnancy blocking in rodents: adaptive advantages for females. Am Nat 118:361–371
138. Rülicke T, Guncz N, Wedekind C (2006) Early maternal investment in mice: no evidence for compatible-genes sexual selection despite hybrid vigor. J Evol Biol 19:922–928
139. Becker SD, Hurst JL (2008) Pregnancy block from a female perspective. Chem Signal 11: 141–150
140. Zipple MN, Roberts EK, Alberts SC, Beehner JC (2019) Male-mediated prenatal loss: functions and mechanisms. Evol Anthropol 28: 114–125
141. Blaffer-Hrdy S (1979) Infanticide among animals: a review, classification, and examination of reproductive strategies among females. Ethol Sociobiol 1:13–40
142. deCatanzaro D (2015) Sex steroids as pheromones in mammals: the exceptional role of estradiol. Horm Behav 68:103–116

Chapter 6

Open-Source JL Olfactometer for Awake Behaving Recording of Brain Activity for Mice Engaged in Olfactory Tasks

Nicole Arevalo, Laetitia Merle, Arianna Gentile-Polese, Andrew Moran, Andrew Parra, Michael Hall, Justin Losacco, Ming Ma, Connor McCullough, Barish Ozbay, Daniel Ramirez-Gordillo, Jose Riguero, Fabio Simoes-de Souza, Kira Steinke, Ryan Williamson, and Diego Restrepo

Abstract

Automated olfactometers are used to study brain function in rodents engaged in olfactory behavioral tasks. In this chapter, we describe the JL olfactometer, an open-source olfactometer developed by our laboratory to record neural activity through electrodes or microscopic imaging in mice engaged in a variety of olfactory behavioral tasks. The JL olfactometer monitors the licks the mouse makes on a lick spout located within an odorant port and controls odorant application and delivery of water reward through the opening and closing of valves. Neural activity is recorded in head-fixed mice using multiphoton microscopy or neuropixels and in freely behaving mice using miniature microscopes, tetrodes, or multielectrode arrays. Mouse movement is recorded with a video camera. An INTAN RHD2000 evaluation system monitors the activity of the JL olfactometer and records other inputs such as sniff recording and signals synchronizing the video camera and other instruments such as the multiphoton microscope. The design is open source and the MATLAB code can be modified to suit research goals. We discuss the use of olfactometry in the study of odorant perception in health and disease.

Key words Olfaction, Go-no go, Associative learning, Olfactometry, Mouse behavior, Multiphoton microscopy, Neuropixels, Tetrodes

1 Introduction

Brain mechanisms involved in olfactory decision-making are fascinating because of the large number of degrees of freedom used by this sensory system for the representation of the odorant

Nicole Arevalo and Laetitia Merle are co-first authors

Raúl G. Paredes et al. (eds.), *Animal Models of Reproductive Behavior*, Neuromethods, vol. 200,
https://doi.org/10.1007/978-1-0716-3234-5_6,

stimulus that produces a unique odor percept [1–3]. Each of ~2500 glomeruli in the olfactory bulb of mice is innervated by olfactory sensory neurons expressing the same olfactory receptor [4]. Even synthetic monomolecular odorants stimulate a large fraction of ~1100 olfactory receptor in mouse [5, 6]. Thus, each receptor has a broad receptive field, responding to multiple odorants sharing chemical features such as an aldehyde moiety. Because of this broad receptive field, stimulation with single volatile odorants activates multiple glomeruli in the olfactory bulb resulting in a high dimensional input [7]. In addition, because of the chromatographic arrival of the odorant to the olfactory receptors located in cilia protruding into the mucus [8] and spatial arrangement of olfactory receptors expression in the epithelium [9, 10], temporal arrival of the stimulus to the olfactory epithelium, gated by sniffing may contribute information on the identity of the odorant. Furthermore, natural stimuli, such as conspecific urine, are complex mixtures of hundreds of monomolecular odorants [11] that elicit discrete patterns of glomerular activation [12]. Thus, the olfactory system must contend with a complex spatiotemporal pattern of glomerular activation in order to discriminate between odorant stimuli. In order to cope with stimulus complexity, the pyramidal cells in the piriform cortex combine and temporally gate information relayed by mitral/tufted cells innervating different glomeruli [13–15]. Furthermore, mitral/tufted cells innervate other brain areas such as the lateral entorhinal cortex and the olfactory tubercle [16, 17]. Importantly, given the complexity of the stimulus and its behavioral relevance, neural processing in these circuits is modulated based on the context of odorant presentation [18, 19].

One approach to understand the neural basis of olfactory decision-making is to study animal behavior in response to odorants. A technique that has received substantial attention due to the ability to study learning of odorant discrimination and detection is automated olfactometry. In an automated olfactometer, the animal responds to defined concentrations of odorants diluted in incoming air with a motor output such as lever pressing or spout licking. This allows examining neural activity and behavioral responses on a trial-to-trial basis. An early report describing olfactometry for studies of animal behavior was published by Pfafmann and co-workers who showed that rats could learn to detect odorants in 20–40 trials in a manually controlled olfactometer [20]. This publication was followed by studies with automated olfactometers by Braun and Marcus [21] to demonstrate generalization to new odorants and by Marshall and Moulton to show that canines can detect odorants at substantially lower concentrations than humans [22]. Finally, behavioral training in an olfactometer was combined with the electrical recording of neural activity by Di Prisco and Freeman to characterize differential EEG responses during odorant discrimination trials in rabbits [23].

Here we describe the JL olfactometer, a custom-made open-source automated olfactometer designed to run olfactory behavioral tasks while monitoring neural activity in the brain of mice by recording the extracellular potential with multisite electrodes or by imaging of fluorescence neural activity sensors through multiphoton microscopy. The JL olfactometer described here is based on an automated olfactometer developed by Bodyak and Slotnick to study odorant detection and discrimination in mice [24] (also *see* [25]) (we refer to this olfactometer as the Bodyak and Slotnick olfactometer). The JL olfactometer can also be used during the modulation of neural activity through optogenetics [26]. This instrument is a lower-footprint modification of an olfactometer described in a previous publication [27] (Fig. 1). The olfactometer is designed to train mice in a go-no-go odorant discrimination task where a thirsty mouse learns to lick on a spout to receive a water reward when the rewarded odorant (S+) is being presented (Fig. 1c). The mouse does not receive a reward when presented with the unrewarded odorant (S−). The absence of reinforcement on the presentation of the S− stimulus quickly results in response inhibition in the presence of the non-rewarded odor. The JL olfactometer has been modified for other behavioral tasks described below.

The olfactometer is controlled by custom MATLAB code (MATLAB 2015b 32-bit run under Windows 7–11) that drives a Measurement Computing USB DIO96H/50 board to control the airflow and water valves (Fig. 1a). This board also records information on licks that is used for decision-making on trial initiation and delivery of the water reward. An INTAN RHD2000 evaluation system is used to record electrical activity through tetrodes or silicon electrodes. The INTAN RHD2000 board also records metadata sent by the olfactometer as an 8-bit digit for each of the behavioral trials with information on items such as which odorant was applied, when the odorant was presented, and when the mouse was rewarded (see examples of metadata in Fig. 2c(i), d(i)). The INTAN board also records lick and sniff signals for offline analysis (Fig. 2c(ii), (iii), d(ii), (iii) are lick recordings). The movement of the mouse is recorded by one or two FLIR cameras at 60 frames per second and the synchronization signal of the camera is recorded by the INTAN board. In addition, when the olfactometer is used along with confocal or multiphoton microscopy, the INTAN board records the per frame synchronization signal sent out by the microscope and the olfactometer can be triggered by the microscope system to start the trials. Finally, the olfactometer controls a laser to stimulate opsins for optogenetic modulation of neural activity during the trials. This awake-behaving recording olfactometer setup is placed on a cart to allow moving it in and out of the microscope rooms if necessary.

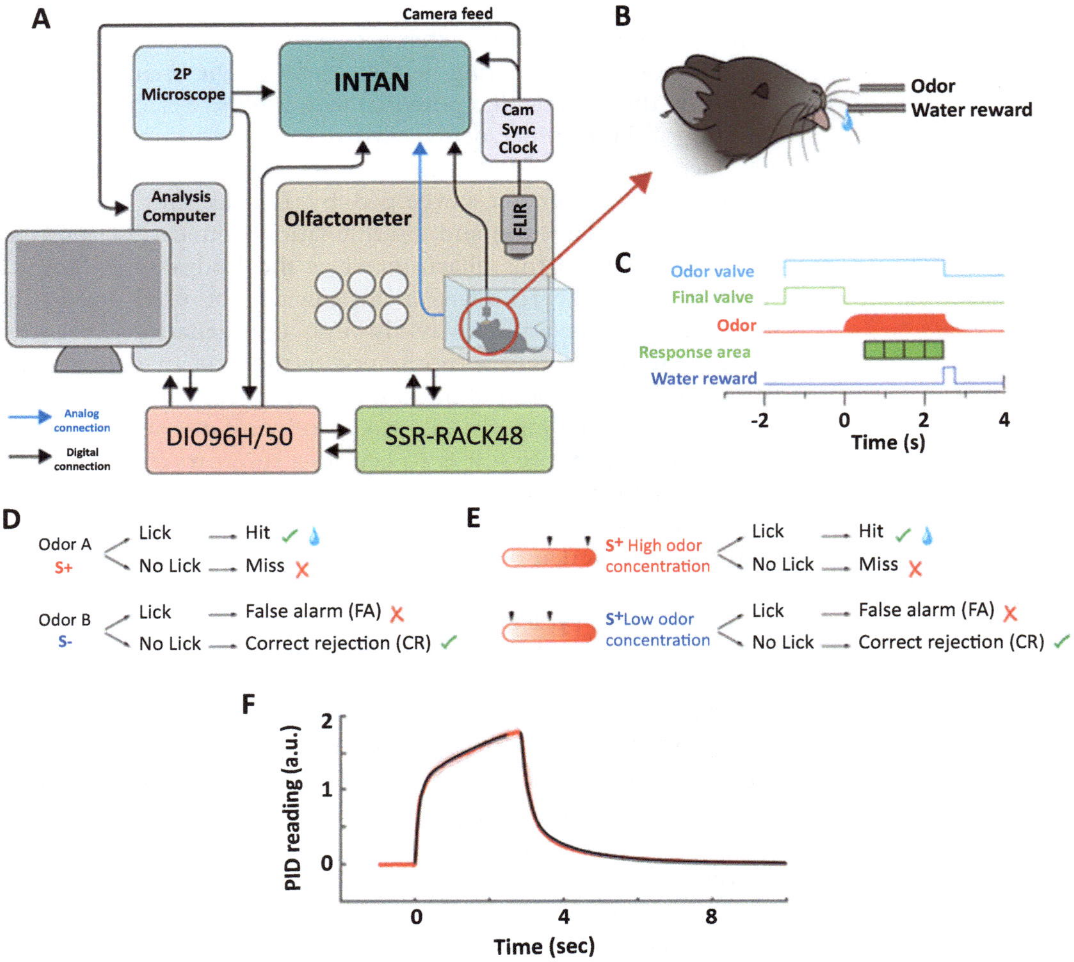

Fig. 1 Diagrams for the JL olfactometer and go-no-go task. (**a**) Block diagram for the JL olfactometer. The air and water valves in the olfactometer are controlled by MATLAB code in the computer through the DIO96H/50 interface. The computer also receives information on licks through digital input ports in this interface (examples of the digital input are shown in Fig. 2c(i), d(i)). An INTAN RHD 2000 evaluation board is used to record: (1) TTL pulses sent out per image from the computer controlling the multiphoton microscope and from the FLIR camera filming the mouse (used for synchronizing image timepoints to behavior metadata for data analysis). (2) Multielectrode extracellular potentials recorded from tetrodes or multielectrode arrays through an INTAN RHD amplifier chip. (3) Synchronization TTL pulses sent out by a Neuropixels setup. (4) Analog voltage output from an amplifier conveying sniff time course. (5) Digital metadata sent by the olfactometer computer through the digital output of the DIO96H/50 board with information on odorant stimulus and behavioral outcome of the trials. (**b**) Mouse self-initiates trials by licking on the water spout. Odorants and water are delivered according to the timeline in B and the decision tree in **c**. (**c**) Timeline for a single go-no-go trial. When the animal enters the port and licks on the spout the airflow is diverted by turning the final valve output to the exhaust and the odorant valve is turned on. At this time, the odor builds up to steady-state concentration flowing away from the animal for 1–1.5 s. At time 0, the final valve airflow is turned toward the odor port for 2.5 s resulting in odor onset 100 ms after the valve is actuated. In order to trigger a water reward for the rewarded odorant, the mouse must lick at least once during each 0.5 s block for four blocks in the response area. If the stimulus is rewarded and the animal licked appropriately, a water reward is delivered. (**d**) Per trial behavioral outcome for the two odorant go-no-go task. Green check mark means the correct decision

2 Materials

2.1 Design of the JL Olfactometer

Here, we provide a summary of the design of the JL olfactometer. In this presentation of the design, we describe the functioning of the olfactometer in reference to the go-no-go task where the mouse differentiates between two different odorants (Fig. 1c, d) [27, 28]. However, the code and the chamber of the olfactometer have been modified to run other behavioral tasks, such as examining mice engaged in a go-no-go task discriminating between different concentrations of the same odorant (Fig. 1e) [29] or odor-shock conditioning [30]. The parts used to assemble the JL olfactometer are listed in Tables 1 and 2, and a detailed description of how to assemble the olfactometer is provided in our GitHub repository (GitHub/restrepd/JLolfactometer). We describe two versions of the automated olfactometer, both having smaller footprints than the olfactometer described in [27]. The "mini" JL olfactometer (Table 1) has a footprint of 22 1/4″ wide × 16″ tall and 9 1/4″ deep and can be placed on a lab cart to move it in and out of microscope rooms and the "2-channel" JL olfactometer (Table 2) has a smaller footprint of 19″ wide × 12.5″ tall and 8.5″ deep designed to be used in a BSL2+ animal facility where the olfactometer has to be placed within a biosafety cabinet for studies with virally infected mice. The Bodyak and Slotnick olfactometer and JL olfactometer have been used by our group in a series of studies of olfactory discrimination, discrimination of stimulus timing, decision-making, learning, and odor-shock conditioning [18, 26, 28, 30–36].

Fig. 1 (continued) was made. Red "X" mark indicates an incorrect decision. Water reward is represented by the water droplet symbol in the case of a hit. (**e**) Per trial behavioral outcome for a go-no-go task where the mouse discriminates between two odorant concentration ranges. The odorant was presented as air equilibrated with six different liquid dilutions (cliq) in mineral oil (10%, 3.33%, 1%, 0.33%, 0.1%, and 0.033%). For the session shown here, the three higher concentrations are the S+ and the three lower concentrations are the S− (the rewarded concentration range is switched for other sessions). When the animal licks at least once in each of the response areas when it is presented with the rewarded odorant the trial is a Hit and the mouse receives a water reward. If the mouse does not lick during one of the response areas it does not receive water (Miss). On the other hand, if the mouse refrains from licking in one of the response areas when presented with the S− stimulus the trial is a correct rejection (CR), and if the mouse licks in all response areas when presented the S− odorant the trial is a false alarm (FA). (**f**) A miniPID (Aurora Scientific) photoionization detector was used to measure the concentration of odorant at the odor spout of the olfactometer. The odorant was air equilibrated with Sharpie ® permanent marker black ink. The red line is the concentration of the odorant in arbitrary units bounded by the bootstrapped confidence interval ($n = 10$ traces). Time was aligned to the onset of the increase in odorant concentration that took place 104.5 + 7 ms (mean + SD, $n = 10$) after the final valve was diverted to deliver the odorant to the odorant spout. The black lines are regression fits of exponential increase and decrease lines with time constants of 0.12 and 2.16 s (increase, two exponential terms) and 0.19 and 1.28 s (decrease, two exponential terms). (**a**–**d**, **f** are modified from [18])

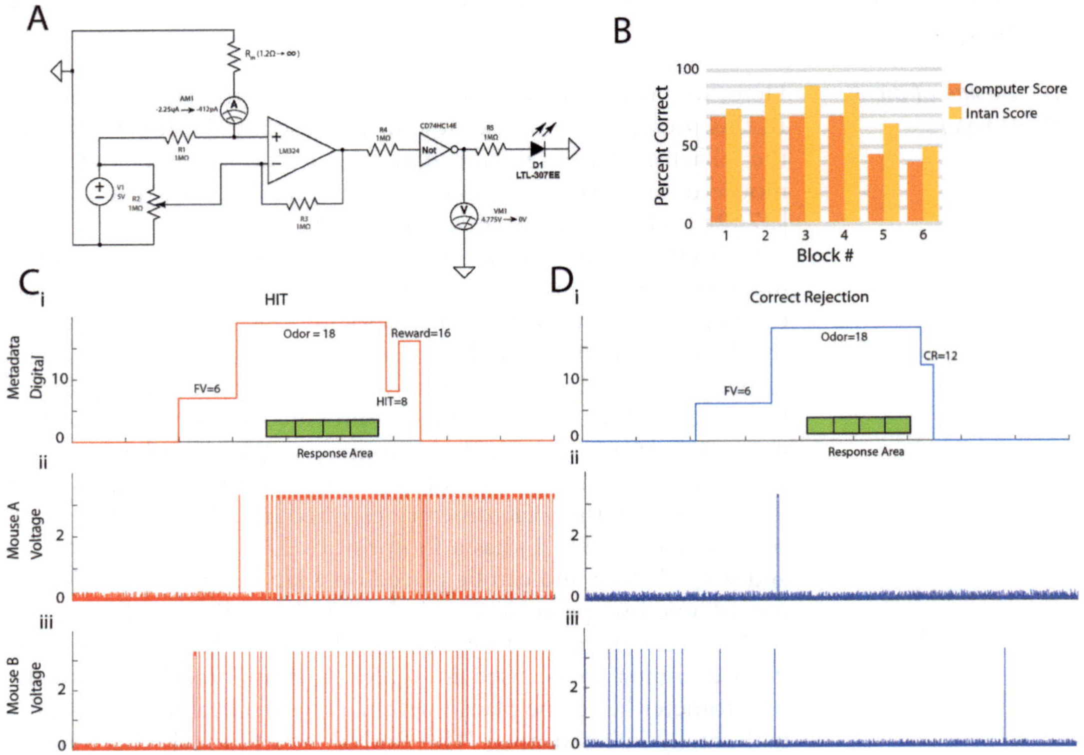

Fig. 2 Detection of licks by the JL olfactometer. (**a**) Lick-detection circuit. The output of the lick-detection circuit changes from 0 to 5 V when the animal licks thereby changing the Rm resistor from infinite (no lick) to ~1.2 Mohms. The signal recorded by the INTAN is capped to 3.3 V because of a limit of the analog input from 0 to 3.3 V. (**b**) Percent correct scoring for the "brief licking" mouse whose licks records are shown in c(iii), d (iii). Percent correct is scored by either the MATLAB computer code or by post hoc scoring from the INTAN recording from the lick circuit. (**c**, **d**) Metadata output from the JL olfactometer control computer (i) and lick traces (ii, iii). The lick traces were recorded from two different mice (ii, iii). **c** are recordings from a Hit trial (**c**(i) metadata, **c**(ii), (iii) lick traces) and **d** are recordings from a correct rejection (CR) trial (**d**(i) metadata, **d**(ii), (iii) lick traces). The numbers defining the events for the metadata are arbitrary (see e.g., in Ci that final valve, FV, is represented by a "6" digital output). The digital metadata output for a Hit is: 6 for final valve diversion, 18 for odor on, 8 for a Hit score, and 16 for water reward. The digital metadata output for a CR is: 6 for final valve diversion, 18 for odor on, and 12 for CR

Mini JL olfactometer. The mini JL olfactometer controls airflow and water delivery valves through a USB-DIO96H/50 Measurement Computing board connected with an SSR-RACK48 (Fig. 1a). The SSR-RACK48 has quad solid-state relay (SSR) modules (1781-OC5Q) to drive the valves with 24 V DC through pins 1–24 (ports A to C) and quad input modules (1781-IB5Q) for TTL input for lick detection through pins 27–34. In the go-no-go task, the mouse starts the trial by licking on the water delivery spout. The odor trial then proceeds by diverting the airflow from the chamber to the exhaust with the final valve diverter and opening the odor valve for 1–1.5 s after the first lick (final valve period, Fig. 1c). During this final valve period, the odorant concentration

Table 1
Parts for the automated Mini JL olfactometer

Item	Details	Manufacturer	Part number	Web site
Interface board	48 channel solid-state relay (OPYO22, Gordos) mounting and interface rack	Measurement Computing	SSR-Rack 48	www.mccdaq.com
Solid-state relay module (output relay for the odorant and diverter solenoid valves)	Quad, VDC output, 5–200 VDC, 3.5 A	Measurement Computing	SSR-4-ODC05A	www.mccdaq.com
Solid-state relay module (input relay for the lick detection and recording)	Quad VDC input, 3–32 VDC	Measurement Computing	SSR-4-IDC-05	www.mccdaq.com
Diverter solenoid valve	Diverts air flow (Series 3: moderate PEEK, EPDM diaphragm, 3-way valve, without bleach, 0.078″ orifice, 24 DC voltage, 1/8: connection barbs)	Parker	003-0258-900	www.ph.parker.com
Water solenoid valve	Engages water system (Series 3: 2-way normally closed valve, 1/8″ barb connection)	Parker	003-0257-900	www.ph.parker.com
Odorant valve	Normally closed 24 V pinch valve	NResearch	161P012	www.nresearch.com
Capacitative lick sensor	Capacitative touch sensor	Sparkfun	AT42QT1010	www.sparkfun.com
RHD USB interface board	USB interface board for neural recording	Intan Technologies	C3100	www.intantech.com

Table 2
Parts for the automated 2-channel JL olfactometer

Item	Details	Manufacturer	Part number	Web site
Interface board	Solid-state I/O device with 8 channels, mounting and interface rack	Measurement Computing	USB-SSR08	www.mccdaq.com
I/O module relay output for the odorant and diverter solenoid valves	Solid-state relay module, single, DC switch, 3–60 VDC @ 3.5 A	Measurement Computing	SSR-ODC-05	www.mccdaq.com
I/O module relay input for the lick detection and recording	Solid-state relay module, single, DC sense, 3–32 VDC	Measurement Computing	SSR-IDC-05	www.mccdaq.com
Diverter solenoid valve	Diverts air flow (Series 3: moderate PEEK, EPDM diaphragm, 3-way valve, without bleach, 0.078″ orifice, 24 DC voltage, 1/8: connection barbs)	Parker	003-0258-900	www.ph.parker.com
Water solenoid valve	Engages water system (Series 3: 2-way normally closed valve, 1/8″ barb connection)	Parker	003-0257-900	www.ph.parker.com
Odorant valve	Normally closed 24 V pinch valve	NResearch	161P012	www.nresearch.com
Capacitative lick sensor	Capacitative touch sensor	Sparkfun	AT42QT1010	www.sparkfun.com
RHD USB interface board	USB interface board for neural recording	Intan Technologies	C3100	www.intantech.com

reaches a final stable level in the background airflow. At the end of the final valve period, the final valve diverter shifts the odorized background air back to the chamber and the 2.5 s odor period starts. For each trial, the olfactometer delivers either the S+ (rewarded when the animal licks) or S− odorant with a pseudorandom Fellow series for S+ and S− where the same odorant is not delivered in more than three consecutive trials. When the S+ stimulus is presented, a water reinforcement is delivered if the mouse meets the criterion response requirement of licking at least once in four 0.5 s periods (0.5–2.5 s after odorant addition, see response area in Fig. 1c). The volume of the water reward can be tittered between 0 and 50 μL by changing the time the water valve is open. In addition, the olfactometer turns on a laser for optogenetic stimulation at specific times during the trials. The USB-DIO96H/50 board in the JL olfactometer is controlled by MATLAB code deposited in https://github.com/restrepd/dropc.

In order to deliver odorant during the trials air is delivered to the odorant sampling port at a flow rate of 2 L/min through a pipe that is connected to a diverter valve (referred to as the final valve, FV). When a trial starts the airflow is diverted by the FV to the exhaust for a random time between 1 and 1.5 s and at the same time 50 mL/min odorized air is delivered into the 2 L/min airflow by opening normally closed pinch valves (the odorant valves) to deliver bubbles of air into an odorant dissolved in odorless mineral oil and to deliver the odorized air into the 2 L/min mainstream airflow. After equilibration of the odorant in the mainstream 2 L/min airflow for 1–1.5 s, the diverter valve turns the airflow back into the sampling odor port resulting in an increase in odor concentration described by two exponentials with time constants of 0.12 and 2.16 s (Fig. 1f) [18]. After an odorant delivery period of 2.5 s, the odorant valves are closed and the odorant concentration at the odor spout decreases exponentially with time constants of 0.19 and 1.28 s. The licks are detected by a custom lick sensor detecting changes in resistance between the lick spout and the grounded metal floor of the chamber (described below) whose output is read by a DC input module that relays the information to the DIO96H/50 board. The circuit is designed to deliver a small amount of current from the mouse's tongue to the floor (2.2 μA) so that the mouse does not sense the current flow (Fig. 2a). The odorant valves and the water delivery valve are controlled by the DIO96H/50 board through solid-state relay modules.

The olfactometer DIO96H/50 board sends metadata information on items such as timing of FV, odor on and odor off, odorant chosen for each trial, water delivery, and behavioral result of the trial to the INTAN RHD2000 evaluation system through the 3.3 V digital input lines (Fig. 2c(i), d(i)). In addition, the INTAN RHD2000 evaluation system records licks, sniffing, and laser on-and-off times through the analog inputs [31]. For tetrode recording experiments, the INTAN RHD2000 evaluation system records extracellular potential through an amplifier chip (RHD2216 for four tetrodes). For multiphoton imaging sessions, the INTAN records camera synchronization signals through the analog input port [36]. Finally, for some olfactometers we record mouse movement from the top at 60 fps using a FLIR camera (BFS-U3-16S2C-CS) with a C Series fixed focal length lens (30 cm above the floor, focal length: 8.5 mm, aperture: f/1.3, requires CS-to-C mount adapter) and USB 3.1 interfacing through custom python code (https://github.com/wryanw/FLIR_JLolfactometer).

2-channel JL olfactometer. The 2-channel JL olfactometer has a further reduced footprint (19″ wide × 12.5″ tall and 8.5″ deep), and all the valves and electronics are within a plexiglass enclosure to allow behavioral assessment of olfactory associative learning in mice infected with BSL2+ biosafety level viruses such as herpes simplex virus 1 (HSV-1). The enclosure ensures that the virus does not reach the valves and electronics that would be difficult to decontaminate. As such, the 2-channel JL olfactometer allows for storage in a cabinet in the confined animal BSL2+ space. The 2-channel olfactometer is controlled by a Measurement Computing USB-SSR08 board that controls airflow valves for two odorants (S+ and S−) to allow performing regular go-no-go associative learning tasks.

3 Methods

3.1 Installing the Software

The software interfaces with the USB-DIO96H/50 board through the MATLAB 2015a data acquisition toolbox in a computer running the Windows 10 operating system. In order to enable the communication with the board, download MCC DAQ Software 6.73 from https://www.mccdaq.com/Software-Downloads and run install to ensure that the computer is interfacing with the board. Then start MATLAB as an administrator and register the mcc driver interface by running daqregister('mcc'). Change the board number in dropcInitializePortsNow.m

handles.dio = digitalio('mcc',1); %(1 or 0 depending on the computer)

Finally, the test begin.m and dropcspm.m by performing a dry run where the user "responds" to each trial by connecting the electrical loop between the lick spout and the metal floor of the chamber that is grounded. Once the user checks the JL olfactometer then they proceed to train the mouse to lick on the spout by running begin.m in stage 1 where the mouse receives an immediate water reward after licking and stage 2 where the final valve period becomes progressively longer. For the 2-channel JL olfactometer follow the same procedure and use begin_2-channel.m and dropcspm_2-channel.m.

3.2 Lick Detection

Accurate evaluation of the behavior of an animal engaged in the go-no-go odorant discrimination associative learning task relies on quantifying its response to the odorants by real time monitoring of its licking behavior. The program must evaluate licking by the animal to decide whether it has responded appropriately to the rewarded odorant and deliver the water reward. The system must capture the animals' licks in real time without missing brief licking. Before trial initiation, the dropcspm program runs a while loop of code that monitors the output of a lick detection circuit until it detects a lick. The circuit, shown in Fig. 2a, performs a binary detection (lick/no lick) of changes in resistance when the animal tongue contacts a sensor. The design utilizes a variable force switch for a force sensor (variable resistor) from https://www.sparkfun.com/products/9375. Here, the Rm resistor in Fig. 2a changes in resistance measured between the lick spout and the floor of the chamber from ~1.2 MΩ for a 3-month C57BL/6 when the mouse's tongue is touching the lick spout to an infinite resistance when the mouse is not licking. A hex inverter circuit is used to output a binary high versus low voltage output. The output changes from high (5 V) when the mouse licks to low when the mouse does not lick (0 V). The change in voltage is detected by the computer through one of the digital inputs in the SSR-RACK48.

Figure 2c, d show the lick signal recorded with the INTAN board with a 20 kHz sampling rate for Hit and correct rejection (CR) trials recorded from two mice (in the INTAN the lick signal shows as 0 to full range 3.3 V). As shown in Fig. 2c, when the stimulus is the rewarded (S+) odorant, the mouse must meet the response criterion. On the other hand, when the odorant is an unrewarded odorant, the mouse performs a correct rejection (CR) if it refrains from licking in at least one of the 0.5 s segments. With a conditional MATLAB loop running at a rate of ~100 Hz, the computer can detect the licks correctly for mice that perform "long" licks of duration >10 ms (Fig. 2c(ii)). However, there are some mice whose licks are brief (Fig. 2c(iii)) and the computer fails to detect the voltage change for a subset of the licks missing some licks and scoring some Hits as Miss trials (Fig. 2b).

To avoid this problem, a 16 MHz Arduino UNO board was added to the olfactometer recording system allowing evaluation of licks at ~8 kHz, fast enough to detect short licks. In a program called dropcspm_UNO.m, the Arduino circuit is triggered when the trial starts and it reports back to the computer through the digital inputs of the SSR-RACK48 board whether the mouse responded in the response area.

3.3 Go-No-Go and Go-Go Associative Learning in Head-Fixed Animals

We have used the JL olfactometer to perform the go-no-go task with head-fixed mice to perform multiphoton imaging while the animal responds to the odorants [36]. In this case, the tubes carrying the airflow and reward water are connected to a cone placed in front of the head-fixed mouse. In this head-fixed associative learning task, the mice were habituated to the stage for one to two sessions before training starts to reduce stress. Begin_hf.m was first used to train animals to obtain water in both S+ and S− trials. After the animal became familiar with the begin behavior paradigm, dropcspm_hf was used to train the mice in the go-no-go task. The water-deprived mice self-initiated the trial by licking on the water port. In S+ trials, the mice needed to lick at least once in two 2 s lick segments to obtain a water reward. In S− trials, the mice need to refrain licking one of the two 2 s segments to avoid a longer inter-trial interval. The animal's behavior performance was evaluated in a sliding window of 20 trials, and the calculated value was assigned to the last trial in the window. The percent correct value represents the percent of trials in which the animal successfully performed appropriate actions, and the animal was considered proficient if percent correct performance is above 80%. In go-go training sessions, the mice obtained a water reward 70% of the time in both "S+" and "S−" trials.

Utilizing the go-no-go paradigm, the mitral/tufted cells in the olfactory bulb of pcdh21-cre mice were imaged with two-photon microscopy during the associated learning. Figure 3 shows examples of the response of recorded neurons with calcium traces and behavior percent correct data demonstrating the learning process.

3.4 Go-No-Go and Olfactory Threshold with the 2-Channel JL Olfactometer

The small footprint "2-channel" JL olfactometer was designed to be used in a BSL2+ facility with little storage space and to be run within a biosafety cabinet to allow safe assessment of olfactory function of human alphaherpesvirus 1 (HSV-1)-infected mice (Fig. 4). The valves are controlled and the licks are detected with a Measurement Computing USB-SSR08 board. The design is similar to the mini olfactometer with the difference that only two odorants can be used. The USB-SSR08 controls the final valve, odorant valves, and water valve through output relays and registers the licks detected by the lick circuit through input relays. The computer interfaces with the USB-SSR08 through the data acquisition toolbox of Matlab 2015a using dropcbegin_2-channel.m and dropcspm_2-channel.m. It is recommended to switch the connections between vials and valves every day: on day 1, the vial containing S+ is connected to valve 1 and the vial containing S− is connected to valve 2 on day 1; on day 2, the vial containing S+ is connected to valve 2 and the vial containing S− is connected to valve 1, etc. This way, if there is any difference in noise or airflow between valve 1 and valve 2, mice will not associate those non-olfactory cues with S+ or S−. Importantly, while the mice learn the

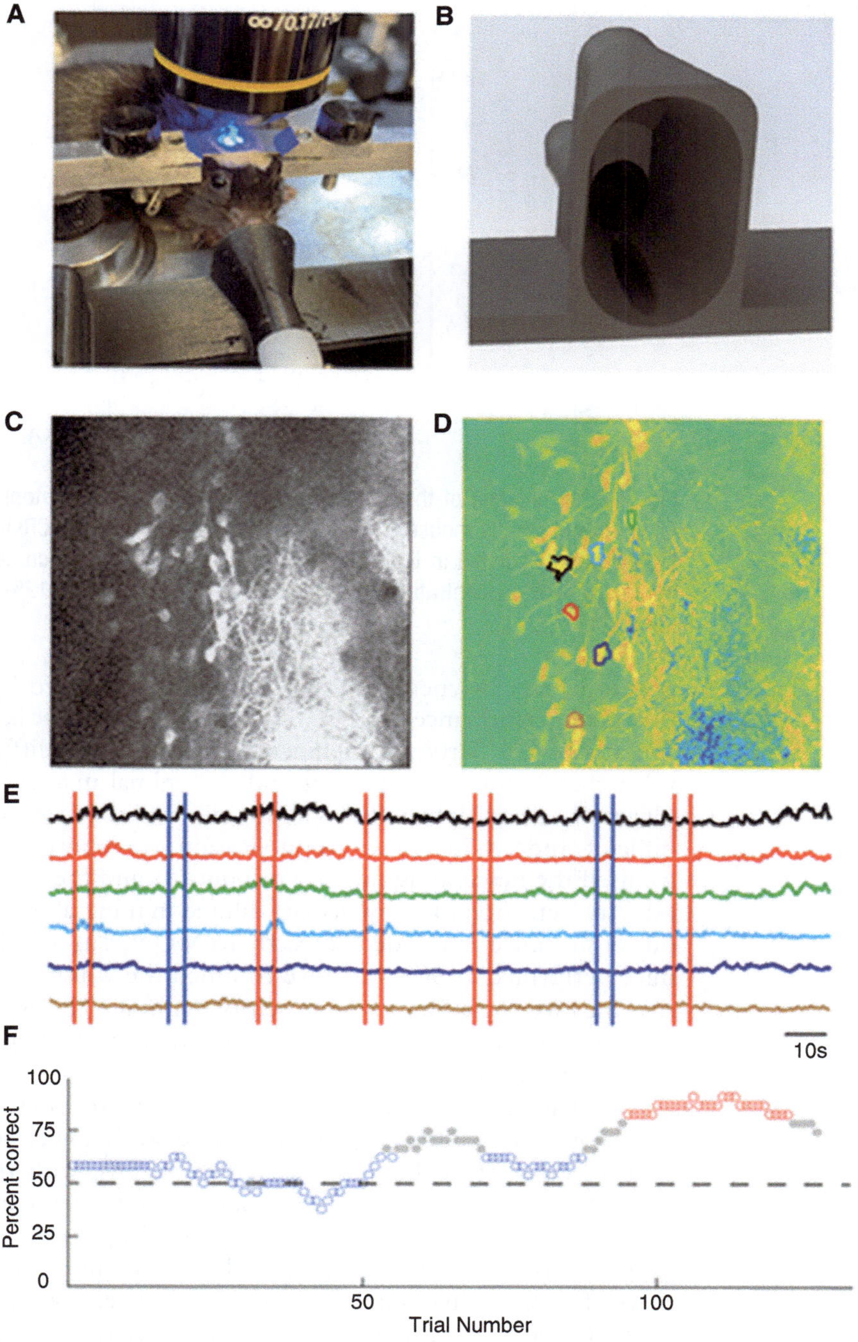

Fig. 3 Head-fixed imaging of mitral/tufted cell activity. (**a**) Head-fixed mouse undergoing the go-no-go task while imaging neural activity in the multiphoton microscope. The odorants were isoamyl acetate (1% dilution in mineral oil) and mineral oil. (**b**) Rendering of the odorant delivery cone. (**b**, **c**) The raw fluorescence intensity (**b**) and dF/F image (**c**) of olfactory bulb mitral/tufted cells imaged with the multiphoton microscope. AAV1-Syn-Flex-GCaMP6f virus was injected into the main olfactory bulb of Pcdh21-Cre mice to visualize and monitor the mitral/tufted cell activity in associative learning. (**d**) The corresponding calcium traces of ROIs in D. The red vertical lines indicate the start and end of S+ odorant application, and the blue vertical lines represent S− trials. (**e**) The learning process of the animal in the head-fixed go/no-go training task. The blue circles represent the trials with percent correct below 65% and red circles show the proficient level with percent correct above 80%. (**f**) Percent correct for the behavioral outcome as a function of trial number

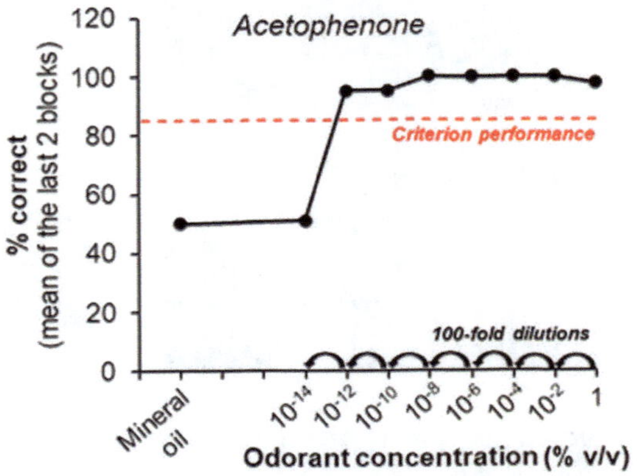

Fig. 4 Determination of the odorant detection threshold for acetophenone in a 6-month-old female mouse. The percent correct (100 * (Hits + CRs)/(number of trials)) was calculated in the last two blocks of 20 trials in each session. The minimum odor concentration detected by this mouse was 10–12% (dilution in mineral oil)

go-no-go task by cueing on the odorant as evidenced by the fact that their performance drops to 50% when the odor cue is removed, somatosensory (airflow differences) and auditory (differences in valve clicks) stimuli are present and the animal may cue on non-olfactory stimuli when the odorant discrimination task is made difficult. Indeed, in a previous publication, we reported that when we used the descending method of limits to find the threshold for odorant detection for ethyl acetate diluted in mineral oil (MO) two out of five mice responded correctly to MO versus MO [37]. The mice performed at 50% when we trimmed the whiskers indicating that they were cueing on somatosensory stimuli. This is the reason why in every set of experiments we test that the mice perform at random when the odorants are removed.

We have used the 2-channel olfactometer to assess olfactory detection in mice trained to detect successively lower concentrations of odorant in the go-no-go task [24, 38], with the aim of comparing odor sensitivity before and after viral infection with HSV-1. Mice have been trained on the dropcbegin_2-channel program using 1% acetophenone (diluted in mineral oil) until successful program completion (percent correct ≥80% for two 20 trial blocks) for 2 consecutive days (typically 5–8 days). After 1 week of rest, mice were then trained on the dropcspm_2-channel with 1% acetophenone as the rewarded stimulus (S+) and mineral oil as the non-rewarded stimulus. Mice became proficient in 1–3 days, performing ~5 blocks (= 100 trials) per day. The criterion performance was defined as reaching at least 80% of correct responses in at least two consecutive blocks of 20 trials. Once the criterion performance was reached, mice were tested with decreasing acetophenone concentration (only one concentration per day), until failure to reach 80% correct responses. The threshold detection for a specific odorant is the last odorant concentration for which criterion performance was achieved before failure. At the end of the entire session, mice were also submitted to a test to determine whether the mouse is cueing exclusively on olfactory stimuli, using a blank (mineral oil) as S+ and S− to verify that mice were not using nonolfactory cues to perform the task. Any mouse reaching the criterion performance during the cheating test should be excluded from the study.

3.5 Working Memory

We implemented a delayed non-match to sample (DNMS) odor task similar to that described in [39, 40] by adding an additional valve to the mini JL olfactometer to maintain constant airflow at 2 L/min and clear residual odorant. Training on the delayed non-match to sample odor task is accomplished in a series of three training phases. In the first phase, the mouse is trained to lick in the presence of the preselected odorant (isoamyl acetate) using the dropcbegin.m program. The purpose of this program is to train the thirsty mouse to lick in the presence of an odorant. After completion of the beginning phase, the mouse is ready for the other two phases of the training. We wrote two MATLAB programs, an intermediate phase 2 where the first odorant is fixed in the DNMS task (dropcspm_WM_AB_DNMS.m, 2 choice DNMS task, phase 2) and the actual DNMS task (dropcspm_WMNP.m, 4 choice DNMS task, phase 3) (Fig. 5a). The code for both programs can be found at https://github.com/restrepd/dropc.

Here, we describe how the two- and three-choice DNMS tasks work highlighting the differences. First, when the mouse licks at the lick spout the computer activates two diverter valves and the odorant valve for 1–1.5 s to equilibrate the odorant in the delivery line while continuing delivery of 2 L/min odorant-free air to the odor spout. After 1–1.5 s, the valves divert the odorant-equilibrated air to the odor port for 1 s. After the exposure to the odorant, the diverter valves turn to deliver odorant-free air, clear the first odorant, and equilibrate the odorant line with the second odorant for a delay period chosen by the user (2–6 s). After the delay period, the animal receives the second odorant for 1 s and has to lick at least once for a subsequent 0.5 s period if the two odorants were different to obtain a water reward (non-match). If the two odorants were the same the animal has to refrain from licking to avoid a 10-s delay in the start of the next trial (*see* Fig. 5). The performance of the mouse is determined by calculating the correct response during the presentations of the odorants in 20 trial blocks where, in random order, for 10 of the trials the first odorant is the same as the second odorant and for 10 of the trials the first odorants is different than the second odorant. See Fig. 5, for example, of a graph for a mouse during naïve and proficient.

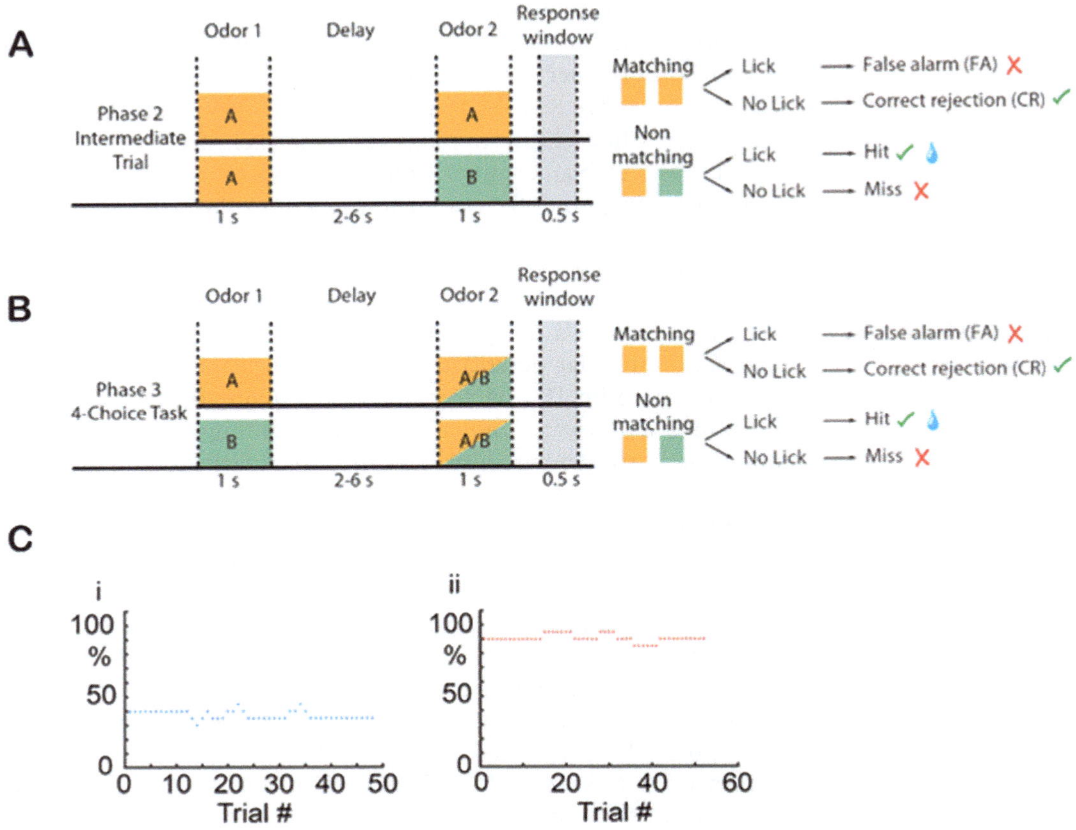

Fig. 5 Delayed non-match to sample (DNMS) working memory behavioral task. (**a**) Phase 2, 2-choice task. In this task, odorant A is delivered for 1 s for every trial. The second odorant can be the same as the first one (A, match) or different (B, non-match). The user can choose the delay time between the first and second odorant. After the selected time the second odorant is released for 1 s. After the end of the second odorant release period, the mouse makes the decision to lick in the 0.5 s response period. When the mouse licks for the non-match case it receives a water reward (Hit). (**b**) Phase 3, 4-choice task. This is similar to the 2-choice task with the exception the first odorant is selected randomly as odorant A or odorant B. (**c**) Example performance for a mouse that is first during the first 4-choice session (i) and after the mouse has learned to respond for non-match (ii)

In the intermediate phase 2 program or 2-choice program, the first odorant is always the same, and the second odorant can be the same as the first one or different (Fig. 5). For the phase 3 program, 4-choice case, the first odorant and second odorants are chosen randomly (Fig. 5). We recommend to first train the animal in the 2-choice program, alternating the first odorant, before moving on the mice to the 4-choice task. Mice can be trained twice per day in the morning and evening depending on the individual time and schedule or once a day. The user can determine which odorant to use, we had success with ethyl acetate versus 2-pentanone.

4 Discussion

Here we describe a custom-made olfactometer, the JL olfactometer, which can be used to flexibly assess olfactory function and learning in a variety of behavioral olfactory tasks. Olfactometers have provided the means to investigate olfactory abilities as well as the underlying brain activity while performing a behavioral task in live animals. Various aspects of olfaction and olfactory abilities can be studied such as odor detection, odor discrimination, working memory, odor-shock conditioning, and odor memory [24, 27, 30, 38, 40]. Olfactometers enable the investigation of olfactory abilities per se but also the learning and memory processes in olfactory-associated tasks. In contrast to olfactory tests that use simple methods such as investigation time, habituation, or time to find a hidden odorous object, olfactometry requires a special apparatus and subject training. However, olfactometers allow for precise control of the concentration and temporal presentation of multiple stimuli over literally hundreds of trials in sessions that provide quantitative, parametric measures of stimulus sampling and the acquisition, discrimination, and memory for odors.

Of particular interest is the sensitivity of mice to social odors such as urine. Mice have an exquisite ability to detect differences in urine (odor types) to differentiate, for example, urine from conspecifics with different major histocompatibility genotypes or urine from virally infected animals and controls [41–43]. This ability is used by the animals for behaviors such as mating choice and social preference. Thus, urine is an attractive or repulsive scent depending on the sex and species of the animal tested versus the animal from which the urine was collected, and it is clearly not a neutral odorant. Olfactometers are useful for the study of discrimination of urine odors. Finally, other fundamental reasons for the advantage of using automated olfactometry over simple test of investigation (e.g., odor preference and habituation) are discussed by refs. [44, 45].

The design of our mini JL olfactometer is based on the olfactometer described by Slotnick and Restrepo [27] that was available commercially from Knosys olfactometers, Inc. We have made some modifications to the design: (1) We use different rapid switch (5 ms) valves (Table 1). We are not aware whether the odorant onset time differs from that of the Knosys olfactometer. (2) We use a different lick circuit. We chose to use this circuit because it has a potentiometer that allows to adjust the sensitivity of the circuit. (3) The most useful feature compared to the commercial olfactometer is that we write custom software allowing us to add capabilities such as sending digital metadata to be recorded by an external ADC board (INTAN in our case) and sending triggers out for synchronization with other instruments (e.g., turning the laser on). (4) The lick output of the JL olfactometer is recorded in real time by the INTAN board allowing analysis of lick-referenced changes in LFP or GCaMP fluorescence.

Using an automated liquid dilution 8-channel olfactometer, mice have been shown to be able to learn to distinguish between urinary odors collected from gonadally intact males and estrous females, or from gonadally intact versus castrated males in a food-motivated olfactory discrimination task [46]. Interestingly, females learned this discrimination significantly less rapidly than males, demonstrating a strong sex dimorphism in olfactory discrimination ability, with males performing better than females. In a thirst-motivated olfactory discrimination task, male and female performance for discriminating between male and female uring was diminished after gonadectomy. The task performance could be restored by administering testosterone propionate in castrated males or estradiol benzoate in ovariectomized females, highlighting the importance of gonadal hormones for olfactory behavior [47].

Another benefit of using operant-olfactometer-based tasks is the diminished requirement of motor activity and the absence of reliance on visual stimuli. As a result, olfaction can be tested in mouse models of neurological disorders with motor or visual deficits [48, 49]. Finally, rodents learn readily when olfactory versus nonolfactory learning tasks are used. For example, the success of rodent match and non-match-to-sample studies stand in sharp contrast to the failure to demonstrate match-to-match and non--match-to-sample learning in rats when nonchemical sense stimuli have been used [39, 40, 50]. The wide range of possible customizations, and the possibility of recording or manipulating neural activity while performing an olfactory task makes the olfactometer an ideal testing apparatus for investigating olfactory behavior in rodents.

Acknowledgments

This research was supported by an Administrative Supplement S1 to NIH UF1 NS116241 (to DRG), NIH R01 DC000566, NIH UF1 NS116241, and NSF BCS-1926676 (to DR).

References

1. Yeshurun Y, Sobel N (2009) An odor is not worth a thousand words: from multidimensional odors to unidimensional odor objects. Annu Rev Psychol 61:219–241
2. Buck LB (2005) Unraveling the sense of smell (Nobel lecture). Angew Chem Int Ed Engl 44: 6128–6140
3. Spors H, Wachowiak M, Cohen LB, Friedrich RW (2006) Temporal dynamics and latency patterns of receptor neuron input to the olfactory bulb. J Neurosci 26:1247–1259
4. Feinstein P, Bozza T, Rodriguez I, Vassalli A, Mombaerts P (2004) Axon guidance of mouse olfactory sensory neurons by odorant receptors

and the beta2 adrenergic receptor. Cell 117: 833–846

5. Soucy ER, Albeanu DF, Fantana AL, Murthy VN, Meister M (2009) Precision and diversity in an odor map on the olfactory bulb. Nat Neurosci 12:210–220
6. Spors H et al (2012) Illuminating vertebrate olfactory processing. J Neurosci 32:14102–14108
7. Wachowiak M, Shipley MT (2006) Coding and synaptic processing of sensory information in the glomerular layer of the olfactory bulb. Semin Cell Dev Biol 17:411–423
8. Mozell MM (1970) Evidence for a chromatographic model of olfaction. J Gen Physiol 56:46–63
9. Ressler KJ, Sullivan SL, Buck LB (1994) Information coding in the olfactory system: evidence for a stereotyped and highly organized epitope map in the olfactory bulb. Cell 79: 1245–1255
10. Vassar R et al (1994) Topographic organization of sensory projections to the olfactory bulb. Cell 79:981–991
11. Kwak J, Willse A, Preti G, Yamazaki K, Beauchamp GK (2010) In search of the chemical basis for MHC odourtypes. Proc Biol Sci 277: 2417–2425
12. Schaefer ML, Yamazaki K, Osada K, Restrepo D, Beauchamp GK (2002) Olfactory fingerprints for major histocompatibility complex-determined body odors II: relationship among odor maps, genetics, odor composition, and behavior. J Neurosci 22:9513–9521
13. Bolding KA, Nagappan S, Han B-X, Wang F, Franks KM (2019) Robust odor coding across states in piriform cortex requires recurrent circuitry: evidence for pattern completion in an associative network. bioRxiv:694331
14. Bolding KA, Franks KM (2018) Recurrent cortical circuits implement concentration-invariant odor coding. Science 361:eaat6904
15. Bolding KA, Franks KM (2017) Complementary codes for odor identity and intensity in olfactory cortex. elife 6:e22630
16. Gadziola MA et al (2020) A neural system that represents the association of odors with rewarded outcomes and promotes behavioral engagement. Cell Rep 32:107919
17. Carlson KS, Gadziola MA, Dauster ES, Wesson DW (2018) Selective attention controls olfactory decisions and the neural encoding of odors. Curr Biol 28:2195–2205 e2194
18. Losacco J, Ramirez-Gordillo D, Gilmer J, Restrepo D (2020) Learning improves decoding of odor identity with phase-referenced oscillations in the olfactory bulb. elife 9: e52583
19. Wu A, Yu B, Komiyama T (2020) Plasticity in olfactory bulb circuits. Curr Opin Neurobiol 64:17–23
20. Pfaffmann C, Goff WR, Bare JK (1958) An olfactometer for the rat. Science 128:1007–1008
21. Jay Braun J, Marcus J (1969) Stimulus generalization among odorants by rats. Physiol Behav 4:245–248
22. Marshall DA, Moulton DG (1981) Olfactory sensitivity to α-ionone in humans and dogs. Chem Senses 6:53–61
23. Di Prisco GV, Freeman WJ (1985) Odor-related bulbar EEG spatial pattern analysis during appetitive conditioning in rabbits. Behav Neurosci 99:964–978
24. Bodyak N, Slotnick B (1999) Performance of mice in an automated olfactometer: odor detection, discrimination and odor memory. Chem Senses 24:637–645
25. Slotnick BM, Schellinck HM (2001) Behavioral methods in olfactory research with rodents. In: Simon SA, Nicolelis MA (eds) Methods in chemosensory research. CRC Press, Boca Raton, pp 552–595. https://doi.org/10.1201/9781420038729
26. Li A, Gire DH, Bozza T, Restrepo D (2014) Precise detection of direct glomerular input duration by the olfactory bulb. J Neurosci 34: 16058–16064
27. Slotnick BM, Restrepo D (2005) Olfactometry with mice. In: Crawley JN et al (eds) Current protocols in neuroscience. John Wiley and Sons, Inc, New York, pp 1–24
28. Doucette W, Restrepo D (2008) Profound context-dependent plasticity of mitral cell responses in olfactory bulb. PLoS Biol 6:e258
29. Losacco J, George NM, Hiratani N, Restrepo D (2020) The olfactory bulb facilitates use of category bounds for classification of odorants in different intensity groups. Front Cell Neurosci 14:430
30. Svalina MN et al (2022) Basolateral amygdala hyperexcitability is associated with precocious developmental emergence of fear-learning in Fragile X Syndrome. J Neurosci. https://doi.org/10.1523/JNEUROSCI.1776-21.2022
31. Li A, Gire DH, Restrepo D (2015) Υ spike-field coherence in a population of olfactory bulb neurons differentiates between odors irrespective of associated outcome. J Neurosci 35: 5808–5822
32. Gire DH, Whitesell JD, Doucette W, Restrepo D (2013) Information for decision-making and

stimulus identification is multiplexed in sensory cortex. Nat Neurosci 16:991–993

33. Doucette W et al (2011) Associative cortex features in the first olfactory brain relay station. Neuron 69:1176–1187
34. Ramirez-Gordillo D, Parra AA, Bayer KU, Restrepo D (2021) Hippocampal-prefrontal theta coupling develops as mice become proficient in associative odorant discrimination learning. bioRxiv 09.28.462143, 2021.2009.2028.462143
35. Ramirez-Gordillo D, Ma M, Restrepo D (2018) Precision of classification of odorant value by the power of olfactory bulb oscillations is altered by optogenetic silencing of local adrenergic innervation. Front Cell Neurosci 12:48
36. Ma M et al (2020) Molecular layer interneurons in the cerebellum encode for valence in associative learning. Nat Commun 11:4217
37. Clevenger AC, Restrepo D (2006) Evaluation of the validity of a maximum likelihood adaptive staircase procedure for measurement of olfactory detection threshold in mice. Chem Senses 31:9–26
38. Williams E, Dewan A (2020) Olfactory detection thresholds for primary aliphatic alcohols in mice. Chem Senses. https://doi.org/10.1093/chemse/bjaa045
39. Otto T, Eichenbaum H (1992) Neuronal activity in the hippocampus during delayed non-match to sample performance in rats: evidence for hippocampal processing in recognition memory. Hippocampus 2:323–334
40. Liu D et al (2014) Medial prefrontal activity during delay period contributes to learning of a working memory task. Science 346:458–463
41. Beauchamp GK, Yamazaki K (2003) Chemical signalling in mice. Biochem Soc Trans 31:147–151
42. Yamaguchi M et al (1981) Distinctive urinary odors governed by the major histocompatibility locus of the mouse. Proc Natl Acad Sci 78:5817
43. Kwak J et al (2008) Genetically-based olfactory signatures persist despite dietary variation. PLoS One 3:e3591
44. Slotnick B, Coppola DM (2015) Odor-cued taste avoidance: a simple and robust test of mouse olfaction. Chem Senses 40:269–278
45. Dewan A et al (2018) Single olfactory receptors set odor detection thresholds. Nat Commun 9:2887
46. Wesson DW, Keller M, Douhard Q, Baum MJ, Bakker J (2006) Enhanced urinary odor discrimination in female aromatase knockout (ArKO) mice. Horm Behav 49:580–586
47. Kunkhyen T et al (2018) Gonadal hormones, but not sex, affect the acquisition and maintenance of a Go/No-Go odor discrimination task in mice. Horm Behav 100:12–19
48. Roddick KM, Roberts AD, Schellinck HM, Brown RE (2016) Sex and genotype differences in odor detection in the 3xTg-AD and 5XFAD mouse models of Alzheimer's disease at 6 months of age. Chem Senses 41:433–440
49. Roddick KM, Schellinck HM, Brown RE (2014) Olfactory delayed matching to sample performance in mice: sex differences in the 5XFAD mouse model of Alzheimer's disease. Behav Brain Res 270:165–170
50. Lu X-CM, Slotnick BM, Silberberg AM (1993) Odor matching and odor memory in the rat. Physiol Behav 53:795–804

Chapter 7

Resting-State Functional Magnetic Resonance Imaging as a Method for the Study of Social Behavior in a Rodent Model

M. Fernanda López-Gutiérrez, Juan J. Ortiz, Wendy Portillo, and Sarael Alcauter

Abstract

The advancement of technology has allowed the implementation of novel techniques that have become very helpful in targeting questions that otherwise would be difficult to answer due to methodological limitations. In this chapter, we describe how magnetic resonance imaging (MRI) can be used in small animals, specifically the prairie vole (*Microtus ochrogaster*), to understand the relationship between functional networks in the brain and the social behaviors observed in this rodent model. The type of MRI data we focus on is resting-state functional magnetic resonance imaging (rsfMRI), which detects signals derived from blood oxygen-level changes during a resting task. Processed rsfMRI data allow the analysis of functional brain networks that can be subsequently correlated with particular behaviors assessed on the subject model. Moreover, by implementing this method with a longitudinal design, it enables describing and understanding the contextual and temporal dynamics of neural circuits in a non-invasive fashion. A methodological overview and requirements of subjects, equipment, anesthesia protocols, and data analysis are given so the reader can clearly outline how to design an experimental protocol around this particular technique.

Key words Resting-state magnetic resonance imaging, Functional connectivity, Prairie vole, Socio-sexual behavior, Brain networks

1 Introduction

Magnetic resonance imaging (MRI) is a complex imaging method that allows the exploration of the brain and the body in vivo. It is based on the resonance effect of the atomic nuclei when immersed in a magnetic field, particularly the hydrogen nucleus ^{1}H, highly abundant in biological systems. When immersed in a magnetic field, the hydrogen atoms will absorb energy from electromagnetic waves of specific frequencies depending on the magnitude of the field. Specifically, the resonance frequency of ^{1}H is directly proportional to the magnetic field, being at a rate of 42.58 MHz per tesla,

Raúl G. Paredes et al. (eds.), *Animal Models of Reproductive Behavior*, Neuromethods, vol. 200,
https://doi.org/10.1007/978-1-0716-3234-5_7,

meaning that for each tesla (T) of magnetic field, the electromagnetic waves need to have a frequency of 42.58 MHz to make the ^{1}H to resonate (absorb energy). In simple terms, after these nuclei absorb energy, they will emit it back and this signal will be captured by a coil placed as close as possible to the imaging sample. Differences in tissue properties (molecular structure, motility, diffusion), among other factors, will result in varying energy re-emission rates. By manipulating the parameters of the imaging sequences, MRI can be used to generate a variety of tissue contrasts to study the brain structure, chemistry, and function.

MRI uses electromagnetic radiation in the same spectrum of the radio waves, the electromagnetic waves used by radio stations. Compared to X-ray and high-energy gamma photons used in nuclear medicine, radio waves are non-ionizing and are not associated with the unwanted associated effects of the ionizing energy. Although there are some reports of acute and local changes in the permeability of some tissues, the technique is safe when taking into consideration the limitations of introducing metal, particularly but not limited to ferromagnetic materials, and electronic devices into large magnetic fields. The feasibility of imaging the whole brain (or large portions of it), with a variety of tissue contrasts, including dynamic studies that provide valuable functional information, in vivo and safely, has positioned the MRI as a valuable tool for longitudinal studies exploring brain–behavior associations. In particular, MRI allows the detection of blood oxygen level-dependent (BOLD) signals that are indirectly related to the activity of neuron populations [1]. This signal may be detected also in the resting state, aiming to explore the spontaneous activity and how it covariates (functional connectivity; [2]) along the brain, allowing the detection of functionally connected regions or resting-state brain networks [3]. Therefore, the use of neuroimaging techniques is an attractive alternative for a non-invasive longitudinal exploration of brain networks, which are often involved in the expression of complex behaviors.

In this chapter, we will describe how this method was useful in studying the relationship between brain functional networks and the social behavior of a rodent model. The model in question was the prairie vole (*Microtus ochrogaster*), a Cricetid rodent that is characterized for being highly social and being capable of forming pair bonds, which can be defined as long-lasting, strong social relationships between individuals in a breeding pair in a monogamous species [2]. If voles pair bond, they will display affiliative behavior toward their partners and will exhibit selective aggression toward stranger voles. These behavioral changes can be observed after 24 h of cohabitation with mating, but long-term brain plasticity processes are reportedly involved in pair bond maintenance [3]. Behaviors like the mentioned above have been previously suggested to be modulated by the social decision-making brain network [4], a theoretical network constituted by a set of brain regions that are hypothesized to interact in order to enable the expression of social behaviors in vertebrates. Pharmacological studies have suggested that several brain regions comprising this putative network are involved in pair bond formation of the prairie vole, including structures associated with sensory, reward, and salience processing [5]. Accordingly, a previous study in the prairie vole showed that the functional interactions of a corticostriatal circuit were associated with social bonding [6]. However, the reported electrophysiological technique was methodologically limited and could not determine if the activity of other brain regions was implicated in the studied behavior. Thus, in order to assess the involvement of other brain regions, a less invasive technique that could be applied in a longitudinal manner would be more appropriate to increase the understanding of multiple functional interactions. Therefore, the forthcoming procedures were part of a study [7] with the objective of acquiring magnetic resonance imaging data that, once processed and analyzed, provided information about the brain functional interactions and potential functional networks involved in the process of pair bond formation and social bonding behaviors in the prairie vole.

2 General Overview of Subjects and Equipment

2.1 Subjects

Having in mind that MRI as a technique has a non-invasive nature over subjects and facilitates the obtention of longitudinal data, the experimental design should ideally make use of such advantage. The usual number of subjects for rodent MRI studies is above 10 per group or treatment [8–10] but higher numbers may be needed depending on the properties being explored. Brain imaging may be used to estimate a small number of specific properties to be compared between groups, for example, the volume of the brain. However, it may also be used to search for voxels (the minimum imaging unit or volume element) with significant differences between groups. These two cases represent the extremes of the type of analysis based on the number of properties to be compared. If several tests are performed (for example one for each voxel), then the statistical tests need to correct for the multiple comparisons being performed, diminishing the statistical power. The ideal strategy to determine the optimal number of subjects is to estimate the sample size based on the expected effect size and statistical power, which may be easier to estimate for sets of specific properties, but gets more complicated to estimate for multivoxel patterns when exploring at the voxel level in large portions of the brain.

The only considerable limitation over subject selection and type of procedures employed together with MRI studies is that they should not have attached or implanted any magnetic object in any part of the body, since it may not only cause injury, but it may damage the equipment or at the least distort the acquired images. MR-compatible alternatives for magnetic objects (cannulae, staples, ID tags) should be considered if necessary. These alternatives are often manufactured with non-magnetic materials, such as plastic for ID tags, or with paramagnetic metals like titanium in the case of screws or cannulae.

2.2 Equipment

2.2.1 MRI Scanner, Coil, and Image Acquisition Sequences

An MRI scanner with the proper equipment and the assistance of a certified technician should be available. MRI scanners are manufactured with diverse specifications, and it is important to verify that the selected animal model is adequate enough for the available MRI machine. Prior testing is recommended to assess if the size and shape of the animal allow proper acquisition of the image sequences required. Rodent MRI acquisition is optimal in specialized scanners for small animals that usually have strong magnetic fields (7 T or above) and a reduced bore diameter. This is necessary because the average rodent brain size and volume are comparatively small, and a high signal-to-noise ratio (SNR) is required to obtain high-quality imaging [11]. To maintain such quality, animals are commonly placed on a scanner bed or adapter that restrains their movement during image acquisition inside the magnet bore. Also, a compatible MR coil that can be adequately placed over the subject's skull is necessary to obtain good quality brain images. While most MRI scanners have software with pre-installed imaging acquisition sequences, it is highly recommended to seek the assistance of trained personnel to test and adjust the sequence protocol according to the experimenter's objectives.

The MRI equipment used in the method described here is a Bruker PharmaScan 70/16US, 7 T magnetic resonance scanner (Bruker, Ettlingen, Germany) with an MRI mouse CryoProbe transmit/receive surface coil (Bruker, Ettlingen, Germany). The software used to perform all imaging protocols was Paravision-6 (Bruker, Ettlingen, Germany).

2.2.2 Anesthesia and Physiological Monitoring

In social behavior studies, subjects are usually scanned alive, i.e. in vivo; thus, it is relevant to ensure physical and physiological stability during image acquisition.

One way to minimize subject motion is through the administration of anesthesia. It is important to consider that as minimal a state of sedation may be, it may alter functional MRI (fMRI) blood oxygen level-dependent (BOLD) response and brain functional connectivity, factors that have to be taken into account depending on the experiment's design. However, various anesthesia protocols have been compared and optimized by several research groups to obtain the best images possible according to the model and the type of MRI study [8, 12]. Gas anesthesia (e.g. isoflurane with air mixture), subcutaneous injection, intravenous infusion (e.g. medetomidine), and combinations between them are common choices (*see* **Note 1**).

Monitoring of physiological readings is necessary to keep track of the animal's state inside the magnet bore and is not only useful in the detection of anomalies that may occur during image acquisition (e.g. animal waking up or moving, abrupt or irregular heart rate or respiration), but also in registering and coordinating physiological readings with MR pulse sequences if necessary. Typical measures are respiration rate, heart rate, temperature, and blood oxygen saturation (particularly for BOLD image sequences). This is achieved by additional MR-compatible equipment and its availability should be consulted with the technician. The MR-compatible monitoring system used in this method records respiration rate, oximetry, heart rate, and temperature (*Model 1030*, SA Instruments Inc.; NY, USA). In addition, a circulating water heating pad within the scanner bed was used to maintain subject body temperature, considering sedation and the environment inside the magnet bore may lower body temperature level.

It cannot be emphasized enough that the experimenter has to do as much as possible to minimize any factor that can influence image quality. Mild-to-severe image artifacts may derive from subject motion, thus making sure physiological readings are stable in live subjects is critical to acquire good data.

2.2.3 Requirements for Data Analysis

After image acquisition, data need to be prepared for analysis. Imaging data pre-processing and subsequent analysis require specific software, and depending on the type of imaging sequences and animal model chosen, additional or different software tools may be available. The most popular MRI analysis software is license-free or open source, such as FMRIB's Software Libraries [13], Statistical Parametric Mapping [14], Advanced Normalization Tools [15] or Freesurfer [16], their installation having specific PC minimum requirements. Additionally, software for statistical computing such as R [17] or MATLAB (MathWorks, Natick, MA) may also be needed depending on the kind of data analyzed (*see* **Note 2**).

Most types of MRI analysis will require the use of an anatomical template. These templates are usually constructed from high-resolution anatomical MRI scans of several subjects, generating a representative brain of the model. Some animal models may not have a standard template available. If that is the case, generating one for the model used in the study should be considered. This method took advantage of previous work in which an anatomical prairie vole brain template was created [18].

3 Methodological Proposal

3.1 Experimental Design

Since the aim of the study was to explore changes in brain functional networks, a resting-state functional magnetic resonance imaging (rsfMRI) protocol was chosen. This type of task-free fMRI sequence allows the analysis of low-frequency fluctuations (<0.1 Hz) of the BOLD signal characteristic of fMRI. The acquired images can then be analyzed for functional connectivity by the estimation of the correlation of rsfMRI signals from anatomically separated regions of the brain [19]. These signals have shown to correlate with neuronal activity [20] and also change under brain plasticity processes [21]. In addition to rsfMRI, anatomical images of the whole brain are required for subsequent image processing and analysis.

Thus, the experimental design was centered on obtaining rsfMRI and anatomical images of male and female prairie voles before and after pair bonding. Male ($n = 16$) and female ($n = 16$) prairie voles were scanned at baseline, 24 h, and 2 weeks after the onset of cohabitation (*see* **Note 3**). After baseline image acquisition, male and female voles were randomly assigned as pairs and were placed for cohabitation to promote pair bond formation. The first 6 h of cohabitation were video-recorded to analyze sociosexual behavior. To evaluate pair bonding, vole pairs were subjected to a partner preference test (PPT) [22] between 48 and 72 h after the onset of cohabitation (Fig. 1).

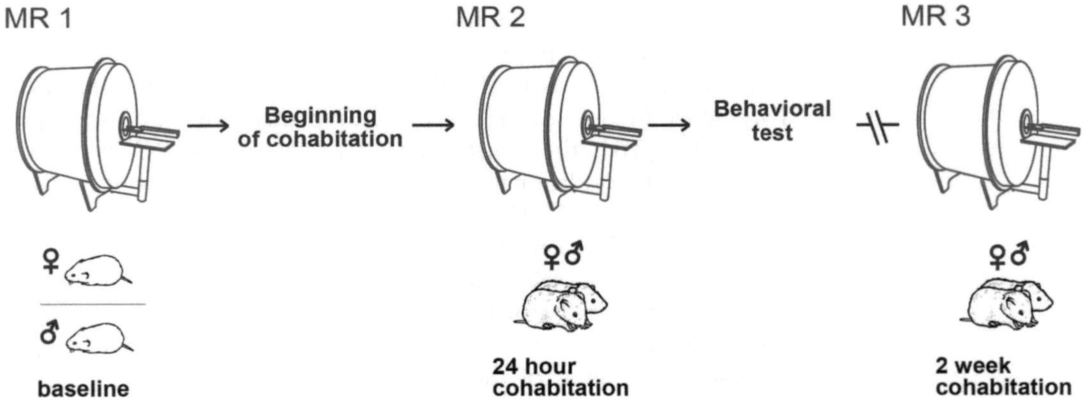

Fig. 1 Example of an experimental design for the study of social behavior using magnetic resonance Imaging. The experiment's main objective was centered on obtaining functional (rsfMRI) and anatomical images of male and female prairie voles before and after pair bonding. Male and female prairie voles were scanned at baseline, 24 h, and 2 weeks after the onset of cohabitation. Between scanning sessions, behavioral recording and testing allowed characterization of social interactions and correlate them with plastic changes in brain functional networks detected through rsfMRI signal analysis as cohabitation progressed

3.2 Prairie Vole Preparation and Base Conditions

Animals used in the study ($n = 32$) were housed in same-sex cages in a room with controlled temperature (23 °C) and light (14:10 light–dark cycle), with ad libitum water and food consisting of sunflower seeds, oat, and rabbit diet HF-5326 (LabDiet, St. Louis, MO). Voles were previously weaned at 21 days.

Female voles are induced ovulators and in natural settings will display sexual receptivity several hours after being exposed to sex-related male olfactory cues [22, 23]. To ensure stable and equal hormonal levels, all female subjects were ovariectomized bilaterally 2 weeks before cohabitation. After surgery recovery and 4 days before cohabitation, silastic capsules filled with estradiol benzoate (E2B; Sigma–Aldrich, St. Louis, MO) dissolved in corn oil (0.5 mg/mL of E2B) were implanted subcutaneously and remained as such for the entire duration of the experiment. Surgical, experimental, and maintenance procedures were carried out in accordance with the local bioethics committee.

3.3 Subcutaneous Anesthesia Preparation

For this protocol of MRI acquisition, a previously standardized combination of isoflurane gas anesthesia (Sofloran; PiSA, Mexico) with s.c. dexmedetomidine solution administration (Dexdomitor; Zoetis, Mexico) was selected given an adequate response in prairie voles (*see* **Note 4**). Subcutaneous anesthesia was prepared before each MRI scanning session and on the same day of use. To obtain the selected dose of 0.05 mg/kg for s.c. administration, stock dexmedetomidine was diluted with saline solution to 1:10 (*see* **Note 5**). Atipamezole (Antisedan; Zoetis, Mexico), which reverses dexmedetomidine/medetomidine anesthetic effects, may be optionally administered to allow fast recovery after sedation and it is highly recommended to do so. Antisedan for use in prairie voles was also diluted with saline solution to 1:10 for the selected dose of 0.5 mg/kg. After use, leftovers should be properly discarded.

3.4 MR Equipment and Subject Preparation

The MR scanner, coil, and accessories for physiological monitoring were checked and prepared before their use by the technician and trained personnel. Once the coil and scanner bed were ready for use, the circulating water heating pad, the MR-compatible pneumatic pillow sensor, and the infrared pulse-oximeter (SA Instruments Inc., NY, USA) were carefully placed and fixed into the scanner bed. Each animal was individually taken to the MR room only right before their image acquisition to avoid unnecessary stress and stimuli exposure. Since animals require to be induced into a state of anesthesia to be positioned in the scanner bed, each vole was weighed and placed inside a gas anesthesia induction cage. The dexmedetomidine solution volume for the subcutaneous anesthesia dose was calculated and prepared according to each subject's weight. For gas anesthesia induction, animals were exposed to isoflurane at 3% concentration in an oxygen/air mixture. Once sedated, voles were transferred from the induction cage to the scanner bed, immobilizing the head with a bite bar, and gas anesthesia was resumed at 2% concentration through the scanner bed anesthesia administration system. All monitoring devices were accordingly placed on the animal to begin physiological recording (*see* **Note 6**). When the animal and recording devices were securely placed, the dexmedetomidine solution was injected subcutaneously by a single bolus in the nape region. After bolus injection, waiting at least 5 min before lowering isoflurane anesthesia to a sedation maintenance level of 0.5%. If physiological readings were stable, that is approximately between 15 and 20 min after bolus injection, the animal was ready for image acquisition and within the scanner bed the vole was placed inside the magnet bore.

3.5 Data Acquisition

Localizer sequences were run in the MR PC-console acquisition software to adjust the subject's physical position in relation to the magnet field. MR functional images must be obtained when physiological readings are stable, and in this protocol, rsfMRI were acquired between 20 and 30 min from bolus administration. Field maps were acquired to optimize local field homogeneity in an ellipsoid placed over the brain and skull. rsfMRI was obtained by a spin-echo echo-planar imaging (SE-EPI) sequence: repetition time (TR) = 2000 ms, echo time (TE) = 19 ms, field of view (FOV) = 18 × 16 mm^2, flip angle (FA) = 90°, matrix dimensions = 108 × 96, slice thickness of 0.7 mm, obtaining 25 slices with a thickness of 0.7 mm of in-plane voxel dimensions of 0.167 × 0.167 mm^2, acquiring a total of 305 (10 min and 10 s) volumes. Other sequence details include EPI bandwidth = 288,461.5 Hz, pi pulse (refocusing pulse) duration = 1.5455 ms with 2200.0 Hz bandwidth, and pi/2 (excitation pulse) = 1.9091 ms and 2200.0 Hz, respectively.

An anatomical image for each subject was also obtained. The sequence selected was a spin-echo rapid acquisition with refocused echoes (Turbo-RARE) with parameters: TE = 38 ms, TR = 1800 ms, number of averages (NA) = 2, RARE factor = 16, FOV = 18 × 20 mm^2, matrix dimensions = 144 × 160, and slice thickness of 0.125 mm with isometric voxels of 0.125 × 0.125 × 0.125 mm^3.

After image acquisition and quality verification, isoflurane gas administration ceased and after gently removing subjects from the scanner bed, a previously prepared dose of antisedan solution was injected subcutaneously and subjects were monitored until recovery and returned to their home cage (*see* **Note 7**).

3.6 Data Analysis

3.6.1 Data Pre-Processing

Data pre-processing through this method generally involves removal of confounding sources from the rsfMRI signal such as physiological noise and motion, and aligning functional data with anatomical data to perform group-wise analysis. To do so, the first five volumes of each subject's functional series were removed to address initial signal instability. The new first volume was then used as the reference volume to determine the deformable transformation for the corresponding anatomical image. Slice-timing and motion corrections were also applied. The transformation to the anatomical image was combined with a non-linear transformation to the prairie vole brain template. Each functional image series was then warped to the brain template and resampled to a voxel dimension of $0.16 \times 0.16 \times 0.7\ mm^3$. The first five eigenvectors within the combined non-gray matter mask were calculated to reduce physiological sources of noise [24]. Six motion parameters and the latter eigenvectors were then regressed from the functional series. In order to retain frequencies of interest, a band-pass filter of 0.01 and 0.1 Hz was applied [25]. A box kernel with size of three-by-three-by-three voxels was used for spatial smoothing. All pre-processing stages were performed with FMRIB's Software Libraries [13] except for image registration, for which Advanced Normalization Tools [15] was employed.

3.6.2 Functional Connectivity Analysis

Brain functional connectivity can be analyzed with multiple techniques and the use of more than one method for one study is also applicable. Common approaches are seed-based connectivity analysis, group independent component analysis (gICA), and fractional amplitude of low-frequency fluctuation (fALFF), among others. In this case, Network-Based Statistics (NBS) was chosen as the most adequate method given the interest in exploring longitudinal changes in multiple brain regions that would potentially constitute one or more networks.

Based on previous studies, regions of interest were chosen according to their relevance for prairie vole social behavior. These regions were the anterior cingulate cortex (ACC), anterior olfactory nucleus (AON), basolateral amygdala (BLA), bed nucleus of the stria terminalis (BNST), lateral septum (LS), medial amygdala (MeA), main olfactory bulb (MOB), medial prefrontal cortex (mPFC), nucleus accumbens (NAcc), retrosplenial cortex (RSC), paraventricular nucleus of the hypothalamus (PVN), ventral pallidum (VP), ventral tegmental area (VTA), dentate gyrus (DG), dorsal hippocampus (dHC), and ventral hippocampus (vHC). The latter were defined by hand within the anatomical prairie vole brain template, selecting the minimum possible number of voxels that covered each region using FSL, and with visual guidance of the Allen Mouse Brain Atlas [26]. MATLAB (MathWorks, Natick, MA) was used to obtain functional connectivity matrices. Using the previously pre-processed functional datasets, the average time series were extracted for each region. Then, partial-correlation estimates were calculated for all possible pairs of regions and correlation values were z-transformed (Fisher's z).

In order to analyze behavioral correlates with functional connectivity data, the NBS Toolbox [27] was used to estimate statistical significance of putative networks by comparison to a null distribution generated by 5000 permutations of the original data.

For the assessment of significant longitudinal changes in functional networks, linear mixed models analysis via the Network-Based R-statistics (NBR) package [28] was applied. Linear mixed models were fitted for each connection, in which sex (male, female) was taken as a fixed variable and scanning session (baseline, 24 h, and 2 weeks after the onset of cohabitation) was taken as random and intercept variables. Networks were identified as sets of connections with significant sex, session, or interaction effects ($p < 0.05$ at the connection level), and corrected at the cluster level by comparing their added magnitude to a null distribution of the largest clusters in the permutated data (5000 permutations).

The models applied through the NBS toolbox revealed significant correlations between behavioral observations and network components of several brain regions.

The first NBS analysis was used to determine if there was a relationship between baseline functional connectivity and the expression of huddling, a prosocial behavior commonly observed in the prairie vole during the process of social bonding. A network component composed by the DG, ACC, NAcc, PVN, BLA, RSC, VTA, LS, MeA, and the VP was shown to negatively correlate with huddling latency. In other words, the higher the level of baseline functional connectivity between these regions, the quicker the voles would huddle during the first hours of cohabitation [7]. Given the brain regions involved in this network, from which some attributed functions may be social reward and goal-oriented behavior, a possible interpretation of this result is that this network modulates the social reward related to huddling, and prairie voles that huddled quicker may find it a more rewarding experience. Another NBS analysis also revealed that functional connectivity obtained at 24 h of cohabitation correlated with the amount of social interaction during a social test. A significant positive relationship was found between the LS, the NAcc, the mPFC, the MeA, the VP, the MOB the DG, with the amount a vole spent interacting with conspecifics [7]. This network may be associated with the modulation of social approach and avoidance, since it involves regions related to social reward, social memory, and sensory processing. These results show that NBS is a useful method in detecting how interactions between brain regions can be associated with a particular behavior.

As mentioned earlier, the NBR linear mixed models package was used for the longitudinal analysis of the data. The linear mixed-models approach was selected instead of ANOVA because it allowed the analysis of the data in spite of some missing data points from some subjects, which otherwise would have had to be removed entirely. This analysis revealed that cohabitation with mating in male and female prairie voles induced significant short- and long-term changes in functional connectivity. Specifically, a network component constituted by the ACC, BLA, dHC, LS, mPFC, NAcc, vHC, VP, and the VTA underwent significant changes as a result of pair bond formation both at 24 h and 2 weeks after the onset of cohabitation [7]. The changes observed could be related to events of brain plasticity that are necessary for pair bonding, which in turn influence the expression of behaviors associated to the maintenance of the pair bond, such as selective aggression and partner preference. Overall, the data evidenced that cohabitation and pair bonding in the prairie vole induced short- and long-term changes in brain functional connectivity, suggesting that prairie vole sociosexual behaviors are modulated by the interaction of multiple brain regions. In addition, correlations were found between the functional connectivity of specific networks and the display of a particular behavior, showing that this method can be useful in studying how brain interactions influence behavior.

As a concluding remark, MRI is a powerful tool that offers a very diverse range of applications, and once the experimenter has been familiarized with the equipment and the mentioned methods, we encourage considering additional implementation of other MRI data acquisition methods. As an example, the use of other MRI pulse sequences, such as sequences designed for diffusion-weighted imaging or MRI spectroscopy, can expand the experimental possibilities by obtaining data related to the microarchitecture of brain tissue or the concentration of molecules of interest, respectively. Thus, the use of several kinds of MRI sequences is advantageous in maximizing the amount and variety of data obtained from experimental subjects, which can be very valuable for answering diverse questions that are related to neurobiology and behavior.

4 Notes

1. Another alternative for subject motion minimization is by training, in which animals are habituated to the whole process of image acquisition, that is to be restrained to the scanner bed and the exposure to loud noises and vibrations consequence of MR image acquisition. While it poses the advantage of avoiding the side effects of sedation, higher stress is reported, special devices have to be designed for that specific purpose and training may take from several days to weeks, which has to be considered for the design of the experiment and the number of subjects to be trained.
2. Many other MRI analysis pipelines and package software are freely available on the web since researchers in the neuroimaging community tend to share both data and analysis tools. It is important to note that most of the latter tools are state-of-the-art software and may require the installing and managing of particular software environments and use of the command line in Unix-based operating systems.

3. Including a group not subjected to social behavior protocols in an MRI experimental design is desirable as a control group, enabling the differentiation of observed changes resulting from social behavior from any changes resulting from the MRI acquisition protocol per se.
4. Anesthesia dose standardization in the selected animal model is strictly necessary before running the experimental protocol. Dose–response-type tests in MR settings are highly recommended in order to optimize the minimum sedation level that inhibits subject motion, but maintains stable physiological readings (heart rate, respiration rate) and enables reliable fMRI BOLD response and functional connectivity.
5. Most reviewed anesthesia protocols in rats use intravenous infusion of dexmedetomidine after bolus injection for fMRI acquisition, which should be considered if this animal model is chosen. Previous anesthesia protocol standardization for this experiment allowed considering only the use of a single bolus of dexmedetomidine.
6. From that moment onward, the experimenter must be attentive to the subject's physiological readings until the end of MR image acquisition.
7. If by any reason the images acquired do not reach the desired quality, it may be worthwhile to repeat image acquisition if the subject's physiological readings remain stable.

Acknowledgments

We thank Jessica Norris for her editing. This work was supported by grants UNAM-DGAPA-PAPIIT IN-212219 to Sarael Alcauter and IN-208221 to Wendy Portillo and CF-2023-G-206 to Wendy Portillo. María Fernanda López-Gutiérrez is a doctoral student from the Programa de Doctorado en Ciencias Biomédicas, Universidad Nacional Autónoma de México (UNAM), and has received CONACyT fellowship (2020-00002602NACF-17340, cvu600922).

References

1. Logothetis NK (2003) The underpinnings of the BOLD functional magnetic resonance imaging signal. J Neurosci 23:3963–3971
2. Friston KJ (2011) Functional and effective connectivity: a review. Brain Connect 1:13–36
3. Fox MD, Snyder AZ, Vincent JL, Corbetta M et al (2005) The human brain is intrinsically organized into dynamic, anticorrelated functional networks. Proc Natl Acad Sci U S A 102:9673–9678
4. O'Connell LA, Hofmann HA (2011) The Vertebrate mesolimbic reward system and social behavior network: a comparative synthesis. J Comp Neurol 519:3599–3639
5. Walum H, Young LJ (2018) The neural mechanisms and circuitry of the pair bond. Nat Rev Neurosci 19:643–654

6. Amadei EA, Johnson ZV, Jun Kwon Y, Shpiner AC et al (2017) Dynamic corticostriatal activity biases social bonding in monogamous female prairie voles. Nature 546:297–301
7. López-Gutiérrez MF, Gracia-Tabuenca Z, Ortiz JJ, Camacho FJ et al (2021) Brain functional networks associated with social bonding in monogamous voles. eLife 10:e55081
8. Grandjean J, Schroeter A, Batata I, Rudin M (2014) Optimization of anesthesia protocol for resting-state fMRI in mice based on differential effects of anesthetics on functional connectivity patterns. Neuroimage 102(Pt 2):838–847
9. Bajic D, Craig MM, Borsook D, Becerra L (2016) Probing intrinsic resting-state networks in the infant rat brain. Front Behav Neurosci 10:1–13
10. Christiaen E, Goossens M-G, Raedt R, Descamps B et al (2019) Alterations in the functional brain network in a rat model of epileptogenesis: a longitudinal resting state fMRI study. Neuroimage 202:116144
11. Hoyer C, Gass N, Weber-Fahr W, Sartorius A (2014) Advantages and challenges of small animal magnetic resonance imaging as a translational tool. Neuropsychobiology 69:187–201
12. Paasonen J, Stenroos P, Salo RA, Kiviniemi V et al (2018) Functional connectivity under six anesthesia protocols and the awake condition in rat brain. Neuroimage 172:9–20
13. Jenkinson M, Beckmann CF, Behrens TEJ, Woolrich MW et al (2012) FSL. Neuroimage 62:782–790
14. Statistical Parametric Mapping (n.d.) Statistical parametric mapping. https://www.fil.ion.ucl.ac.uk/spm/
15. Avants BB, Tustison NJ, Song G, Cook PA et al (2011) A reproducible evaluation of ANTs similarity metric performance in brain image registration. Neuroimage 54:2033–2044
16. Dale AM, Fischl B, Sereno MI (1999) Cortical surface-based analysis: I. Segmentation and surface reconstruction. Neuroimage 9:179–194
17. R Core Team (2020) R: a language and environment for statistical computing. R Foundation for Statistical Computing, Vienna
18. Ortiz JJ, Portillo W, Paredes RG, Young LJ et al (2018) Resting state brain networks in the prairie vole. Sci Rep 8:1231
19. Friston KJ, Frith CD, Liddle PF, Frackowiak RSJ (1993) Functional connectivity: the principal-component analysis of large (PET) data sets. J Cereb Blood Flow Metab 13:5–14
20. Mateo C, Knutsen PM, Tsai PS, Shih AY et al (2017) Entrainment of arteriole vasomotor fluctuations by neural activity is a basis of blood-oxygenation-level-dependent "resting-state" connectivity. Neuron 96:1–13
21. Nasrallah FA, To XV, Chen DY, Routtenberg A et al (2016) Functional connectivity MRI tracks memory networks after maze learning in rodents. Neuroimage 127:196–202
22. Williams JR, Catania KC, Carter CS (1992) Development of partner preferences in female prairie voles (Microtus ochrogaster): the role of social and sexual experience. Horm Behav 26: 339–349
23. Taylor SA, Salo AL, Dewsbury DA (1992) Estrus induction in four species of voles (Microtus). J Comp Psychol 106:366–373
24. Behzadi Y, Restom K, Liau J, Liu TT (2007) A component based noise correction method (CompCor) for BOLD and perfusion based fMRI. Neuroimage 37:90–101
25. Gorges M, Roselli F, Müller HP, Ludolph AC et al (2017) Functional connectivity mapping in the animal model: principles and applications of resting-state fMRI. Front Neurol 8:1–14
26. Lein ES, Hawrylycz MJ, Ao N, Ayres M et al (2007) Genome-wide atlas of gene expression in the adult mouse brain. Nature 445:168–176
27. Zalesky A, Fornito A, Bullmore ET (2010) Network-based statistic: identifying differences in brain networks. Neuroimage 53:1197–1207
28. Gracia-Tabuenca Z, Alcauter S (2020) NBR: network-based R-statistics for (unbalanced) longitudinal samples. bioRxiv. https://doi.org/10.1101/2020.11.07.373019

Chapter 8

Evaluation of Sexual Behavior in Laboratory vs Seminatural Conditions

Xi Chu and Anders Ågmo

Abstract

Sexual behavior is, by necessity, sexually dimorphic. Males transfer sperm to females, whereas females receive sperm from males. Discussions of sex differences in copulatory behavior are consequently trivial. However, the behaviors associated with copulation, for example mate choice or postcopulatory reactions, may well be similar in males and females. Such differences, even subtle, are far easier to observe in seminatural environments than in the standard laboratory cage. We will present examples of the use of seminatural environments in insects and rodents. Even though most studies of insect sexual behavior are performed in relatively simple laboratory procedures, there are also some studies performed in natural or seminatural conditions. We briefly describe the most common procedures used and mention the main results. It is noteworthy that insect studies focus on sexual approach behaviors, particularly the role of visual and olfactory stimuli in mate location. The actual copulatory behavior, i.e., how gametes are transferred from one individual to another, seems to be of less interest.

The sexual behavior of rats has traditionally been studied in heterosexual pairs, despite the fact that they often copulate in groups. Nevertheless, data obtained in the simplified environment have advanced knowledge of the endocrine and neurobiological control of sex behavior in a quite spectacular way. The understanding of the dynamics of the sexual interaction and the possible function of the many peculiarities of rat sexual behavior has not advanced to a similar degree. Studies in seminatural environments may provide valuable data concerning sociosexual interactions and how such interactions are modified by contextual events. Furthermore, observations made in an environment, which incorporates the basic features of rats' natural habitat, offer some external validity. This is of importance when we want to generalize our results to contexts outside the laboratory, and it becomes of paramount important when we want to make inferences about behavior in other species, for example the human. We offer here a detailed description of an environment designed for studies of group-living rats, with notes on the observation procedure and the analysis of the large quantity of data generated in the environment.

Key words Sexual behavior, Insects, Rodents, Representative design, Social interaction, Olfaction

1 Introduction

The vast majority of animals reproduce sexually. With the exception of self-fertilizing hermaphrodites, and perhaps animals with external fertilization, this obliges animals from the simplest to the most

Raúl G. Paredes et al. (eds.), *Animal Models of Reproductive Behavior*, Neuromethods, vol. 200,
https://doi.org/10.1007/978-1-0716-3234-5_8,

complex species to engage in some kind of sexual behavior. Consequently, sexual behavior has been studied in several classes of invertebrates [1] as well as in many chordates. Whereas many invertebrates have quite modest nervous systems, the insect brain processes sensory information, integrates different sensory modalities, and controls motor systems in a way similar to the vertebrate brain [2]. In contrast to their complex behavior, the insect brain is small and simple compared to that of mammals. Therefore, insects may be used as model organisms for studying the neural bases of behavior. However, insects are often overlooked in reviews of sexual behavior. We find it important to draw attention to the valuable data obtained in these animals. Consequently, we devote the first part of this chapter to a synopsis of procedures employed in research on insect sexual behavior. The second part of this chapter is dedicated to a rather detailed presentation of a seminatural environment used in studies of rat sociosexual interactions.

In all species with internal fertilization, for example mammals, gametes need to be transferred from one individual to another. Males produce sperm and females are equipped with eggs. For some reason, nature has arranged that sperm must be deposited in the interior of the female reproductive organs before fertilization can occur. This is a necessity because of the dimorphic reproductive organs. The fact that the male expels gametes from his body whereas the females receive them in hers has made male and female sexual behavior entirely different. In fact, sexual behavior is inevitably dimorphic. At the same time, it must be remembered that males of most species are perfectly able to display typically female behavior patterns, and females can easily display male sexual behavior patterns. In rats, males now and then display lordosis and occasionally also paracopulatory behaviors. Female rats frequently mount other rats, and sometimes they can show behavior most similar to the male behavioral pattern associated with vaginal penetration and even ejaculation. The difference is that the likelihood for a male to display a female pattern is far lower than the likelihood that he displays a male pattern. The inverse is true for females, i.e., the likelihood for a female to display male behavior patterns is far lower than the likelihood of displaying female patterns.

Even though male and female reproductive behaviors, i.e., copulatory behavior and behaviors associated with parental care, are dimorphic by necessity, there is no a priori reason to believe that behaviors unrelated to reproduction are. Even though the frequency of many behaviors, for instance acts of aggression, may differ between males and females, such quantitative differences are not inevitable. To the contrary, only males can transfer sperm and only females can lactate. Seminatural environments offer excellent opportunity for observing a vast spectrum of non-sexual, non-dimorphic social behaviors as well as emotional reactions to aversive as well as attractive stimuli. We will briefly provide an example of this at the end of this chapter.

2 Examples of Insect Model Organisms

Several insect species have been used in research studies for different purposes. Fruit flies (e.g., *Drosophila melanogaster*), with a lifespan well-suited to laboratory research, are small, abundant, and easy to manipulate. Currently studied by over 1800 labs around the world, they are the most powerful model insect in neuroscience for understanding multisensory signal processing in courtship (e.g., [3, 4]), odor memory, and synaptic plasticity [reviewed by 5], as well as neurodegenerative diseases, like Parkinson's, Alzheimer's, and Huntington's disease [reviewed by 6]. The majority of observations of sociosexual behaviors in fruit flies are performed in varied laboratory conditions, while only few studies systematically document fruit fly sexual behaviors in the natural habitat. Field observations demonstrated that the mating territory of each fly was rather small, and a number of flies could copulate in close proximity [7]. After orienting and approaching a potential mate, the male engaged in an approximate 2.5-min species-specific courtship before attempting to mount and copulate [7, 8]. During copulation, a third fruit fly might approach the mating pair, and according to the field work conducted by Dukas [9], a male fly was more likely to display approach to mating pairs than was a female fly. Multiple copulation with different sex partner is common in fruit flies. In addition to fruit flies, lepidoperan species are also used model insects in pheromone tracing, e.g., silk moth (*Bombyx mori*) [10], as well as a group of sympatric heliothis moths that use the same-sex pheromone components (airborne chemicals) in different ratios for reproductive isolation (e.g., [11]). Species-specific sex pheromones plume released by the sexually mature female is first detected by olfactory sensory neurons located on the male antenna, and the male will demonstrate an upwind, zigzag flight [12] to trace the origin of these pheromone signals. Multiple mating to different partners is common in heliothis moths, e.g., female tobacco budworm moths, *Heliothis virescens* [13], and female and male cotton bollworm moths, *Helicoverpa armigera* [14].

3 Multisensory Processing in Insects' Mate Choice and Mate Searching and Its Importance in Experimental Methods Selection

Orientation to a proper mate is vastly important in reproductive success. Insects make good subjects for the study of discerning mechanisms and processes governing mate search. Many insects rely on simultaneous sensory signals in more than one modality to trace the potential mate partner. Within their small but complex nervous system, the insect first encodes the external information into electric signals via different sensory organs and then computes this information into different strategical plans. Based on the tracing distance, we find two types of reproductive searching, the short-distance and long-range mate tracing. Both short- or long mate-tracing activities are multimodal navigation, which is one of the fundamental activities for insects to locate the source of an odor that may indicate food, a mating partner, proper oviposition site, or aggregation into swarms [15, 16]. The short-distance exploration of the local environment for mates usually has a small traveling distance. For instance, male fruit flies orient to female at a distance of 1.5–2.5 mm [17]. Such tracing activity is based on the female mating signals combining olfactory, visual, and wing-vibratory information [18]. Generally, male flies preferentially mate with larger, more productive females [8, 19]. Impaired sensory perception can affect mate choice in fruit flies, particularly in females. It appeared that female mate choice persists in a rational, transitivity manner, only when one of the sensory cues, i.e., either visual or olfactory modality, is impaired, but not when both sensations are impaired [20].

The long-range tracing, on the other hand, occurs mainly in moth species, in which the males are extraordinarily good at finding the conspecific females from a distance of more than 1 km [21]. With a supreme sensitivity, male moths can detect a single molecule of female-produced pheromone [22]. However, sensing attractive odors alone is not enough to result in approaching. Flying moths rely profoundly on the visual system when tracing an odor source [reviewed by 23]. Without visual feedback, moths find it very difficult to track airborne pheromone plumes successfully [24]. Contrary to our intuition, no odor-mediated behavior in moths is based on *chemotaxis*, i.e., there is no direct navigating response to odor concentration gradients. Moths rather steer against the odor source in correspondence with optomotor *anemotaxis* (*anemo*: wind, taxis: directed movement), which requires olfactory, visual, and mechanosensory feedback [25].

Although insects may not necessarily utilize all components of a multisensory cue, more ecologically relevant laboratory assays would greatly advance our understanding of the mechanisms of sexual-behavior-related orientation. It is worth mentioning that selection of experimental method depends on the perspective of the individual research question. In the studies examining neurophysiological properties of individual sensation (e.g., olfaction, vision, audition, etc.), providing a single-component cue is highly desired. For instance, the experiments reporting the mapping of the pheromone pathways were conducted in such controlled manner, exposing the male moths to the female-produced pheromone puff and minimizing the interference of foreign emissions and other factors such as visual/auditory stimuli as much as possible (e.g., [26, 27]). However, in complex behavioral studies, such as exploration of the mate-tracing flight pattern, to focus on only one sensory cue while neglecting the involvement of multisensory perception may lead to results with less external validity.

4 Experimental Methods Used in Laboratory

4.1 Small Behavior Chamber

The shape and size of the chamber used for studying insect sexual behavior are diverse, but they are generally small, suited for subjects like fruit flies for a short period of behavioral assay. The acoustic behavior chamber (Fig. 1a) used in studying auditory experience and sexual response in fruit fly was 50 mm long × 10 mm wide × 6 mm high [28]. The courtship chamber (Fig. 1b) used to examine the choice between feeding and sexual behaviors in fruit flies was a petri dish measuring 2 cm wide [29]. The number of available sexual partners in such behavioral tests is one in most of the cases. Thus, the observations of the behavioral patterns in a pair of flies or a few flies in this kind of chamber/apparatus usually are far less representative to the wild fruit flies living in their natural conditions.

A Y-tube or T-maze bioassay apparatus is also a commonly used behavior chamber in olfactometer experiments. As shown in 1C, the Y-tube used in Biasazin et al. [30] was intended to test the detection of food volatiles in fruit flies. It was made of a borosilicate

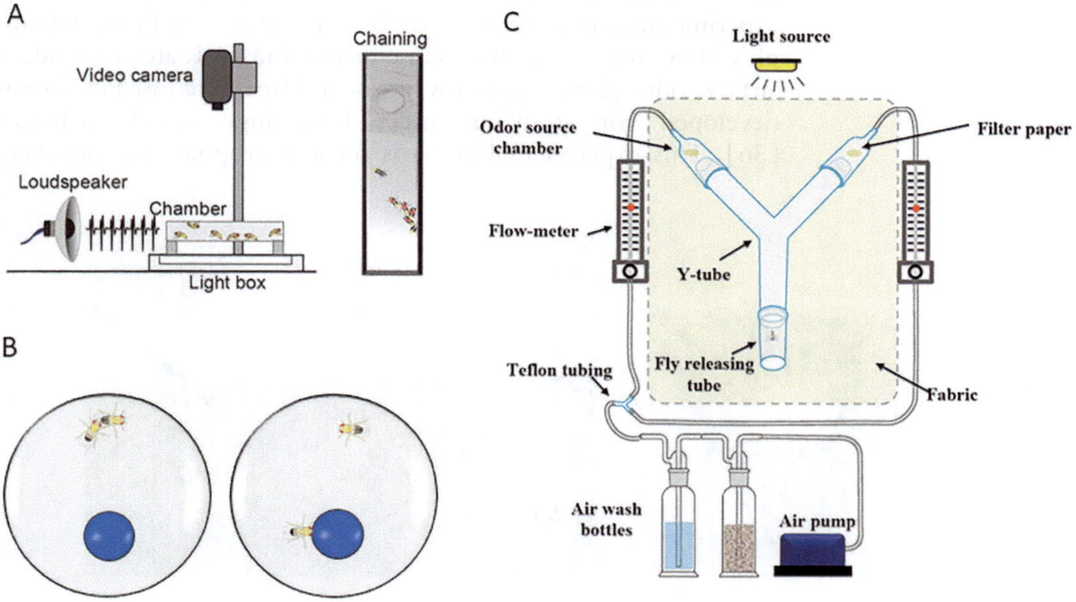

Fig. 1 Small behavior chambers. (**a**) Acoustic behavior chamber. (Reproduced with permission (CC BY 4.0) from Fig. 2 in [28]). (**b**) Example of courtship chamber, in which a male fruit fly is allowed to choose between feeding source (blue circle) and courting a receptive female. (**c**) Y-tube bi-choice chamber. (Reproduced with permission (CC BY 4.0) from Fig. 1 in [30])

glass tube with an inner diameter of 3.1 cm. The length of the upstream arms was 16 cm each and the common arm was 14 cm long. The upstream end consisted of glass odor source chambers (8.5 cm and 2 cm id) containing a stimulus. Such apparatus with different size can be used in preference tests with other volatile organic compounds, e.g., sex pheromones, in fruit flies and moths [31, 32].

4.2 Physiological Recordings in Individual Insects

For studies focusing on understanding the physiological mechanism of sexual attractiveness, observing the response to sensory cues on a neuronal level is broadly used. The most relevant stimuli in these experiments are a group of olfactory cues functioning as sexual attractant or repellents. Sex pheromones, produced by a conspecific female, serve as the main sexual attractant, whereas sex antagonists, produced by a sympatric heterospecific female, serve as repellents. Extracellular recording and intracellular recording/staining techniques allow to obtain the instant electrophysiological activity of individual neuron across different period around the stimulation window (e.g., [26, 27, 33]). Another method, the calcium-imaging experiment, is to monitor the change of cellular calcium signal in a group of neurons (e.g., [34, 35]). In these assays exploring the odor perception, the subject is normally immobilized, and the stimuli are closely applied. However, mate searching is not based on olfaction exclusively. Flying or walking is commonly occurring simultaneously with olfaction perception. A modern experimental arena designed to allow the insect free flying in complex 3D virtual reality environments (MultiMoVR arena including airflow, odor plumes, skyglow, grass, and trees, etc., *see* Fig. 2) was developed for studying insect long-range search behavior [36]. This experimental arena is most appropriate for behaving

Fig. 2 MultiMoVR Arena. (**a**) The MultiMoVR arena is a prism-shaped arena composed of three monitors, measured 32 cm wide and 60 cm high. The behaving apply fly is placed in the center of the arena. (**b**) The insect is surrounded by a dynamic seminatural scenery according to its flying activity. Directional wind and odor are provided as well. (The figure is reproduced with permission (CC BY-NC-ND 4.0) from Movie S1 in [36])

insects (particularly flying), yet it has not been used in studies of insect mate search strategy. This kind of methods are highly inspiring and they have huge potential to impact behavioral neuroscience.

4.3 Observing Apparatus Used in a Population of Insects

In many species, the territorial area of each insect in the wild is rather small, and multiple mating is common. Thus, the representative value of examining a pair's behavior is lower compared to analyzing a group of insects. A population observation cage was designed to observe sexual selection and sexual isolation in a such context [37–39]. In the sexual isolation studies [37, 38], a total of 24 pairs (12 virgin females and males from one population and 12 virgin females and males from a second one) were used for random mating experiments for 90 min. Group copulation tests involving multiple pairs were also used in evaluating mating preference. In Korol et al. [39], 40 flies (10 pairs from one location and 10 pairs from a second location) were introduced simultaneously into a mating chamber and were observed for 60 min. As soon as a pair started copulating, the mated types were recorded before the pair was removed from the chamber [39]. A population cage has also been used in testing the tracing of sexual signals emitted by the opposite sex in the laboratory. Keys, Mills [32] investigated the attractiveness of female sex attractant (odor) in highly dense male angoumois grain moths (*Sitotroga cerealella*) population. The cage used in this study contained 600 individuals simultaneously, having a size of 63.5 cm long × 63.5 cm wide × 63.5 cm high, in which three traps with female-produced attractant were hang 20 cm from the top in a 30 cm triangular pattern in this cage and the number of trapped males was monitored.

4.4 Seminatural Environment with Host Plants

In many of the experimental methods we mentioned above, the subject insects are either restrained or confined to a very small space with unnaturally strong experimental stimuli, such that the subject could not avoid perceiving them. Additional to these unwanted scenarios, host plants were seldom provided. The chemical, visual, and tactile cues from the host plants are specifically important for some sexual behaviors, e.g., mate choice and female ovipositing activity. Only a few studies used ecological enriched laboratory assays, e.g., seminatural environment with host plants [40, 41]. In a study by Arita and Kaneshiro [40], multiple male fruit flies were placed into a 23 cm long × 23 cm wide × 23 cm high glass cage with a potted Syngonium plant for a 24-h period to acclimate to the cage conditions, before a virgin female was placed into the cage with the two males, and her choice of mate was recorded. A simplified version of seminatural environment was used in an experiment to test female egg-laying preference in gravid female hawk moth, *Manduca sexta* [41]. In that experiment, two species of nonflowering host plants of similar size were placed at the upwind end of a

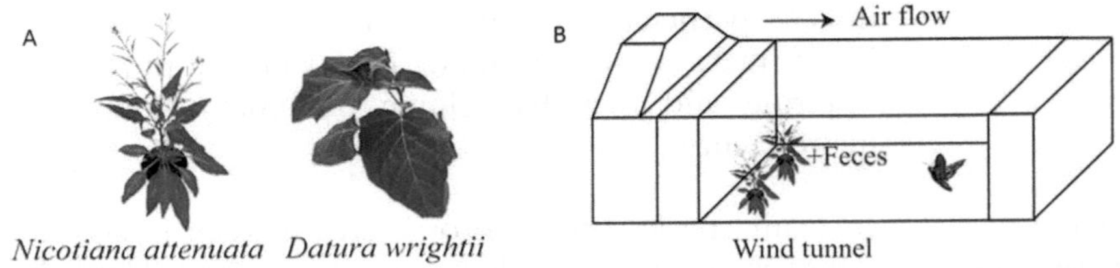

Fig. 3 Example of seminatural environment with host plant. (**a**) Two host plants of *M. sexta* coyote tobacco (*N. attenuata*, Left) and jimson weed (*D. wrightii*, Right). (**b**) Schematic drawing of the behavioral assay to test the oviposition deterring effect of larval feces. (The figure is reproduced with permission (CC BY-NC-ND 4.0) from Fig. 1 in [41])

transparent wind tunnel (220 cm long × 90 cm wide × 90 cm high, *see* Fig. 3), and the oviposition deterrent effect of feces from conspecific larvae to the females was compared by the number of eggs between two plants.

5 Insect Conclusion

This short overview of some of the procedures used for studying insect sexual behavior should illustrate that the main theme of research in this area concerns mate localization and mate choice. The stimuli, olfactory and visual, activating sexual approach behaviors have been studied in some detail, and the neurophysiological activities generated by these stimuli are starting to be understood. The sexual acts themselves have been far less studied. The motor patterns displayed during copulation and their sensorimotor control have not been as carefully described as the sexual approach behaviors. An important issue that has not been sufficiently addressed is whether the observations made in insects in any way contribute to the understanding of mammalian, including human, sexual behavior.

6 Why Study Rodent Sexual Behavior?

Among rodent behavior patterns, sexual behaviors are exceptional for several reasons. To justify this assertion, we can use the female rat as an example. Intact females usually reject mating attempts from males or other females with a series of rather stereotyped behaviors. For a short period every 4 or 5 days, depending on the length of the estrous cycle, the female's behavior is drastically changed. Instead of rejecting males' mounting attempts, the female actively incites males to mount with a few behavior patterns named paracopulatory (formerly proceptive) behaviors. When the male mounts, the female arches her back in a concave fashion, extends the hind legs, and moves the tail to one side. Thereby the vaginal opening is made accessible for the male's erect penis, and vaginal penetration may occur. Neither this response to a mount, lordosis, nor the paracopulatory behavior are displayed outside this short period in the cycle. This short period, called proestrus, is preceded by increasing blood concentration of estradiol and coincides with the LH surge, ovulation, and a rise in progesterone. The female rat sexual behavior is, in fact, strictly dependent on ovarian hormones. No other known behavior shows this extremely rigid dependence on hormones. In male rodents, the sexual behavior is also dependent on gonadal hormones, but this dependence is not so immediate. Castration of a male rat will lead to a gradual decline in sexual behavior during a few weeks until the behavior disappears. It can be restored by the administration of testicular hormones, but restoration is gradual and needs 1 or 2 weeks. Nevertheless, it can be maintained that male sexual behavior is as dependent on gonadal hormones as female behavior is, although on a longer time scale.

Not only is rodent sexual behavior strictly hormone dependent, it is also highly stereotyped. In females, paracopulatory behavior and lordosis can be considered as simple reflexes, produced by stimuli from the male. The fundamental unit of male behavior, the mount accompanied by anteroposterior pelvic thrusting, is also highly stereotyped and can be considered a reflex activated by tactile stimulation of the preputial and perineal areas by the female's back.

Sexual behaviors are biologically significant, yet they do not depend on organismic need as eating and drinking do. Abstinence from sex has no deleterious consequence for the individual. Furthermore, the behavior is unlearned. Most males and females will engage in sexual interaction when given the opportunity to do so, without any prior training.

Because of the reasons outlined above, sexual behavior is a perfect choice for those interested in studying the neurobiology of hormone-dependent modifications of behavior. There is no mammalian, hormone-dependent behavior pattern better described than lordosis in female rats. The sensory pathways from the skin on the flanks to the hypothalamus as well as the motor output to the muscles in the back responsible for the lordosis posture have been carefully mapped, and the molecular actions of estradiol in the crucial hypothalamic neurons are beginning to be elucidated [42–44]. Knowledge of the neural control of mounting and the ensuing penile insertion lags seriously behind, but progress has certainly been made ([45–47], *see* also [48] for a review).

7 The Role of Representative Design in the Studies of Rat Sexual Behavior

The knowledge summarized above has been obtained in very simple behavioral procedures. One male and one female are put together in a cage, and the experimenter just observes what they are doing. This procedure is straightforward and efficient for obtaining data in a short time. It has proved invaluable in neurobiological studies of sexual behavior. Now and then scientists are interested in making generalizations from their data to contexts outside of the laboratory. Some want to speculate about the adaptive value of some behavioral element, and others may want to generalize their rat data to the human, i.e., use rat sexual behavior as a model for human sexual behavior. Then the simple behavioral procedures may not be appropriate. Enclosing two individuals of opposite sex in a barren arena dramatically limits the behavioral repertoire. Furthermore, since the subjects cannot escape from each other, and since few alternative behaviors are possible, they are almost forced to engage in sexual activities. The situation is entirely different in rats' natural habitat. According to observational data, a sexually receptive female attracts a pack of males, all trying to copulate with her. The female can control copulation by escaping into a burrow or elsewhere [49, 50]. The natural environment also offers possibilities for other behavior patterns, like foraging, or non-sexual social interactions with other individuals appearing in the rat's home range. The behavioral repertoire is extremely rich in this environment. This means that motivational factors determine which of all possible behaviors the rat will choose to perform. The fact that rats copulate in groups rather than in couples changes the dynamic of the sexual interaction, and since alternative behaviors are available, motivation to have sex relative to all other motives becomes of paramount importance.

Whenever we wish to make assumptions about the function or adaptive value of a behavior, we need to assure that we study the behavior in an experimental setup as similar to the natural habitat as possible. This was pointed out many years ago [51–53], and experimental designs incorporating these features are called representative. Only a representative design allows for generalizations from our results to the world outside of the laboratory. If we want to generalize our findings to another species in addition, it is likely that representative design, in the brunswikian sense, becomes still more important, as discussed elsewhere [54, 55].

Here, we need to mention that a variant of the standard observation setup allows the female to escape from the male, thereby acquiring control of the pace of sexual interactions [56, 57]. The mating arena is simply divided into parts connected with small holes. The small female can pass through the holes and is therefore free to move about the entire cage. The larger male cannot penetrate the holes and is, consequently, confined to one part. This variant, usually called paced mating, has been rather popular for many years. It approaches the natural situation in the way that the female can escape from the male, but the behavioral repertoire is still quite limited. Furthermore, a fundamental element of the natural context, group mating, is entirely absent.

In an effort to study rat sexual behavior with a representative design in the laboratory, we have adapted a seminatural environment, originally described by McClintock [58–60]. Rats live in male–female groups in a large environment consisting of a burrow system similar to a real burrow, attached to a large open field with some objects available. Thus, the main features of rats' natural habitat (group living, a burrow, and a large open space) are included, making it possible to claim that this procedure satisfies the brunswikian criteria for a representative design. Detailed descriptions of sexual interactions in this context [55, 61] have revealed several important differences from the standard small cage. The relationship between sexual behavior patterns and fertility has also been analyzed [62]. We believe that data obtained in this environment can be generalized to the world outside the laboratory. Whether it also can be generalized to other species remains an open question, although there are some indications that it actually can. However, the validity of such generalizations remains an open question.

In the following, we will give a detailed description of the experimental setup as well as the procedure for performing experiments and analyzing the enormous amount of data obtained.

8 Materials

8.1 Apparatus

A drawing of the environment is provided in Fig. 4. The burrow is constructed on sheets of dark gray unplasticized polyvinylchloride. The distribution of the tunnels as well as of the nest boxes is based on descriptions of wild rat burrows [50]. The size of the tunnels and boxes, for example, is similar to those found in a real rat burrow. They are all built of sheet steel, covered with black plastic on the inside, and fixed to the polyvinylchloride base. During experiments, Plexiglas fixed to the boxes and tunnels with building tape covers the entire area. The floor is covered with an approximately 2-cm-thick layer of aspen wood shavings (Tapvei, Harjumaa, Estonia). Nesting material in the form of 6 pieces of square mats of non-woven hemp fibers (5 × 5 cm, 0.5 cm thick; Happi mat, Datesend) is put in each nest box. The burrow connects to an open area via 4 openings (8 × 8 cm). Twelve small wooden sticks (2 × 2 cm, 10 cm long, Tapvei) are randomly put on the floor of the open area. Three plastic shelters of red polycarbonate (15 × 16.5 cm, height 8.5 cm; Datesend, Manchester, UK) are also provided. The

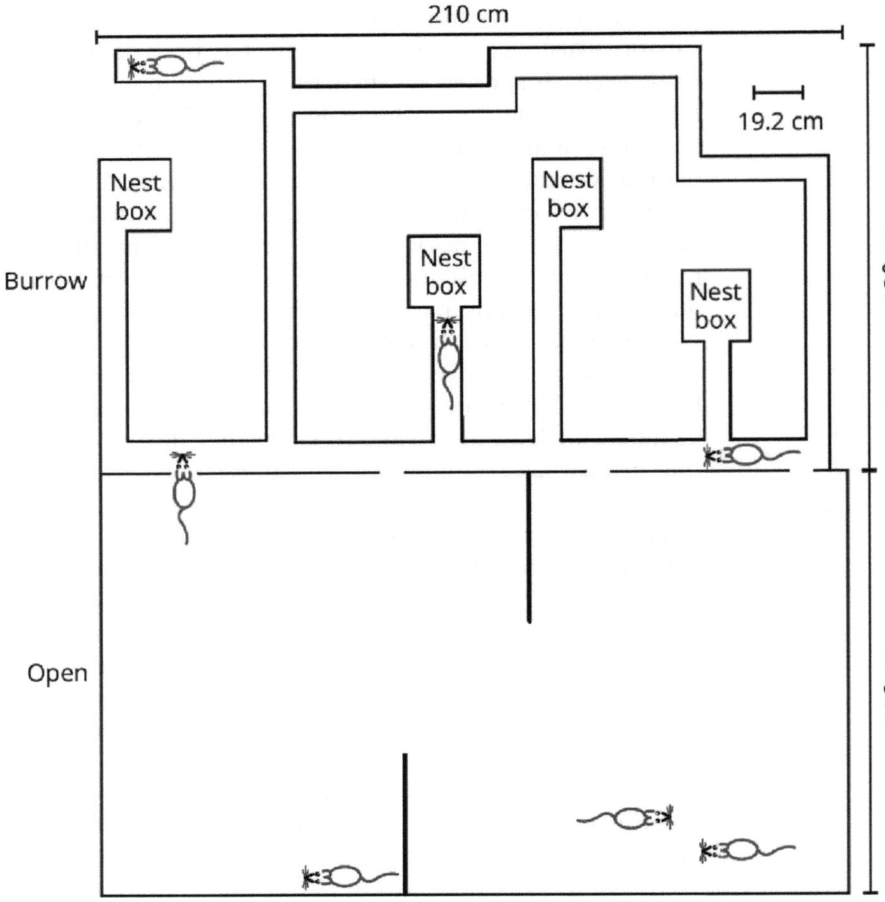

Fig. 4 The seminatural environment. The thick black lines extending from the walls in the open area are 40 cm high, black dividers about 40 cm long. The purpose with them is to enhance complexity in the open area

walls surrounding the open area are 75 cm high, which is enough to ensure that the rats are unable to escape from the environment even though no cover is used. Like in the burrow, the floor is covered by about 2 cm of aspen wood shavings. Four 500-mL drinking bottles hang on the exterior wall, with the spout extending 5 cm through a hole about 10 cm above the floor on the inside.

The burrow and the open area are separated by a wall of heavy dark cloth extending from floor to ceiling and from wall to wall. This makes it possible to maintain the burrow in complete darkness even though the open area has a 12:12-h light/dark cycle. The light needed for video recording comes from two infrared lamps (850 nm; model Sal-60, New Surway Digital Technology (Shenzhen), Guangdong, P.R. China) located on opposing walls of the room. In the open area, the transition from light to darkness and vice versa is made gradually during 30 min at the beginning and end of the light phase. Light intensity is 180 lx at floor level during daytime and 1 lx during nighttime.

One digital video camera (Basler acA2000-50gc, Ahrensburg, Germany) is located about 2 m above the burrow and another camera is fixed above the open area. Both are connected to a computer equipped with Noldus Media Recorder 2.5 (Noldus, Wageningen, the Netherlands). The animal quarters' ventilation system produces a constant background noise of about 40 dB.

8.2 Animals

We have constantly used Wistar Han IGS rats from Charles River (Sulzfeld, Germany). Other strains and other providers would probably work equally well. Rats are ordered as adults (250 g for females, 300 g for males) and kept in the animal quarters for at least 2 weeks before experiments. They are housed under the light/dark schedule to be used during experiments in a room with constant temperature (21 ± 1 °C) and controlled relative humidity (55 ± 10%). Same-sex pairs are cohabitating in Makrolon® IV cages, with food (RM1, Special Diets Services, Witham, UK) and tap water constantly available.

In some experiments, the rats are gonadectomized or given intracerebral injections, or whatever treatment may be required. This is always done during the 2-week period between arrival and start of experiments, allowing the animals sufficient time for complete recovery.

8.3 Procedure

8.3.1 General Considerations

The groups used in the environment consist always of three males and four females. This sex ratio (57% females) is similar to what has been reported for adult rats observed in the wild [63]. The population density in the environment is 1.4 rat per square meter. This should be considered a very high density in the wild. Practical considerations do not allow for maintaining a population density more similar to the wild condition. In experiments requiring treatment of some kind, e.g., drug administration or manipulations of gene expression, a reasonable number of subjects in each treatment is necessary. This means that several groups must be run in the environment before a reasonable number of rats has received each experimental treatment. The smaller the number of subjects in each group, the larger the number of replications needed. The number of replications becomes important because of the large size of the environment and the consequent need for substantial lab space combined with the length of the procedure.

8.3.2 Preparation of the Animals

When seven animals are to be observed simultaneously in a large and rather complex environment, easy identification of each of them is fundamental. Therefore, we shave different areas of the back of the animals. A rectangle, about 2 × 4 cm, is carefully shaved on the back of the rats about 3 h before introduction into the environment. On one female, the rectangle is close to the tail, in another it is in the middle of the back, and in a third it is close to the neck. The fourth female is not marked on the back. Males are marked exactly as the females. Because of their bigger size, there is no danger of confusion between males and females. The tail is marked with one, two, three, or four transversal, thick black lines. Any permanent marker pen can be used, e.g., Edding 500 (Edding international, Ahrensburg, Germany). The double marking makes it possible to identify the rats on the video record with a certain ease.

8.3.3 Preparing the Environment

The floor needs to be covered with fresh wood shavings, a Plexiglas cover needs to be fixed over the burrow, food (about 2 kg of standard rat pellets) must be put on the floor in one corner of the open area, and water bottles must be filled. The video recording system should be checked and started.

8.3.4 Using the Environment

The rats are introduced in rapid succession over a period shorter than 5 min. Only one animal per cage is used, assuring that all animals are unknown to each other at the beginning of the experiment. When all 7 rats are in the environment, humans leave the room. The rats remain there for 8 days. The only disturbance during that time is a daily, human visit to check the proper functioning of the equipment and that water remains in the bottles. This visit is always made shortly before the end of the light phase of the cycle.

Pilot studies indicated that rats showed elevated locomotor activity when introduced into the environment. Activity was gradually reduced over the first 48 h and remained stable for the rest of the observation period. It also appeared that social interactions became stable after 48–72 h. Nevertheless, we do not start experimental manipulations until day 5 in the environment, and whenever possible, we wait until day 7. This might be an unnecessary precaution, but we want to have the social group solidly established at the moment experiments are performed. In purely observational studies, we use data from any moment of the period, as appropriate. For example, when we wanted to see how devocalization affected the initial social interaction, we observed behavior during the hour following introduction of the animals into the environment [64].

In some experiments, all or some of the animals need to be manipulated (e.g., [65]). Drugs or hormones can be injected whenever appropriate. This requires capture of the animals, which is easily accomplished. During capture, we usually block the entrances to the open area. All animals that are to receive treatment are first captured, then treated, and thereafter reintroduced into the environment at the same time. Injections or other treatments are performed in another room in order to avoid disturbing animals that shall not be treated and reduce the likelihood that possible aversive experiences during treatment become associated with the environment. As far as we have observed, removal of the animals, injection, and subsequent return to the environment have only a short-lasting effect on behavior. Within 15 min of return, the rats have resumed their regular activity.

All kinds of events may be introduced into the environment, something useful when we want to study emotional reactions. Pleasant and aversive odors may be distributed in the burrow or in the open area or both. We use an olfactory stimulation equipment (Olfactory Stimulus Package, Medical Associates, Georgia, VT) connected to nozzles placed on the walls of the tunnels in the burrow and in the open area. Assuring an efficient flow of scented air, the rats can be exposed to any odor. We have used the aversive odor 2,5-dihydro-2,4,5-trimethylthiazoline (TMT; Contech, Delta, BC, Canada), smelling of fox, and the pleasant odor of lavender oil (AromaBio, Lyon, France). Tasty foods, e.g., chocolate tasting pellets (Supreme Mini-Treats 1 g; F05472; Bio Serv, Frenchtown, NJ), can be provided in limited amounts, allowing observation of who has first access and how much each individual consume, among other things. Fear-inducing, sudden events, like intense white noise, can be introduced. This is easily produced by a noise generator (e.g., from Lafayette instruments, Lafayette, IN) and loudspeakers placed in the ceiling above the burrow and open area. A sound intensity of 90 dB is sufficient for producing an immediate fear reaction.

We have used a sequence of aversive and attractive events administered during part of the dark period, trying to mimic the events a rat is likely to encounter during a nightly walk in the home range [66–68]. Only the scientist's creativity imposes limits on the events that can be used.

8.4 Behavioral Observations

All observations of behavior are made from the video records. The records from the burrow and the open area are synchronized and shown on the same screen. Before beginning observation, it is indispensable to establish an ethogram, such as the one shown in Table 1, including all the behavior patterns of interest. In addition to record the frequency and duration of the behavior patterns defined in the ethogram, it can be extremely useful to determine to whom the behavior was directed (in case of social or sexual behavior) and where in the environment it occurred. If social interaction is of interest, it is fundamental to know not only the emitter of the behavior but also the receptor. Behavior patterns are not uniformly distributed in the environment. Therefore, the spatial distribution of behavior can offer important information. In addition to the obvious distinction between burrow and the open area, it can be useful to divide the open area in sectors. Thus, besides recording emitter and receptor, we systematically record location.

Efficient software, fixing behavior patterns in time besides recording the data mentioned in the preceding paragraph, is essential for the recording of behavior. We use the Observer XT, version 12.5 (Noldus, Wageningen, the Netherlands) to that end. The normal procedure for making observation is to observe one rat at a time, during a prefixed time interval. Since there are 7 rats in the

Table 1
Typical ethogram used in the rat seminatural environment

Category	Behavior pattern	Definition
Female sexual behaviors	Lordosis; f	Posture of the female arching her back, exposing her vagina.
	Paracopulatory behaviors; f, d	Approach to a male followed by runaway, often associated with hops, darts, and ear wiggling.
	Rejection; f	Female kicks, boxes, or assumes a belly-up posture.
Prosocial behaviors	Resting with other females; f, d	Rests immobilized in relaxed position at a distance shorter than one rat from one or several females.
	Resting with males; f, d	Rests immobilized in relaxed position at a distance shorter than one rat from one or several males.
	Sniffing other females; f, d	Snout close to a female, sniffing the fur.
	Sniffing males; f, d	Snout close to a male, sniffing the fur.
	Hiding with another rat; f, d	Immobilized in a corner or in a nest box within one body length of the other rat.
Antisocial behaviors	Nose-off male; f, d	The female faces a male, nose to nose, heads up, with or without boxing.
	Nose-off female; f, d	The female faces another female, nose to nose, heads up, with or without boxing.
	Flee from male; f	Escapes from agonistic interaction by running away or simply turning head away from a male.
	Flee from another female; f	Escapes from agonistic interaction by running away or simply turning head away from a female.
Non-social behaviors	Resting alone; f, d	Rests immobilized in relaxed position at a distance longer than one rat to a conspecific.
	Drinking; f, d	Self-explanatory.
	Self-grooming and scratching; f, d	Self-explanatory.
Exploratory behaviors and ambulatory activity	Hide alone; f, d	Immobilized in a corner or nest box at a distance longer than one body length to another rat.
	Grabbing; f	Grabbing food with paws or mouth.
	Eating; f, d	Chews on food pellet.
	Freezing; f, d	Immobilized in rigid position without any movement, including those of vibrissae.
	Startle; f	Sudden reflex contractions of the major muscles of the body, leading to a little jump on the spot.
	Sniffing the floor, f, d	Sniffs the floor material with all four paws on the floor.
	Rearing; f, d	Sniffs the air while standing on the hind legs.
	Transitions; f	Displays a behavior in a zone different from the one in which the previous behavior was displayed.

Both the number of occurrences (frequency) and the duration of each occurrence were registered for most behavior patterns. For some, like lordosis, duration is not always possible to record accurately from the video record. For other behaviors, duration does not offer any useful information beyond that contained in frequency
f frequency, ***d*** duration

environment, this interval must be multiplied by 7 when estimating the time needed for behavioral observation. The amount of time required for data collection is probably the main drawback of this procedure. Furthermore, observers must be trained, and interobserver reliability must be determined if more than person is in charge of the observations.

Although not feasible at present, it is most likely that highly efficient algorithms incorporating deep learning may be able to replace human observers in the near future. Specific body parts and their position relative to each other as well as their movement can already be determined and linked to specific behaviors (e.g., [69]). It should be possible to identify behaviors like grooming, mount, and lordosis, etc., with relative ease. Moreover, it appears that the identity of similar looking animals, like rats or mice, can be preserved even when the animals are interacting closely [70]. When the entire ethogram can be computerized, an enormous amount of information can be extracted with minimal investment of human labor.

8.5 Data Analysis

Classic statistical procedures may be used both for describing and for analyzing the data. Depending on the number of behavior patterns and the complexity of social interactions, such analyses may be cumbersome and time-consuming. Nevertheless, since they produce a lot of easily understandable information, they are certainly most useful. Often, the number of behavior patterns that are of interest calls for a multivariate analysis, particularly if some treatment has been given to the animals. Multivariate analysis can be performed using standard statistical packages such as SPSS (IBM Corp., Armonk, NY, USA), or with the free software R. The latter has the advantage of including non-parametric, multivariate analyses [71], a feature not found in SPSS. It is likely that a few multivariate analyses rather than a large number of univariate tests appear more attractive to statisticians concerned with multiple hypothesis testing.

The Observer software records the behaviors in chronological order from the start of observation. The record is, then, a description of the continuous flow of behavior, in which one behavior pattern always is followed by another. Instead of just counting the frequencies and duration of these behavior patterns, the sequence of behavior can be analyzed. One way to do so is to determine how often one specific behavior pattern is followed by another specific behavior. The most obvious example would be how often an intromission is followed by genital grooming. Another example would be how often nose-off is followed by flight. However, limiting the analysis to adjacent behaviors gives little new information in addition to the obvious. To the contrary, it is most informative to establish a window of 4 or 8 behaviors and determine how often two behaviors occur in the same window. This is a procedure known as co-occurrence analysis. The window moves, by steps of one behavior pattern, over the entire individual record. The frequency of co-occurrence is subjected to a descending hierarchical classification with the purpose of identifying clusters of temporally related behaviors. This classification is based on the probability for an item to be proportionally more present in a cluster than it is in the entire data set, as evaluated by chi-square analysis. Each item is permutated from one cluster to another to test the robustness of the classification, until statistically independent profiles of items appear. The criterion for including elements in their respective cluster is a higher frequency of co-occurrence compared to the average occurrence, as well as an association with the cluster determined by chi-squared values equal to or higher than 3.84. This gives an error margin of 0.05 when df = 1.61. These rather complex statistical procedures have been described elsewhere [72, 73]. Clusters can be interpreted as groups of behaviors significantly more co-occurring together than with items of another cluster. The clusters and their relationship are visualized using the Fruchterman–Reingold algorithm [74]. These calculations are performed with the software IRAMUTEQ (Interface de R pour les Analyses Multidimensionnelles de Textes et de Questionnaires; available for free at http://www.iramuteq.org).

The co-occurrence analysis can be applied to each of the treatments included in an experiment. Even when classical analyses fail to detect significant treatment differences, a co-occurrence analysis can reveal that the behavior patterns are grouped differently under different treatments. An example is shown in Fig. 5. In fact, this analysis makes it possible to detect subtle and unpredicted changes in the structure of behavior, something particularly important when the animals are able to express a substantial part of their behavioral repertoire. Statistics designed for hypothesis testing does exactly what they are intended for, but does not allow for the discovery of unexpected behavioral changes produced by the experimental manipulations or the dynamics of complex social interactions in unmanipulated animals.

9 Seminatural Environments and Sex Differences in Non-Sexual Behaviors

We already pointed out that copulatory behavior by necessity is dimorphic. Search for sex differences in that behavior is, therefore, a triviality. There is no such obligatory sexual dimorphism in non-reproductive behaviors. Although not the theme of the present chapter, we want to briefly mention that seminatural environments can offer an externally valid procedure for analyzing sex differences in all kinds of behaviors and in all kinds of situations. For example, when the mixed-sex groups in the seminatural environment were exposed to a highly aversive stimulus, sudden,

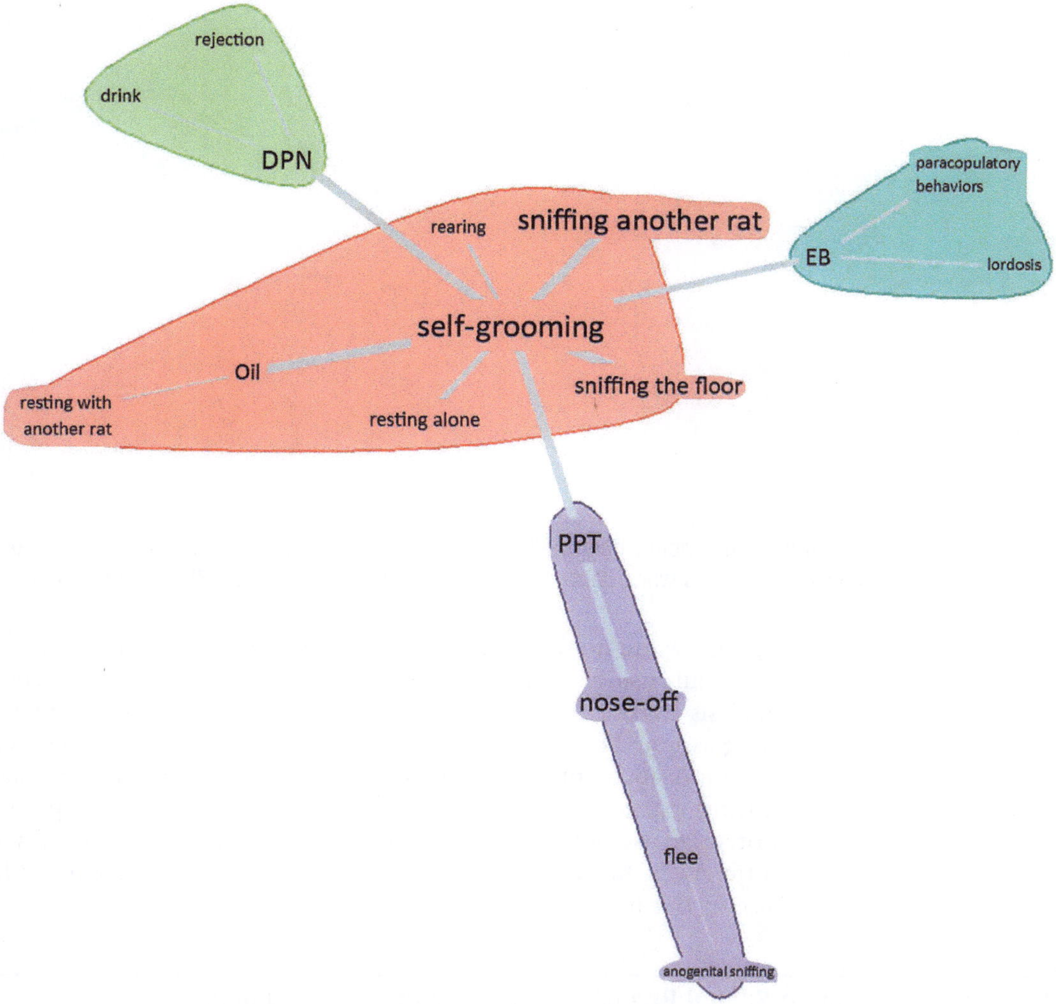

Fig. 5 The graphical representation of the result of a co-occurrence analysis. In this experiment, female rats had been treated with either estradiol benzoate (EB), the estrogen receptor α agonist propylpyrazoletriol (PPT), the estrogen receptor β agonist diarylpropionitrile (DPN) or oil. All subjects received progesterone (1 mg/rat) 48 h after the priming injection and 4 h before the start of observation. Each treatment appeared in a different cluster. The sexual behaviors, lordosis and paracopulatory behavior, were associated with EB treatment. The ERα agonist was related to the antisocial behaviors of nose-off and fleeing, and to a minor degree with anogenital sniffing. The females treated with the ERβ agonist were characterized by the antisocial behavior of rejection and the solitary activity of drinking. Oil treatment was mainly related to the self-centered behavior of grooming, resting, and social and non-social sniffing. Rearing was also typical for these females. Many of the associations between treatment and particular behavior patterns were not detected in the analyses of frequency or duration, revealing the superior power of the co-occurrence analysis. For further details of this experiment, see [66]. (Reproduced with permission)

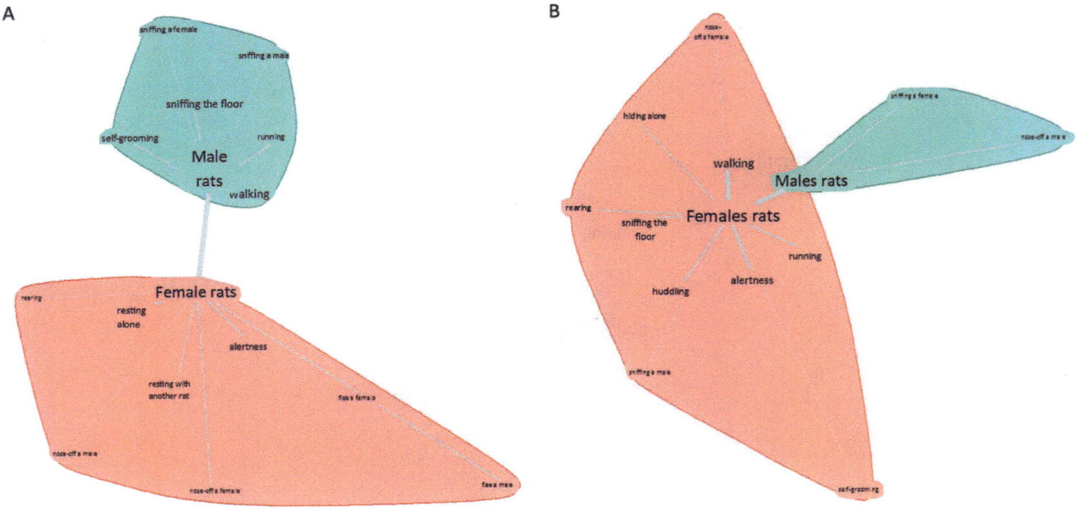

Fig. 6 Co-occurrence analysis of behavior in male and female rats housed in a seminatural environment before (**a**) and during (**b**) exposure to a 90 dB white noise. (Reproduced with permission (CC BY 4.0) from [67])

intense white noise, the behavioral responses of males and females were surprisingly similar [67]. However, when a co-occurrence analysis of the behavioral record was made, males and females were assigned to different clusters. Interestingly, the sex difference was more apparent before presentation of the aversive stimulus than during exposure to it (*see* Fig. 6). This is a modest example of potential uses of seminatural environments in the study of sex differences. At least it shows that representative designs are possible also in this field of research.

10 A Note on Human Sexual Behavior and Seminatural Conditions

In this context, it may be interesting to make a brief comparison between rat and human sexual behavior. Outside the laboratory, human sexual interactions are similar to those of rats in the seminatural environment. Sex mostly occurs in a place which the participants have chosen and which they are free to leave at any moment. Furthermore, the participants have always the possibility to engage in a considerable number of alternative behaviors, meaning that sexual activity is only one possibility among many others. There is, however, one important difference: Humans normally copulate in couples, whereas rats engage in group sex. Even though humans may do the same, it does not seem to be the most common form of sexual interaction. For reasons that are evident, laboratory studies of human sexual interactions are extremely rare. In fact, only the ground-breaking work of Masters and Johnson [75] employed observations of people copulating in the laboratory. Studies before and after have either employed questionnaires or recordings of sexual responses to solitary sex (masturbation). Exceptionally, a partner has been asked to stimulate the genitals of the experimental subject (e.g., [76, 77]). This is probably the closest we have come to a natural sexual interaction in the laboratory. Thus, it can be maintained that most laboratory studies of human sexual responses do not satisfy the criteria for a representative design. The consequences of this are probably far less important than the consequences of absent representative design in rat studies. In the case of rodents, observations must not only be generalized to contexts outside the specific setup used, but also to another species. In humans, generalizations are only made from one context to others within one species.

In one exceptional study, women were asked to record their behavior or genital responses to sexually relevant stimuli at home as well as in the laboratory (e.g., [78]). The former is very close to a representative design. Interestingly, there were important differences between the data obtained in the lab and those obtained at home, showing that even intra-species generalizations from one context to another are not perfect. We know of only one additional study in which genital responses were recorded in a home setting [79]. Unfortunately, no laboratory data were obtained, making comparisons between the two contexts impossible. We feel it is important to draw attention to the fact that excellent representative designs could easily be used in humans. In insect and rodent studies, we are only able to approximate such designs through the use of seminatural environments.

11 Conclusion

Experimental setups incorporating basic characteristics of the experimental subjects' social and physical environment, and permitting the expression of as large as possible part of the natural behavioral repertoire, can be expected to be externally valid. This means that results from such setups can be generalized to conditions outside of the laboratory setting. Whenever the purpose of a study is to elucidate the biological function, adaptive value, or evolution of particular behavior patterns, the use of an externally valid procedure is mandatory. The preceding presentation of this kind of procedures in insects and rodents will hopefully contribute to an enhanced interest in and use of seminatural environments. The development of powerful image analyzing tools may make data acquisition far more time-efficient than at present, and novel statistical approaches may provide the possibility to fully explore the millions of data points generated in complex environments. Perhaps this will contribute to a breakthrough in our understanding of brain–behavior relations.

References

1. Hodgson AN (2011) Reproduction and sex in invertebrates. Encyclopedia of life support systems. UNESCO, Paris
2. Kinoshita M, Homberg U (2017) Insect brains: minute structures controlling complex behaviors. In: Shigeno S, Murakami Y, Nomura T (eds) Brain evolution by design: from neural origin to cognitive architecture. Springer, Tokyo, pp 123–151. https://doi.org/10.1007/978-4-431-56469-0_6
3. Oh SM, Jeong K, Seo JT, Moon SJ (2021) Multisensory interactions regulate feeding behavior in *Drosophila*. Proc Natl Acad Sci U S A 118(7):e2004523118. https://doi.org/10.1073/pnas.2004523118
4. Griffith LC, Ejima A (2009) Multimodal sensory integration of courtship stimulating cues in Drosophila melanogaster. Ann N Y Acad Sci 1170:394–398. https://doi.org/10.1111/j.1749-6632.2009.04367.x
5. Keene AC, Waddell S (2007) Drosophila olfactory memory: single genes to complex neural circuits. Nat Rev Neurosci 8(5):341–354. https://doi.org/10.1038/nrn2098
6. Moloney A, Sattelle DB, Lomas DA, Crowther DC (2010) Alzheimer's disease: insights from Drosophila melanogaster models. Trends Biochem Sci 35(4):228–235. https://doi.org/10.1016/j.tibs.2009.11.004
7. Spieth HT (1974) Courtship behavior in *Drosophila*. Annu Rev Entomol 19(1):385–405
8. Ewing AW (1961) Body size and courtship behaviour in *Drosophila melanogaster*. Anim Behav 9(1):93–99. https://doi.org/10.1016/0003-3472(61)90055-0
9. Dukas R (2020) Natural history of social and sexual behavior in fruit flies. Sci Rep 10(1): 21932. https://doi.org/10.1038/s41598-020-79075-7
10. Takasaki T, Namiki S, Kanzaki R (2012) Use of bilateral information to determine the walking direction during orientation to a pheromone source in the silkmoth *Bombyx mori*. J Comp Physiol A 198(4):295–307
11. Lelito JP, Myrick AJ, Baker TC (2008) Interspecific pheromone plume interference among sympatric heliothinae moths: a wind tunnel test using live, calling females. J Chem Ecol 34(6): 725–733. https://doi.org/10.1007/s10886-008-9475-6
12. Cardé RT, Hagaman TE (1979) Behavioral responses of the gypsy moth 1 in a wind tunnel to air-borne enantiomers of disparlure 2. Environ Entomol 8(3):475–484. https://doi.org/10.1093/ee/8.3.475
13. Lamunyon CW (2000) Sperm storage by females of the polyandrous noctuid moth Heliothis virescens. Anim Behav 59(2): 395–402. https://doi.org/10.1006/anbe.1999.1294
14. Hou M, Sheng C (1999) Fecundity and longevity of *Helicoverpa armigera* (Lepidoptera: Noctuidae): effects of multiple mating. J Econ Entomol 92(3):569–573. https://doi.org/10.1093/jee/92.3.569
15. Shepherd GM (1974) The synaptic organization of the brain: an introduction. Oxford University Press, New York
16. Guo X, Yu Q, Chen D, Wei J, Yang P, Yu J, Wang X, Kang L (2020) 4-Vinylanisole is an aggregation pheromone in locusts. Nature 584(7822):584–588. https://doi.org/10.1038/s41586-020-2610-4
17. Connolly K, Cobb M, Burnet B (1985) Courtship behaviour in the melanogaster species sub-group of Drosophila. Behaviour 95(3–4): 203–230. https://doi.org/10.1163/156853985X00136
18. Halfwerk W, Varkevisser J, Simon R, Mendoza E, Scharff C, Riebel K (2019) Toward testing for multimodal perception of mating signals. Front Ecol Evol 7:124. https://doi.org/10.3389/fevo.2019.00124
19. Byrne PG, Rice WR (2006) Evidence for adaptive male mate choice in the fruit fly Drosophila melanogaster. Proc Royal Soc B: Biol Sci 273(1589):917–922
20. Arbuthnott D, Fedina TY, Pletcher SD, Promislow DEL (2017) Mate choice in fruit flies is rational and adaptive. Nat Commun 8(1): 13953. https://doi.org/10.1038/ncomms13953
21. Priesner E, Witzgall P, Voerman S (1986) Field attraction response of raspberry clearwing moths, *Pennisetia hylaeiformis* Lasp.(Lepidoptera: Sesiidae), to candidate pheromone chemicals. J Appl Entomol 102(1–5):195–210
22. Kaissling K-E, Priesner E (1970) Die Riechschwelle des Seidenspinners. Naturwissenschaften 57(1):23–28. https://doi.org/10.1007/bf00593550
23. Baker TC, Hansson BS (2016) Moth sex pheromone olfaction. In: Allison JD, Cardz RT (eds) Pheromone communication in moths: evolution, behavior, and application. University of California Press, Oakland
24. Willis MA, Avondet JL, Zheng E (2011) The role of vision in odor-plume tracking by walking and flying insects. J Exp Biol 214(24):

4121–4132. https://doi.org/10.1242/jeb.036954
25. Kennedy JS, Marsh D (1974) Pheromone-regulated anemotaxis in flying moths. Science 184(4140):999–1001
26. Chu X, Heinze S, Ian E, Berg BG (2020) A novel major output target for pheromone- sensitive projection neurons in male moths. Front Cell Neurosci 14:147. https://doi.org/10.3389/fncel.2020.00147
27. Kymre JH, Liu X, Ian E, Berge CN, Wang G, Berg BG, Zhao X, Chu X (2021) Distinct protocerebral neuropils associated with attractive and aversive female-produced odorants in the male moth brain. eLife 10:e65683. https://doi.org/10.7554/eLife.65683
28. Li X, Ishimoto H, Kamikouchi A (2018) Auditory experience controls the maturation of song discrimination and sexual response in Drosophila. eLife 7:e34348. https://doi.org/10.7554/eLife.34348
29. Cheriyamkunnel SJ, Rose S, Jacob PF, Blackburn LA, Glasgow S, Moorse J, Winstanley M, Moynihan PJ, Waddell S, Rezaval C (2021) A neuronal mechanism controlling the choice between feeding and sexual behaviors in Drosophila. Curr Biol 31(19):4231–4245. https://doi.org/10.1016/j.cub.2021.07.029
30. Biasazin TD, Chernet HT, Herrera SL, Bengtsson M, Karlsson MF, Lemmen-Lechelt JK, Dekker T (2018) Detection of volatile constituents from food lures by tephritid fruit flies. Insects 9(3):119. https://doi.org/10.3390/insects9030119
31. Ren L, Ma Y, Xie M, Lu Y, Cheng D (2021) Rectal bacteria produce sex pheromones in the male oriental fruit fly. Curr Biol 31(10): 2220–2226.e2224. https://doi.org/10.1016/j.cub.2021.02.046
32. Keys RE, Mills RB (1968) Demonstration and extraction of a sex attractant from female angoumois grain moths 1. J Econ Entomol 61(1):46–49. https://doi.org/10.1093/jee/61.1.46
33. Namiki S, Kanzaki R (2019) Morphology and physiology of olfactory neurons in the lateral protocerebrum of the silkmoth Bombyx mori. Sci Rep 9(1):16604. https://doi.org/10.1038/s41598-019-53318-8
34. Ian E, Kirkerud NH, Galizia CG, Berg BG (2017) Coincidence of pheromone and plant odor leads to sensory plasticity in the heliothine olfactory system. PLoS One 12(5):e0175513
35. Wu H, Xu M, Hou C, Huang L-Q, Dong J-F, Wang C-Z (2015) Specific olfactory neurons and glomeruli are associated to differences in behavioral responses to pheromone components between two Helicoverpa species. Front Behav Neurosci 9:206. https://doi.org/10.3389/fnbeh.2015.00206
36. Kaushik PK, Renz M, Olsson SB (2020) Characterizing long-range search behavior in Diptera using complex 3D virtual environments. Proc Natl Acad Sci U S A 117(22): 12201–12207. https://doi.org/10.1073/pnas.1912124117
37. Koukou K, Pavlikaki H, Kilias G, Werren JH, Bourtzis K, Alahiotis SN (2006) Influence of antibiotic treatment and Wolbachia curing on sexual isolation among *Drosophila melanogaster* cage populations. Evolution 60(1):87–96
38. Ehrman L (1965) Direct observation of sexual isolation between allopatric and between sympatric strains of the different *Drosophila paulistorum* races. Evolution 19:459–464
39. Korol A, Rashkovetsky E, Iliadi K, Michalak P, Ronin Y, Nevo E (2000) Nonrandom mating in *Drosophila melanogaster* laboratory populations derived from closely adjacent ecologically contrasting slopes at "Evolution Canyon". Proc Natl Acad Sci U S A 97(23): 12637–12642. https://doi.org/10.1073/pnas.220041397
40. Arita LH, Kaneshiro KY (1988) Body size and differential mating success between males of two populations of the Mediterranean fruit fly. Pac Sci 42(3–4):173–177
41. Zhang J, Bisch-Knaden S, Fandino RA, Yan S, Obiero GF, Grosse-Wilde E, Hansson BS, Knaden M (2019) The olfactory coreceptor IR8a governs larval feces-mediated competition avoidance in a hawkmoth. Proc Natl Acad Sci U S A 116(43):21828–21833. https://doi.org/10.1073/pnas.1913485116
42. Pfaff D (2017) How the vertebrate brain regulates behavior. Direct from the lab. Harvard University Press, Cambridge
43. Pfaff DW (1999) Drive. Neurobiological and molecular mechanisms of sexual motivation. MIT Press, Cambridge
44. Pfaff DW (1980) Estrogens and brain function: neural analysis of a hormone-controlled mammalian reproductive behavior. Springer, New York
45. Holmes GM, Chapple WD, Leipheimer RE, Sachs BD (1991) Electromyographic analysis of male rat perineal muscles during copulation and reflexive erections. Physiol Behav 49(6): 1235–1246
46. Sachs BD (1982) Role of striatal penile muscles in penile reflexes, copulation, and induction of pregnancy in the rat. J Reprod Fertil 66(2): 433–443

47. Moralí G, Soto MAP, Contreras JL, Arteaga M, González-Vidal MD, Beyer C (2003) Detailed analysis of the male copulatory motor pattern in mammals: hormonal bases. Scand J Psychol 44(3):279–288
48. Hull EM, Rodríguez-Manzo G (2017) Male sexual behavior. In: Pfaff DW, Joëls M (eds) Hormones, brain and behavior, 3rd edn. Academic Press, Oxford, pp 1–57. https://doi.org/10.1016/B978-0-12-803592-4.00001-8
49. Robitaille JA, Bovet J (1976) Field observations on the social behaviour of the Norway rat, Rattus norvegicus (Berkenhout). Biol Behav 1:289–308
50. Calhoun JB (1962) The ecology and sociology of the Norway rat. USPHS Publication No. 1008. US Governnment Printing Office, Washington, DC
51. Brunswik E (1955) Representative design and probabilistic theory in a functional psychology. Psychol Rev 62(3):193–217
52. Petrinovich L (1980) Brunswikian behavioral biology. In: Hammond KR, Wascoe NE (eds) Realizations of Brunswick's representative design, New directions for methodology of social and behavioral science, vol 3. Jossey-Bass, San Francisco, pp 85–93
53. Petrinovich L (1989) Representative design and the quality of generalization. In: Poon LW, Rubin DC, Wilson BA (eds) Everyday cognition in adulthood and late life. Cambridge University Press, New York, pp 11–24
54. Chu X, Ågmo A (2016) Sociosexual interactions in rats: are they relevant for understanding human sexual behavior? Int J Psychol Res 9(2):76–95
55. Chu X, Ågmo A (2015) Sociosexual behaviors of male rats (Rattus norvegicus) in a seminatural environment. J Comp Psychol 129(2): 132–144
56. Peirce JT, Nuttall RL (1961) Self-paced sexual behavior in the female rat. J Comp Physiol Psychol 54(3):310–313
57. Erskine MS, Kornberg E, Cherry JA (1989) Paced copulation in rats – effects of intromission frequency and duration on luteal activation and estrus length. Physiol Behav 45(1): 33–39
58. McClintock MK, Adler NT (1978) The role of the female during copulation in wild and domestic Norway rats (*Rattus norvegicus*). Behaviour 67:67–96
59. McClintock MK, Anisko JJ (1982) Group mating among Norway rats I. Sex differences in the pattern and neuroendocrine consequences of copulation. Anim Behav 30(2):398–409
60. McClintock MK, Anisko JJ, Adler NT (1982) Group mating among Norway rats II. The social dynamics of copulation: competition, cooperation, and mate choice. Anim Behav 30(2):410–425
61. Chu X, Ågmo A (2014) Sociosexual behaviours in cycling, intact female rats (*Rattus norvegicus*) housed in a seminatural environment. Behaviour 151(8):1143–1184
62. Chu X, Guarraci FA, Ågmo A (2015) Sociosexual behaviors and reproductive success of rats (*Rattus norvegicus*) in a seminatural environment. Physiol Behav 151:46–54
63. Schein MW (1950) The relation of sex ratio to physiological age in the wild brown rat. Am Nat 84(819):489–496
64. Heinla I, Chu X, Ågmo A, Snoeren E (2021) Rat ultrasonic vocalizations and novelty-induced social and non-social investigation behavior in a seminatural environment. Physiol Behav 237:113450. https://doi.org/10.1016/j.physbeh.2021.113450
65. Le Moëne O, Hernández-Arteaga E, Chu X, Ågmo A (2020) Rapid changes in sociosexual behaviors around transition to and from behavioral estrus, in female rats housed in a seminatural environment. Behav Process 174:104101. https://doi.org/10.1016/j.beproc.2020.104101
66. Le Moëne O, Ågmo A (2018) Behavioral responses to emotional challenges in female rats living in a seminatural environment: the role of estrogen receptors. Horm Behav 106: 162–177
67. Le Moëne O, Ramírez-Rentería ML, Ågmo A (2020) Male and female immediate fear reaction to white noise in a semi-natural environment: a detailed behavioural analysis of the role of sex and oestrogen receptors. J Neuroendocrinol 32(10):e12902. https://doi.org/10.1111/jne.12902
68. Le Moëne O, Stavarache M, Ogawa S, Musatov S, Ågmo A (2019) Estrogen receptors a and β in the central amygdala and the ventromedial nucleus of the hypothalamus: sociosexual behaviors, fear and arousal in female rats during emotionally challenging events. Behav Brain Res 367:128–142. https://doi.org/10.1016/j.bbr.2019.03.045
69. Mathis A, Schneider S, Lauer J, Mathis MW (2020) A primer on motion capture with deep learning: principles, pitfalls, and perspectives. Neuron 108(1):44–65. https://doi.org/10.1016/j.neuron.2020.09.017
70. Lauer J, Zhou M, Ye S, Menegas W, Nath T, Rahman MM, Di Santo V, Soberanes D, Feng G, Murthy VN, Lauder G, Dulac C,

Mathis MW, Mathis A (2022) Multi-animal pose estimation and tracking with DeepLabCut. Nat Methods 19(4):496–504. https://doi.org/10.1038/s41592-022-01443-0
71. Ellis AR, Burchett WW, Harrar SW, Bathke AC (2017) Nonparametric inference for multivariate data: the R package npmv. J Stat Softw 76(4):1–18. https://doi.org/10.18637/jss.v076.i04
72. Reinert A (1983) Une méthode de classification descendante hiérarchique: application à l'analyse lexicale par contexte. Les Cahiers de l'analyse des Données 8(2):187–198
73. Valax MF, Marine C, Reinert M (1990) Traitement de données structurées par un ordre temporel ou hiérarchique: utilisation dans l'analyse de l'activité. Le Travail Humain 53(1):79–89
74. Fruchterman TMJ, Reingold EM (1991) Graph drawing by force-directed placement. Softw-Pract Exp 21(11):1129–1164. https://doi.org/10.1002/spe.4380211102
75. Masters WH, Johnson VE (1966) Human sexual response. Little, Brown & Co., Boston
76. Georgiadis JR, Kortekaas R, Kuipers R, Nieuwenburg A, Pruim J, Reinders AATS, Holstege G (2006) Regional cerebral blood flow changes associated with clitorally induced orgasm in healthy women. Eur J Neurosci 24(11):3305–3316. https://doi.org/10.1111/j.1460-9568.2006.05206.x
77. Georgiadis JR, Holstege G (2005) Human brain activation during sexual stimulation of the penis. J Comp Neurol 493(1):33–38
78. Bloemers J, Gerritsen J, Bults R, Koppeschaar H, Everaerd W, Olivier B, Tuiten A (2010) Induction of sexual arousal in women under conditions of institutional and ambulatory laboratory circumstances: a comparative study. J Sex Med 7(3):1160–1176. https://doi.org/10.1111/j.1743-6109.2009.01660.x
79. Poels S, Bloemers J, van Rooij K, Goldstein I, Gerritsen J, van Ham D, van Mameren F, Chivers M, Everaerd W, Koppeschaar H, Olivier B, Tuiten A (2013) Toward personalized sexual medicine (part 2): testosterone combined with a PDE5 inhibitor increases sexual satisfaction in women with HSDD and FSAD, and a low sensitive system for sexual cues. J Sex Med 10(3):810–823. https://doi.org/10.1111/j.1743-6109.2012.02983.x

Chapter 9

Sexual Incentive Motivation

Patty T. Huijgens, Roy Heijkoop, and Eelke M. S. Snoeren

Abstract

Motivation can be described as the processes that activate, direct, and determine the persistence of goal-directed behavior. A certain level of sexual motivation is required to activate approach behavior which could potentially lead to copulation and reproduction. Here, we discuss the use of the sexual incentive motivation test to assess the magnitude of sexual motivation in rodents in response to different incentive stimuli. Our goal is to provide the reader with guidelines on how to set up the test and what measures can be considered to study sexual incentive motivation.

Key words Sexual motivation, Incentive motivation, Protocol, Sexual behavior

1 Introduction

1.1 Sexual Incentive Motivation

Motivation can be described as the processes that activate, direct, and determine the persistence of goal-directed behavior. A certain level of sexual motivation is required to activate approach behavior which could potentially lead to copulation and reproduction. As summarized in [1], different models of motivation have been discussed throughout history that mainly focused on the role of motivation for the continuation of the behavioral act. However, sexual motivation can be subdivided into two different components: the incentive motivation and the motivation to continue copulation. Incentive motivation could then be seen as the magnitude of motivation that is present before approach behavior has occurred, and the motivation to continue copulation as the processes needed to maintain the continuation of the behavior until satiety.

As proposed in a model from Ågmo and Bindra [1–3], sexual incentive motivation is based on three important elements: a central motive state, an incentive stimulus, and a certain level of general

Patty T. Huijgens and Roy Heijkoop contributed equally with all other contributors.

Raúl G. Paredes et al. (eds.), *Animal Models of Reproductive Behavior*, Neuromethods, vol. 200,
https://doi.org/10.1007/978-1-0716-3234-5_9,

arousal. The central motive state was defined as "a hypothetical set of neural processes that promotes goal-directed actions in relation to particular classes of incentive stimuli" [2]. This internal state of the animal, if brought to a certain level, can activate sexual incentive motivation and trigger the individual to take action. However, the internal state by itself will not induce approach behavior. For that, an incentive stimulus needs to be present, resulting in a positive effect on the central motive state. The incentive stimulus may be any stimulus with hedonic properties. With regard to sexual motivation, the incentive stimulus could for instance be the presence of an attractive partner. Last, but not least, a level of general arousal is also required to activate the central motive state. This means that the individual should be ready for the perception of an incentive stimulus. In a state of sleep, for instance, general arousal is absent, but even when awake, different factors can influence general arousal, such as the hormonal status of the individual. When all elements are present and cause an enhancement of the central motive state, a level of sexual incentive motivation can be reached which may induce approach behavior toward the mate partner.

As already mentioned, gonadal hormones play an important role in the determination of sexual incentive motivation. Not only are hormones necessary for the appropriate internal state leading to sexual incentive motivation, and do they influence general arousal, potential partners will also only have an incentive value when they are in the appropriate hormonal state. A female rat, for instance, will approach an intact male more frequently when she is sexually receptive, i.e., on the night of the transition from proestrus to estrus, shortly after estrogen and progesterone levels have peaked [4, 5]. Similarly, only receptive females are normally sexually approached by intact males [6]. Castrated male rats and non-receptive females are not a sexual incentive to their conspecifics and will also not sexually approach a potential mate [6]. Thus, the appropriate hormonal status that drives sexual motivation also causes animals to emit certain stimulatory signals making them sexually attractive. In rodents, the secretion of specific odors and pheromones is an example of such a signal that is important for the incentive value of males and females. Gonadally intact male rats approach the odor of sexually receptive females more than they approach non-receptive females [7–10], while anosmic males lose their capacity to distinguish between females in estrus or non-estrus [11]. Male odors also seem to have the same effect on female rats, i.e., receptive females prefer odors of intact males over odors of castrated males, and exposure to male urine can affect subsequent copulatory behavior [7, 12].

Olfactory stimuli are not the only factors determining the incentive value of a stimulus. As approach behavior to a potential mate can be induced by distant stimuli, other potential sensory modalities involved in mate choice include audition (e.g., via ultrasonic vocalizations) or vision (the sight of a potential mate). In rats, visual stimuli are not as important, since males approach females even in complete darkness [13, 14]. The role of auditory stimuli is potentially more complex. Although rats emit 50 kHz ultrasonic vocalizations (USVs) in the presence of a sexual partner and during copulation [15–17], we have shown that these USVs do not have any incentive value for male and female rats, i.e., playback of USVs alone does not induce any preference, and devocalized potential mates have the same incentive value as intact mates [9, 18]. Interestingly, a study in which the different modalities and the combination of these modalities in inducing sexual incentive value was studied showed that there is a synergistic function for the different modalities in increasing the incentive value of an individual and thus the induction of approach behavior toward a potential mate [19]. Specifically, the presence of a single-sensory sexual stimulus with one modality (olfaction, vision, or "others," but not audition) was sufficient to induce approach, but not above approach toward a social stimulus animal. A multisensory stimulus of multiple modalities was necessary to induce incentive sexual motivation (i.e., a preference over a social stimulus) [19].

Another important fact is that sexual incentive motivation, at least in rats, is unconditioned, meaning that sexual experience is not necessary for sexual motivation to be induced by sexual incentives. In fact, sexual experience does not even modify sexual incentive motivation in either male or female rats [6, 20]. This is remarkable because gain of sexual experience reduces mount/intromission latency and ejaculation latency in a copulation test in males [21] and influences paced mating behavior in females (e.g., return latencies after intromission and ejaculation) [22]. Urine of a receptive female alone can elicit approach behavior in gonadally intact male rats, even when still sexually naïve. However, when bedding of a receptive female is used as olfactory stimulus (meaning a mix of urine with other odors), sexual motivation is induced in sexually experienced, but not in sexually naïve animals [6]. This illustrates an important point; whereas sexual motivation itself is innate and unconditioned, it seems that a learning component through rewarding sexual experience leads to a greater capability of perceiving sexual incentives during future encounters under complex conditions with several (non-sexual) stimuli present.

After the induction of sexual motivation, approach behavior could follow leading to the start of copulation. Sexual behavior as a whole can thus be divided into different phases: the introductory phase, copulatory phase, and executive phase (ejaculation or orgasm) [23, 24]. It is important to note that these different phases may be driven by different neurobiological processes, at least in rats [25]. Different subregions of brain areas are activated during the different phases [26], suggesting that different mechanisms are involved in the different parts of sexual behavior. The dissociation between motivation, copulation, and ejaculation is also evident in lesion studies showing, e.g., that motivation or copulation can be abolished while ejaculation is stimulated [27], or chemogenetic studies in which incentive motivation remained unaffected while ejaculation was disrupted [28]. Furthermore, the relative magnitude of sexual incentive motivation (measured as relative preference for a sexual vs. a social incentive) does not correlate with several outcome parameters of copulation assessment in either male or female rats. For example, the latency to mount and number of paracopulatory behaviors, which are often taken as measures of sexual motivation in copulation tests, as well as number of mounts and intromissions, ejaculation latency, and duration of post-ejaculatory interval in males, and percentage of exits and return latency after mounts and intromissions in females, do not correlate with sexual incentive motivation measured in a separate test [24, 29]. The lack of relationships between these measures can then be attributed to the different elements of sexual motivation, with incentive motivation on the one side and motivation to continue copulation on the other. Together, the lack of a correlation between motivation and copulatory measures as well as the ability of physiological manipulations to affect motivation but not copulation stress the relevance for why it is important to study the different phases of sexual behavior (incentive motivation and copulation) separately with the appropriate tests (reviewed in [24]).

1.2 Assessing Sexual Incentive Motivation

Several paradigms have been used to study sexual motivation (reviewed in [24]). Importantly, not all paradigms measure the same aspects of sexual motivation. Briefly, it is important to distinguish between paradigms that measure incentive sexual motivation versus paradigms that actually measure (1) motivation to continue copulation, (2) the willingness to work or overcome adverse events in order to access a mate, or (3) reward anticipation. In contrast to sexual incentive motivation, these measures are dependent on previous sexual experience and heavily influenced by the rewarding properties of copulation.

In our lab, we assess sexual motivation in rats with the sexual incentive motivation test (SIM test). In the SIM test, a subject animal is tracked for 10 min during exploration of a large arena to which two stimulus boxes are attached that house either a sexual incentive or a social incentive. The time spent in the vicinity of the sexual incentive, relative to the time spent in the vicinity of the social incentive, is a measure for the sexual incentive motivation of the subject rat. Measuring sexual motivation above and beyond motivation for social interaction in the same test is a clear advantage of the SIM test setup over other tests in which only one incentive is used. As rats are highly social animals, any sexual incentive animal is inherently also a social incentive. Since tests of sexual motivation often neglect this aspect, this needs to be taken into account when assessing the motivation to approach a sexual incentive.

Other advantages of the SIM test include the fact that no physical interaction between subject and stimulus animals is necessary and that the outcome measure for magnitude of sexual motivation is not dependent on operant responses, nor on sexual experience, and less sensitive to variations in motor function than other measures (e.g., operant responses). Even if ambulatory activity is affected by certain treatments, it is first and foremost easy to assess this by looking at distance traveled, velocity, and immobility during the test, but it would also most likely not affect the outcome of the test because the subject animal will still spend relatively more time in the vicinity of the sexual incentive compared to the social incentive. This relative preference has been observed in the face of both enhanced ambulatory activity (e.g., upon amphetamine administration) as well as decreased ambulatory activity (e.g., upon cis(Z)-flupenthixol administration) [30].

1.3 Sex Differences

Sexual behavior in general is very different in males and females. Male rats, for example, copulate in mount bouts in which they show patterns of mounts, intromissions, and ejaculations [24, 31], while females show paracopulatory behaviors and lordosis [24]. The setup of a copulation test is often also different for male and female rats, as the sex that is under investigation should be allowed to control the pace of copulation [24]. The different types of behaviors and different testing setup make it difficult to compare results between the different sexes. Still, by comparing the behaviors in the right phases, it remains possible to draw parallels [23].

In contrast to the copulation test, the SIM test is conducted in exactly the same way for both sexes and does not need to be modified for sex differences. Both males and females show similar levels of sexual incentive motivation, despite the fact that different gonadal hormones are involved. As long as the rats are sexually mature and in the appropriate hormonal state, male and female rats spend similar amounts of time with a sexual incentive and a social incentive in the SIM test, resulting in equal levels of sexual incentive motivation (ratio typically around 0.7) [20]. It is then also expected that certain manipulations will have similar effects in males and females on sexual incentive motivation. For example, both male and female rats lose sexual incentive motivation when they have engaged in copulation with extensive stimulation (4 h of copulation up to exhaustion in males, 3 ejaculations received in females) [20]. These similarities between males and females, as well as the aforementioned overlapping importance of hormonal status, the reliance on the sensory modality of olfaction, the role of sexual experience, and the lack of a correlation between incentive motivation and copulation parameters, make the SIM test a very suitable test to study sex differences.

2 Materials

2.1 Equipment

The setup for the sexual incentive motivation test consists of the following parts:

- SIM apparatus
- Dim, non-directional light source
- Digital, light-sensitive camera, connected to computer
- Computer with tracking software

2.2 Apparatus and Setup

The SIM apparatus consists of a rectangular arena (50 × 100 cm) covered with dark gray polyvinyl chloride, with short sides that are rounded (Fig. 1). The walls are 60 cm high and made of 0.6 mm thick coated sheet steel. They are coated on the inside with a black polymer, which minimizes light reflections and makes the walls easy to clean. In either of the long walls of the arena, a 23 × 23 cm square opening is routed. On the outside of the openings, two vertical guide rails allow stimulus boxes to slide in place. These boxes (25 × 15 × 25 cm, and closed with a lid on hinges) are made from the same material as the arena and connect with the arena with a wire mesh.

A light-sensitive, digital camera is positioned roughly 200 cm above the SIM apparatus so that the whole apparatus is in view. A 15 Watt (85 lumen), incandescent light source, mounted just above the camera, provides about 5 lux of light in the arena, roughly comparable to light of a full moon. This is just bright enough for proper functioning of the camera and tracking software. To track the subject animal in the SIM arena, we use EthoVision XT16 (Noldus, Wageningen, The Netherlands), but any behavior-tracking software that enables predefining areas within the test arena will work.

Beforehand, we design an experiment in the EthoVision software. For details on how to do this, we refer to the manual provided by Noldus (Noldus, Wageningen, the Netherlands). It is important to classify the arenas of your test setup and define two virtual incentive zones (20 × 30 cm; A4 paper size for convenience) in front of the stimulus box openings in each arena. The subject's position is tracked with a frequency of at least 5 Hz.

2.3 Animals

We have always used adult Wistar rats, but we see no reason why other rat strains, or even other animal species, could not be used in this same test (*see* **Note 7**). Animals are typically kept on a reversed light–dark cycle (11 am–11 pm) and testing is performed during the dark period (typically between 12 and 5 pm). Subject and stimulus animals can undergo a wide variety of treatments for testing purposes before the SIM test. In case the animals undergo

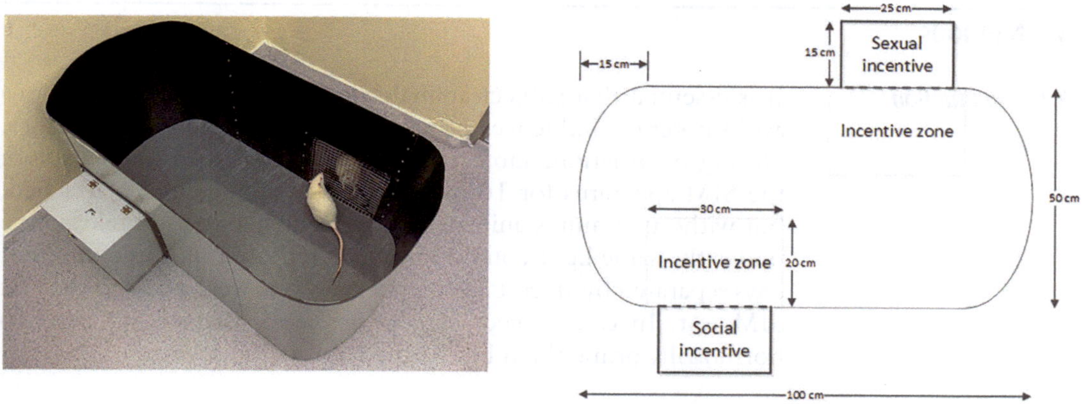

Fig. 1 Photograph (left) and schematic drawing (right) of the sexual incentive motivation test setup

invasive treatment (e.g., stereotaxic brain surgery), a recovery period of at least one week is used.

Most often we use conspecifics as an incentive stimulus (*see* **Note 4**), even though other types of stimuli can be used as well (*see* **Note 6**). If the subject animal is a male, one stimulus box contains an intact male as a social incentive and the other a sexually receptive female as sexual incentive. In order to induce sexual receptivity at the moment of testing, these stimulus females undergo a surgical procedure at least one week prior to the SIM test for ovariectomy and implantation of a Silastic capsule containing 10% 17β-estradiol in cholesterol (see for a protocol: [32], and *see* **Note 4**). Four hours before testing, the females are hormonally primed by subcutaneous injection with progesterone (dissolved in peanut oil, 5 mg/mL, 0.2 mL). In our experience, these females can be used as stimulus at least twice a week for at least 3 months. An ovariectomized female without hormonal treatment (non-receptive female) could also be used as social incentive instead of a male rat for a male subject.

If the subject animal is a female, one can decide whether to study naturally cycling females or ovariectomized and hormonally primed females (also *see* **Note 5**). In case naturally cycling females are used as subjects, we have good experience using an estrous cycle monitor (Muromachi Kikai Co., Tokyo, Japan) to track the estrous cycle of female rats. For a female subject, the sexual incentive is an intact male, and the social incentive is either a castrated male or another female. All stimulus animals are usually sexually experienced.

3 Methods

3.1 Habituation

It is essential that subject animals are familiarized to the arena to avoid potential influences of novelty on the behavioral outcomes. During habituation, subject animals are allowed to freely explore the SIM apparatus for 10 min, with the stimulus boxes attached, but without stimulus animals present. The habituation takes place under the same light conditions and is repeated on three different days separated by at least 24 h, preferably one week prior to the first SIM test. In case subject females are used, there is no need to hormonally prime them before the habituation session.

3.2 Testing

After habituation, the test sessions can start. The SIM test can be repeated as many times as the experiment calls for, with a minimum of 24 h between test sessions. The same type of incentives or a variety in incentives could be used for the different SIM tests, depending on the purpose of the experiment (*see* also **Note 6**). Both, within- and between-group comparisons are possible. When conducting the SIM test in combination with a copulation test, *see* **Notes 1–3** for some considerations.

Before each test session, the incentive animals are placed in their respective stimulus boxes. About 5 min later, the first experimental subject can be introduced into the middle of the arena and the 10 min of observation is started. The experimenter can then leave the room, or stay in the room quietly. At the end of the test, the subject animal and any feces and urine on the floor are removed from the arena. Then, the following subject animal can immediately be introduced.

The position of the incentive animals is semi-randomly changed throughout the experimental session to rule out location bias for the sexual vs. social incentive. This is done by switching the detachable stimulus boxes on the arena while stimulus animals can remain in the boxes. Stimulus animals are often changed between tests to prevent effects of dehydration, food deprivation, and stress of small space confinement. We typically change the stimulus animals after every third SIM test during a testing day. It is essential to place the same type of stimulus animal (social or sexual incentive) in the previously used stimulus box, so that odor contamination is avoided. If a different type of incentive stimulus animal must be placed in a previously used box, careful cleaning is highly recommended between changing the animals.

After the last test of a day, the SIM arena is cleaned with 0.1% acetic acid solution, and the stimulus boxes are washed with washing machine of the animal facility.

3.3 Data Analysis

The tracking software program can determine the following main outcome parameters:

- The time spent in each of the virtual incentive zones (i.e., time spent near the sexual incentive and time spent near the social incentive)
- The latency to enter each of the incentive zones
- The frequency of entering each of the incentive zones
- Mean duration of visits to each of the incentive zones
- The distance moved during the test
- The mean velocity of movement
- The time spent moving

In addition, a preference score can be calculated which is the time spent in the sexual incentive zone divided by the sum of time spent in the sexual and social incentive zones. The time spent in the rest of the arena is thus not included in this calculation. The preference score has a value between 0 and 1 and with 0.5 as no preference for either the social or sexual incentive and above 0.5 to indicate the sexual incentive was efficient to induce sexual motivation. For comparison between incentives, the preference score as well as the time spent in the vicinity of the incentives should differ in order to consider one incentive as different from another. A double criterion is needed in order to avoid false-positive effects. A high preference score could be caused by a very short time spent in the social incentive zone just as well as by a very long time spent in the sexual incentive zone. However, a short time spent in the vicinity of the social incentive zone does not necessarily indicate a superior sexual incentive stimulus. Consequently, the chance of false-positive differences between incentives is reduced by using a combination of both criteria.

Furthermore, the number of entries into each incentive zone as well as the latency to visit and mean duration of each visit can be determined. As indicators of ambulatory activity, we use the total distance moved during the test, the mean velocity of movement while moving, and the time spent moving.

The newest version of EthoVision XT 16 allows for detection of head and tail leading to the exploration of more behavioral outcomes. This gives the possibility to add tracking parameters such as head directions, and distance between nose and predetermined points of interest in the arena (i.e., the wire mesh). In addition, other behavioral observations, such as general mobility, self-grooming, freezing, and rearing, can be made using the video files.

4 Conclusion

Sexual incentive motivation is of the same magnitude in gonadally intact male and receptive female rats and is similarly affected by different types of manipulations. The SIM test is highly suitable to assess sexual incentive motivation and is conducted in the same way for males and females. The many advantages and potentials of the SIM test advocate for this test to be used much more in the study of sexual behavior, for example with different types of incentives (*see* **Note 6**), other animal species (*see* **Note 7**), in translational research (*see* **Note 8**), and in studies with timed interventions or when needing measures to be relatively unpolluted by motor reflexes (*see* **Note 9**).

5 Notes

1. The SIM test can be conducted on the same day as a copulation test. In that case, it is important to perform the SIM test before the copulation test, as prior sexual stimulation can negatively affect sexual incentive motivation [20], as also noted in the introduction. Pre-exposure to sexual odors in the SIM test may have some influence on copulation in the copulation test, but these conditions will be the same for all subject animals (unlike amount of sexual stimulation in the copulation test).
2. It is recommended to not use stimulus animals in the SIM test that have just been used before as stimulus animals in a copulation test. After copulation, they can carry odors from the mate partner, which can mask the odors of the stimulus animal itself.
3. It is possible to modify the stimulus boxes so that the front wire mesh can be removed and the stimulus box opens toward the arena. Now the stimulus animal can enter the arena and studying the transition from motivation to copulation in one single test becomes possible. It should be noted though that if the subject animal is a female, she cannot pace the copulation in this setup. Importantly, this modification has not been tested or taken in use yet.
4. As mentioned in the Subheading 3, different kinds of social stimuli can be used when male or female subjects are tested. In case of a male subject, one can choose between another male rat or a non-receptive female as social stimulus that is paired with a receptive female as sexual stimulus [6]. Ovariectomized females without hormonal priming are usually used as non-receptive female. As receptive female, either an intact female in the appropriate phase of the estrous cycle or an ovariectomized females primed with hormones is used. The primed

ovariectomized female can either be implanted with a 5 mm long Silastic capsule (medical grade Silastic Tubing, 0.0625 in. inner diameter, 0.125 in. outer diameter, Degania Silicone, Degania Bet, Israel) containing 10% 17β-estradiol (Sigma, St. Louis, MO, USA) in cholesterol (Sigma, St. Louis, MO, USA) with medical grade adhesive silicone (Nusil Silicone Technology, Carpinteria, CA USA)-sealed ends (leading to average estradiol serum levels of 75 pg/mL [33]), or primed with a estradiol benzoate 25 μg/rat (in peanut oil) subcutaneous injection 52 h before the test. In both cases, the females should be primed with progesterone 1 mg/rat (in peanut oil) 4 h before testing. Hormone priming protocols may vary in the literature. It is most important that the result is that the female stimulus is sexually receptive, as would be reflected in a lordosis quotient of 100 or more.

5. When testing female subjects, either a castrated male or another female rat can also be used as social incentive stimulus (e.g., [34]), whereas an intact male rat is used as sexual incentive.
6. When studying female subjects, a consideration is whether to study naturally cycling females or hormonally primed ovariectomized females. The advantage of the latter is simplicity, as all females can be tested whenever they are hormonally primed. It has however been shown that the dose of estradiol (in addition to progesterone dose kept constant) affects the preference for an intact male over a castrated male [35] and affects paracopulatory and lordosis behaviors [36]. Therefore, naturally cycling subject females might be the most natural option to not mask possible effects of hormonal manipulations on the assessed sexual incentive motivation. When choosing to study ovariectomized female subjects, it should be considered to lower the dose of estradiol in order to better mimic natural hormone levels (e.g., 18 μg/kg estradiol benzoate [35, 37]). Finally, in female rat copulation studies, intact females are sometimes primed with estradiol only [38]. We have never employed this in the SIM test, but it could be a consideration if surgery is to be avoided. 6. The SIM test is highly suitable to test approach toward any kind of incentive, besides toward behaving animals. We have for example assessed approach behavior toward anesthetized animals, devocalized animals, auditory stimuli (playback of ultrasonic vocalizations), and odors [9, 18, 19]. Odors can for example be introduced to a stimulus box by leaving an animal into the stimulus box for 6 h prior to the test in order to collect urine and feces in the box [9, 19]. Auditory stimuli, on the other hand, can be introduced by placing a loudspeaker in one of the stimulus boxes [9, 18, 19].
7. An advantage of the SIM test is that it should be suitable to be used to test any kind of small animal species. The SIM test has

been successfully used in mice [39], and similar setups have been used for hamsters [40], and prairie voles (although direct contact between animals was also possible in this test) [41].

8. The SIM test can be used as a valid behavioral paradigm to assess sexual motivation in the translational setting. Whereas copulation is quite different in humans compared to rodents, sexual approach is rather similar between these species (for a discussion, *see* [29]). The SIM test is not often used for characterization of effects of pharmacological or other interventions in animals for translational purposes. However, using the SIM test or a very similar setup, it has for example been demonstrated that SSRIs, which can reduce sexual motivation in both human sexes, also reduce sexual incentive motivation in male (fluoxetine; [42]) and female rats (paroxetine; [43]). We advocate for the SIM test to be used in this translational context more often.

9. The SIM test is highly suitable for the study of any kind of intervention, from pharmacology/chemogenetics [28, 44, 45] to optogenetics and photometry. For example, in optogenetic experiments, the setup can easily be exploited for timed intervention when the animal is in any of the zones, even in a "closed loop" with the help of the right tracking software that can be linked to TTL signals. For photometry, the SIM test is especially beneficial because signals will be mostly unpolluted by motor reflexes, in contrast to for example operant response paradigms in which specific motor patterns are required for the performance of the behavior.

References

1. Ågmo A (1999) Sexual motivation--an inquiry into events determining the occurrence of sexual behavior. Behav Brain Res 105(1):129–150
2. Bindra D (1974) A motivational view of learning, performance, and behavior modification. Psychol Rev 81(3):199–213
3. Bindra D (1978) How adaptive-behavior is produced - perceptual-motivational alternative to response-reinforcement. Behav Brain Sci 1(1):41–52. https://doi.org/10.1017/S0140525x00059380
4. Snoeren EMS (2019) Female reproductive behavior. Curr Top Behav Neurosci 43:1–44. https://doi.org/10.1007/7854_2018_68
5. Pfaff DW, Schwartz-Giblin S (1988) The physiology of reproduction. Raven, New York
6. Ågmo A (2003) Unconditioned sexual incentive motivation in the male Norway rat (Rattus norvegicus). J Comp Psychol 117(1):3–14
7. Carr WJ, Loeb LS, Dissinger ML (1965) Responses of rats to sex odors. J Comp Physiol Psychol 59:370–377
8. Lemagnen J (1952) *Les Phenomenes Olfacto-Sexuels Chez Le Rat Blanc. Arch Sci Physiol 6(4):295–331
9. Snoeren EM, Ågmo A (2013) Female ultrasonic vocalizations have no incentive value for male rats. Behav Neurosci 127(3):439–450. https://doi.org/10.1037/a0032027
10. Stern JJ (1970) Responses of male rats to sex odors. Physiol Behav 5(4):519–524
11. Thor DH, Flannelly KJ (1977) Peripheral anosmia and social investigatory behavior of the male rat. Behav Biol 20(1):128–134
12. Geyer LA, Mcintosh TK, Barfield RJ (1978) Effects of ultrasonic vocalizations and males urine on female rat readiness to mate. J Comp Physiol Psychol 92(3):457–462

13. Hetta J, Meyerson BJ (1978) Sexual motivation in male rat - methodological study of sex-specific orientation and effects of gonadal hormones. Acta Physiol Scand 453:5–68
14. Meyerson BJ, Lindstrom LH (1973) Sexual motivation in the female rat. A methodological study applied to the investigation of the effect of estradiol benzoate. Acta Physiol Scand Suppl 389:1–80
15. Barfield RJ, Thomas DA (1986) The role of ultrasonic vocalizations in the regulation of reproduction in rats. Ann N Y Acad Sci 474: 33–43
16. Bialy M, Rydz M, Kaczmarek L (2000) Precontact 50-kHz vocalizations in male rats during acquisition of sexual experience. Behav Neurosci 114(5):983–990
17. Brudzynski SM (2005) Principles of rat communication: quantitative parameters of ultrasonic calls in rats. Behav Genet 35(1):85–92
18. Snoeren EM, Ågmo A (2014) The incentive value of males' 50-kHz ultrasonic vocalizations for female rats (Rattus norvegicus). J Comp Psychol 128(1):40–55. https://doi.org/10.1037/a0033204
19. Ågmo A, Snoeren EM (2017) A cooperative function for multisensory stimuli in the induction of approach behavior of a potential mate. PLoS One 12(3):e0174339. https://doi.org/10.1371/journal.pone.0174339
20. Ågmo A, Turi AL, Ellingsen E, Kaspersen H (2004) Preclinical models of sexual desire: conceptual and behavioral analyses. Pharmacol Biochem Be. 78(3):379–404. https://doi.org/10.1016/j.pbb.2004.04.013
21. Dewsbury DA (1969) Copulatory behaviour of rats (Rattus norvegicus) as a function of prior copulatory experience. Anim Behav 17: 217–223. https://doi.org/10.1016/0003-3472(69)90004-9
22. Meerts SH, Schairer RS, Farry-Thorn ME, Johnson EG, Strnad HK (2014) Previous sexual experience alters the display of paced mating behavior in female rats. Horm Behav 65(5): 497–504. https://doi.org/10.1016/j.yhbeh.2013.12.015
23. Snoeren EM, Veening JG, Olivier B, Oosting RS (2014) Serotonin 1A receptors and sexual behavior in female rats: a review. Pharmacol Biochem Behav 121:43–52. https://doi.org/10.1016/j.pbb.2013.11.017
24. Heijkoop R, Huijgens PT, Snoeren EMS (2018) Assessment of sexual behavior in rats: the potentials and pitfalls. Behav Brain Res 352:70–80. https://doi.org/10.1016/j.bbr.2017.10.029
25. Ågmo A (2002) Copulation-contingent aversive conditioning and sexual incentive motivation in male rats: evidence for a two-stage process of sexual behavior. Physiol Behav 77(2):425–435. https://doi.org/10.1016/S0031-9384(02)00874-0
26. Veening JG, Coolen LM (1998) Neural activation following sexual behavior in the male and female rat brain. Behav Brain Res 92(2): 181–193
27. Kippin TE, Sotiropoulos V, Badih J, Pfaus JG (2004) Opposing roles of the nucleus accumbens and anterior lateral hypothalamic area in the control of sexual behaviour in the male rat. Eur J Neurosci 19(3):698–704
28. Huijgens PT, Heijkoop R, Snoeren EMS (2021) Silencing and stimulating the medial amygdala impairs ejaculation but not sexual incentive motivation in male rats. Behav Brain Res 405:113206. https://doi.org/10.1016/j.bbr.2021.113206
29. Le Moëne O, Ågmo A (2019) Modeling human sexual motivation in Rodents: some caveats. Front Behav Neurosci 13(187). https://doi.org/10.3389/fnbeh.2019.00187
30. Ellingsen E, Ågmo A (2004) Sexual-incentive motivation and paced sexual behavior in female rats after treatment with drugs modifying dopaminergic neurotransmission. Pharmacol Biochem Be. 77(3):431–445. https://doi.org/10.1016/j.pbb.2003.12.008
31. Huijgens PT, Guarraci FA, Olivier JDA, Snoeren EMS (2021) Male rat sexual behavior: insights from inter-copulatory intervals. Behav Process 190:104458. https://doi.org/10.1016/j.beproc.2021.104458
32. Ågmo A (1997) Male rat sexual behavior. Brain Res Brain Res Protoc 1(2):203–209
33. Legan SJ, Allyn Coon G, Karsch FJ (1975) Role of estrogen as initiator of daily LH surges in the Ovariectomized Rat12. Endocrinology 96(1):50–56. https://doi.org/10.1210/endo-96-1-50
34. Chu X, Snoeren E, Södersten P, Ågmo A (2021) Sexual incentive motivation and male and female copulatory behavior in female rats given androgen from postnatal day 20. Physiol Behav 237:113460. https://doi.org/10.1016/j.physbeh.2021.113460
35. Spiteri T, Ågmo A (2006) Modèles précliniques du désir sexuel. Theol Sex 15(4): 241–249. https://doi.org/10.1016/j.sexol.2006.05.001
36. Snoeren EM, Bovens A, Refsgaard LK, Westphal KG, Waldinger MD, Olivier B et al (2011) Combination of testosterone and vardenafil increases female sexual functioning in

sub-primed rats. J Sex Med 8(4):989–1001. https://doi.org/10.1111/j.1743-6109.2010.02177.x

37. Spiteri T, Musatov S, Ogawa S, Ribeiro A, Pfaff DW, Ågmo A (2010) Estrogen-induced sexual incentive motivation, proceptivity and receptivity depend on a functional estrogen receptor α in the ventromedial nucleus of the hypothalamus but not in the Amygdala. Neuroendocrinology 91(2):142–154. https://doi.org/10.1159/000255766
38. Snoeren EMS, Chan JSW, De Jong TR, Waldinger MD, Olivier B, Oosting RS (2011) A new female rat animal model for hypoactive sexual desire disorder; behavioral and pharmacological evidence. J Sex Med 8(1):44–56. https://doi.org/10.1111/j.1743-6109.2010.01998.x
39. Portillo W, Antonio-Cabrera E, Camacho FJ, Díaz NF, Paredes RG (2013) Behavioral characterization of non-copulating male mice. Horm Behav 64(1):70–80. https://doi.org/10.1016/j.yhbeh.2013.05.001
40. Ballard CL, Wood RI (2007) Partner preference in male hamsters: steroids, sexual experience and chemosensory cues. Physiol Behav 91(1):1–8. https://doi.org/10.1016/j.physbeh.2007.01.005
41. Ophir AG, delBarco-Trillo J (2007) Anogenital distance predicts female choice and male potency in prairie voles. Physiol Behav 92(3):533–540. https://doi.org/10.1016/j.physbeh.2007.04.030
42. Kaspersen H, Ågmo A (2012) Paroxetine-induced reduction of sexual incentive motivation in female rats is not modified by 5-HT1B or 5-HT2C antagonists. Psychopharmacology 220(2):269–280. https://doi.org/10.1007/s00213-011-2475-1
43. Matuszcyk JV, Larsson K, Eriksson E (1998) The selective serotonin reuptake inhibitor fluoxetine reduces sexual motivation in male rats. Pharmacol Biochem Be 60(2):527–532. https://doi.org/10.1016/S0091-3057(98)00010-0
44. Snoeren EM, Lehtimaki J, Agmo A (2012) Effect of dexmedetomidine on ejaculatory behavior and sexual motivation in intact male rats. Pharmacol Biochem Behav 103(2):345–352. https://doi.org/10.1016/j.pbb.2012.09.007
45. Chu X, Ågmo A (2016) The adrenergic alpha2-receptor, sexual incentive motivation and copulatory behavior in the male rat. Pharmacol Biochem Behav 144:33–44. https://doi.org/10.1016/j.pbb.2016.02.008

Chapter 10

Pro-Choice: Partner Preference as a Method to Assess Sexual Motivation

Fay A. Guarraci and Sarah H. Meerts

Abstract

The partner preference test has been used to evaluate the neurobiology of sexual motivation in animal models. Despite the variable characteristics of the apparatus used, two key elements are endemic to all versions of this paradigm: approach and choice. The partner preference paradigm relies on approach behavior to indicate motivation to interact with an opposite-sex (sexual stimulus animal) or same-sex (social stimulus animal) conspecific. The partner preference paradigm also requires choice; subjects are free to a make a choice between spending time near or with either of two stimulus animals. In some tests, physical contact is limited whereas other tests allow for full mating interactions. Here, we describe various methodological considerations, highlight critical procedures, and discuss the ways partner preference tests contribute to our overall understanding of sexual function. We focus primarily on the rat as a model organism. In addition, we present new findings that explain apparent sex differences in sexual motivation often observed in partner preference tests. In conclusion, partner preference is a methodology widely used to elucidate the neural, hormonal, and experiential factors underlying sexual motivation in both female and male animals.

Key words Sexual behavior, Lordosis, Mount, Intromission, Ejaculation, Long-Evans rats

1 Introduction

Partner preference tests easily and effectively evaluate sexual motivation in animal models and are a common paradigm used in studies with rats to better understand the motivational components of sexual behavior [1, 2]. The display of copulatory behaviors in male or female animals is preceded by a drive to seek sexual contact. The partner preference paradigm enables scientists to quantify behaviors indicative of sexual drive separately from the act of copulation, which in itself is not sufficient to indicate sexual motivation [3]. Partner preference tests have been used to identify the neural, hormonal, and behavioral factors that contribute to sexual motivation. A key advantage of the partner preference test is that the same protocol can be used with female or male rats, thereby facilitating

Raúl G. Paredes et al. (eds.), *Animal Models of Reproductive Behavior*, Neuromethods, vol. 200,
https://doi.org/10.1007/978-1-0716-3234-5_10,

analysis of sex differences in sexual motivation. A variety of experimental paradigms have been called "partner preference"; however, here we limit our discussion to paradigms involving choice between a potential sexual partner and a potential social partner, not any other choices such as two or more sexual partners or conditioned preferences. When a subject is given a choice between a sexual partner and a social partner, these choice tests are sometimes referred to as incentive motivation tests, sexual preference tests, or partner preference tests; they are essentially the same test, but referred to by different names. In this review, we will discuss the nature and options for subjects and partners (sexual and social stimulus animals), the structural elements of the apparatus used, the multitude of dependent measurements, as well as sex differences that are commonly observed in rats.

2 The Experimental Animals

Partner preference tests are straightforward yet versatile, enabling assessment of sexual motivation in animals under a variety of experimental conditions. Depending on whether the experimental question relates to increased or decreased sexual motivation, the subjects used may be in a state that would produce sexual vigor or in a state of diminished sexual vigor. Male rats that show robust sexual activity are typically gonadally intact, sexually mature (i.e., post-pubertal), and sexually experienced. Whereas prepubertal male rats do not show a sexual preference [4] and gonadectomized male rats show reduced sexual behavior/motivation when tested several weeks following surgery [5]. Furthermore, sexually naïve male rats exhibit reduced preference for a sexual stimulus (c.f., [6]), whereas sexually experienced male rats exhibit distinct preferences for a sexual stimulus [7–13] (c.f [14]). Sexual vigor can be confirmed in males during the partner preference test, if mating is permitted, or after the test, if mating is prevented.

Female rats that show robust sexual receptivity are often gonadectomized to prevent pregnancy but hormone-treated to induce behavioral estrus (i.e., lordosis response to sexual stimulation). Administering hormones to all female subjects holds hormone levels constant across subjects thereby eliminating the influence of natural variation in ovarian hormone secretion. When lack of sexual behavior in female rats is of interest, female subjects are frequently gonadectomized and injected with an oil vehicle to control for handling and stress of injections. Sexual receptivity and sexual motivation are comparable in female rats treated with both estradiol benzoate (EB) and progesterone (P) or in females that are gonadally intact and naturally enter behavioral estrus [15]. In contrast, although sexual receptivity can be induced by 6 consecutive days of 2 μg EB alone, sexual motivation is lower than in rats treated with both EB + P [15]. However, 50 μg of EB alone in gonadally intact females does induce a preference [16, 17]. Hormone priming has the added advantage of providing convenient control of when the female rats can be used for testing, enabling groups of rats to be tested on the same day to reduce potential environmental confounds. When natural variation in circulating hormones is of importance, experimenters can wait for individual intact female rats to enter behavioral estrus. Sexual receptivity can be confirmed during partner preference tests that allow for sexual interaction or following tests that limit physical contact by observing lordosis response to 10 mounts or manual palpation if maintaining sexual naivete is necessary. Using an appropriate subject—male or female rat, maximally sexually motivated or minimally sexually motivated—the role of neural circuits, hormone receptors, and experience can easily be tested by administration of pharmacological agents (e.g., agonists, antagonists, different hormones) or by observing behavior in rats before and after they receive sexual experience.

3 The Stimulus Animals

Choice between a sexual and non-sexual partner is common to all versions of the partner preference test, whether the experimental subject is female or male. Sexually receptive female subjects prefer males over females and sexually experienced male subjects prefer females over males [3, 7, 18]. The sexual partner (i.e., a sexual stimulus animal) is typically a sexually experienced rat that is the opposite sex of the experimental subject and has sufficient circulating gonadal hormones to activate sexual behavior. Therefore, a gonadectomized female stimulus primed with ovarian hormones (EB + P) is commonly used as a sexual stimulus for male subjects. Ovarian hormones are often supraphysiological to ensure sexual receptivity, such as EB (10–25 μg/rat) 48 h (and sometimes again at 24 h) prior to testing, combined with P (500 μg – 2 mg/rat) 4 h prior to testing [1, 19–21]. What matters most is that the female stimulus is sexually receptive. Gonadally intact, sexually mature (i.e., post-pubertal) male rats that have demonstrated reliable copulatory behavior (mounts, intromissions, ejaculations) are the appropriate sexual stimulus for female experimental subjects. Often the stimulus animals are tested immediately prior or at the beginning of the day of testing in a different apparatus to ensure sexual vigor (i.e., achieve an intromission with a stimulus female for the male stimulus animals or display lordosis for the female stimulus animals). To avoid conditioned associations with particular animals, stimulus animals should be unfamiliar to the experimental subjects [22–24]. Using familiar stimulus animals could be a confound when testing subjects repeatedly (i.e., a pre-test and a post-manipulation test). Additionally, placement of the stimulus animals in the apparatus should be random and then alternated to prevent location effects.

The non-sexual social stimulus animal acts as a foil to the sexual stimulus, so should not be sexually active but should provide similar social interaction opportunities. Whether the social stimulus animal is the same or opposite sex of the experimental animal does not affect preference for the sexual stimulus [25, 26]; therefore, the selection of the alternative social stimulus animal should be dictated by the experimental question. For example, if having the two stimulus animals be similar in size is important, then two same-sex stimulus animals would be appropriate [20], whereas opposite-sex stimuli enable investigation of questions regarding sexual orientation [27, 28].

4 The Apparatus

Essential to the apparatus used for partner preference tests is the ability to assess the experimental subject's choice between a sexual and social stimulus. The experimental subject should be able to move freely throughout the apparatus, whereas stimulus animal placement should be restricted and consistent across subjects. The goal is to quantify time spent by the subject in particular locations within the apparatus relative to the stimulus animals; physical proximity to one of the stimulus animals represents motivation to seek contact with that stimulus animal [3].

The nature of the contact permitted between the experimental subject and the stimulus animals affects how interest in a stimulus animal is defined. Restricted access can allow for the exchange of only olfactory, auditory, and visual information whereas unrestricted access can allow for sexual stimulation. We refer to these two phases of the partner preference test as No Contact and Contact, respectively. To restrict physical access to the stimulus animals so no sexual interaction can occur, the stimulus animals can be placed in small wire mesh enclosures which are embedded in the walls of the chamber [see Fig. 1a adapted from [29]] or inside the chamber [see Fig. 1b adapted from [21]]. In some No Contact procedures, stimulus animals are anesthetized to give subjects full access to bodily cues without the behavior of the stimulus animals influencing choice [11, 30]. During a Contact phase of the partner preference test, a tether [see Fig. 1c adapted from [31–34]] can be used to restrict the movement of the stimulus animals and allow the subject to have control of their proximity to the stimulus animals and subsequent sexual contact if they come within reach of the tethered stimulus animals [35]. Alternatively, stimulus animals can be trained to stay behind a barrier in one section of the apparatus with mild punishment (e.g., tapping the animal gently on the nose if it attempts to leave its section), but be permitted to move freely

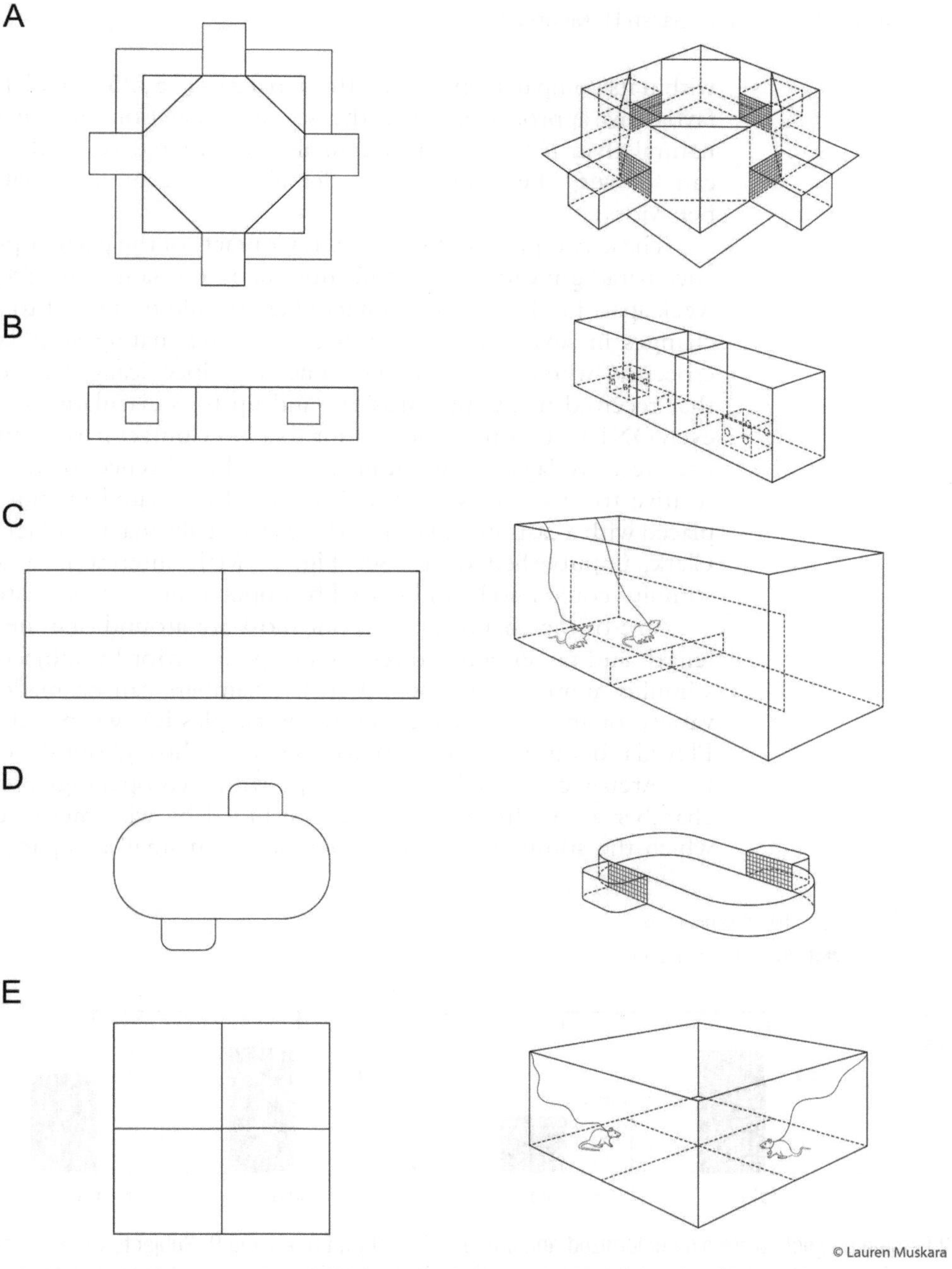

Fig. 1 Artist renderings of each apparatus used by a number of different labs to measure partner preference (Copyright Lauren Muskara, Scientific Illustrator). Line drawings were made to scale and proportional to one another (and proportional to Fig. 3), based on the dimensions and descriptions provided in published articles. (**a**) A dark Plexiglas hexagonal arena with stimulus animals placed in small containers embedded into the walls and covered with mesh [29]. (**b**) A Plexiglas rectangle divided into three compartments, two larger outer compartments and one smaller central area. Each stimulus animal is placed in either outer compartment, but inside a small Plexiglass container with air holes [21]. (**c**) A wooden rectangular arena that is divided in half with tethered stimulus animals at one end [32–34]. (**d**) An elliptical arena with stimulus animals placed in small containers embedded into the walls and covered with mesh [39]. (**e**) A square arena divided into quadrants with stimulus animal tethered in opposite quadrants [19]

within its compartment within the apparatus [15, 25, 36, 37]). We favor this approach because the sexual interaction can be more natural than if the stimulus animals are tethered (e.g., the male can pursue the female; the female can exhibit solicitation behaviors).

These two phases (No Contact, Contact) of the partner preference paradigm can be tested in subjects on the same day [38] or a week apart [15]. The No Contact phase should occur first to avoid changes in sexual motivation after receiving mating stimulation, especially when tested on the same day. Specifically, female rats that received many intromissions and up to 3 ejaculations [1] or even ONLY 10 intromissions prior to a No Contact partner preference test displayed a significantly reduced preference for the male relative to rats that were yoked in time to a mated rat but were placed with a castrated male rat who was sexually inactive (Meerts & Clark, unpublished data). See Fig. 2. Male interest in a sexual stimulus could also be influenced by opportunities to ejaculate [1].

Size of the chambers vary, but most are around one meter in length and tall enough to reduce escape behavior by subjects and stimulus animals (e.g., 30–50 cm). Chambers can be made of a variety of materials (e.g., wood, metal, plastic); we prefer clear Plexiglas because it allows visual observation throughout the chamber. Arenas can be elliptical in shape with two openings into the chamber across from each other, occluded by wire mesh behind which the stimulus animals are contained in small compartments

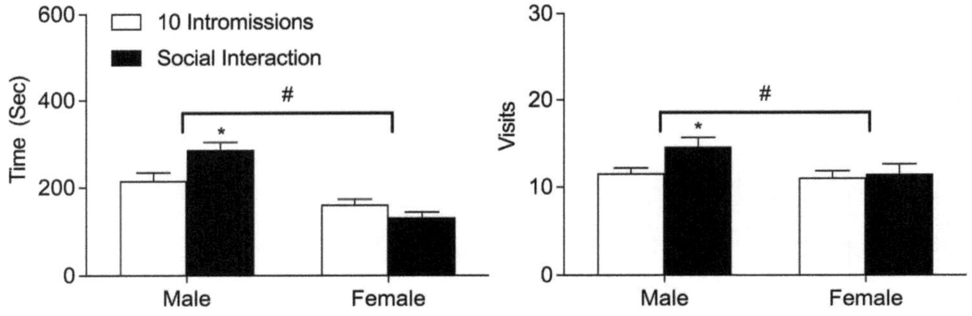

Fig. 2 Female subjects were ovariectomized and primed with 10 μg EB + 1 mg P. Subjects were assigned to receive 10 intromissions from a sexually vigorous male rat (10 Intromissions, $n = 16$) or to spend time with a castrated male rat that showed no sexual behavior (Social Interaction, $n = 16$). Each subject in the Social Interaction group was yoked to a subject in the 10 Intromissions group so that time with the male rat was comparable between the two groups. LEFT: Female rats spent significantly more time with the male stimulus than the female stimulus, independent of group (10 Intromissions or Social Interaction); however, female rats in the Social Interaction group spent significant more time with the male stimulus than female rats in the 10 Intromission group. RIGHT: Female rats made significantly more visits to the male stimulus independent of group (10 Intromissions or Social Interaction); however, female rats in the Social Interaction group made significantly more visits to the male stimulus than female rats tested in the 10 Intromissions group. Data are presented as means (± SEM); n = number of subjects in each group. An asterisk (*) indicates a significant difference between the 10 Intromission and Social Interaction groups

[see Fig. 1d adapted from [39, 40]]. A virtual area adjacent to the wire mesh covered opening is defined as the incentive zone [1]. Another possibility is a square chamber marked by lines on the floor into quadrants with tethered stimulus animals in two opposite quadrants and two empty quadrants [19]; time in each quadrant is recorded [see Fig. 1e adapted from [19]]. Rectangular chambers can also be divided into two long, narrow compartments adjacent to each other with a third smaller compartment located at one end of the chamber [see Fig. 1c adapted from [31]]. Some labs, including our own individual labs, use a rectangular chamber separated into 3 compartments by dividers with holes that permit the experimental animal to move freely throughout the apparatus, while confining one stimulus animal to each outer compartment at opposite ends [see Fig. 3a, b adapted from [15, 25, 36, 37]]. The subject starts in the center compartment and then can choose to enter either of the two outer side compartments. When the experimental animal's entire body is in one of the outer compartments, it is considered to be choosing the stimulus animal present in that compartment. The rectangular apparatus can be used for both phases of the partner preference test, No Contact and Contact, which facilitates comparison of sexual motivation under the two conditions. This apparatus is most similar to the apparatus typically used to measure paced mating behavior in female rats [41]. A critical aspect of each and every apparatus is that the subject has an area where they can retreat and be alone, not near either stimulus animal, otherwise the choice would be artificially forced.

5 The Benefits and Drawbacks of No Contact and Contact Phase of the Partner Preference Test

Both the No Contact and Contact phases of the partner preference test have benefits and drawbacks. Mating is prohibited during the No Contact phase of a partner preference test; therefore, the motivational state of a subject can be assessed independently from the effects of copulatory stimulation. Time in the male compartment is greater for female subjects during the No Contact phase than Contact, because withdrawals from the male exhibited as part of paced mating behavior, reduce time spent with the male stimulus [15, 25, 42]. Thus, it would appear that sexual motivation is reduced by the opportunity to receive mating stimulation when in fact the reduced preference may merely be an artifact of the display of paced mating behavior. Studies have also shown that sexual motivation can be robust during the No Contact phase in non-sexually active, but sexually experienced rats [10, 43]. Furthermore, female rats with lesions of the nucleus accumbens avoid the male when mating is possible, but show robust preference for a sexual partner when physical contact is restricted [19]. Similarly,

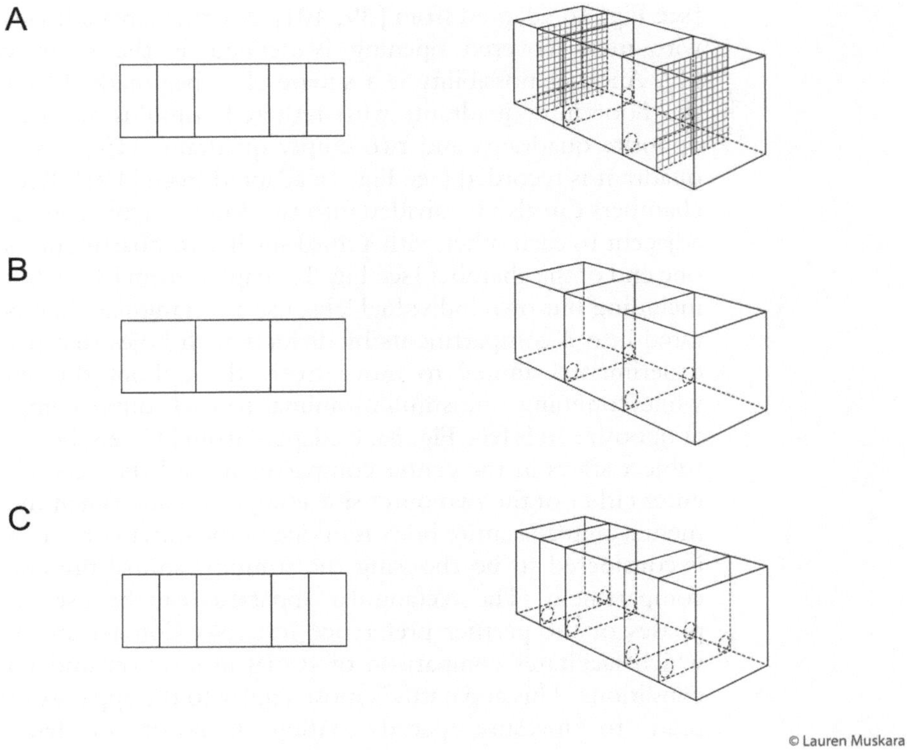

Fig. 3 Artist renderings of each apparatus used in our labs to measure partner preference. Line drawings were made to scale and proportional to renderings depicted in Fig. 1 (Copyright Lauren Muskara, Scientific Illustrator). (**a**) A Plexiglas arena divided into three equal compartments with two clear Plexiglas dividers that have a hole in each bottom corner. The subject is placed in the center compartment and stimulus animals are placed in either of the outer compartments. The outer compartments are further divided in half with a wire mesh divider to limit physical contact between the subject and the stimulus animals from [15, 25, 36, 37]. (**b**) The same Plexiglas arena without the mesh dividers so physical contact is not limited once the subject enters a side compartment. (**c**) The same Plexiglas arena with additional clear Plexiglas dividers subdividing the outer compartments in half to increase control by the subject when approaching the stimulus animals. Entries are counted when the first set of dividers are cleared. Therefore, the stimulus animals have limited mobility within that larger area as each is restricted to the outer most compartments

female rats treated with zaprinast, a phosphodiesterase type-5 inhibitor, exhibit significantly longer contact-return latencies after ejaculation but do not differ from saline controls in their preference for the male stimulus when physical contact is restricted [44]. Motivational states direct an animal to perform particular behaviors to obtain rewards, so understanding whether an experimental manipulation affects an animal's interest in a sexual or social stimulus, without the possibility of mating, is informative.

It is also critical to understand the effect of copulation on sexual motivation. As noted by Henely and colleagues [45, 46], when subjects are only able to exchange visual, auditory, and olfactory information with the stimulus animals, the preference measured cannot be *confirmed* as sexual in nature, because no sexual behaviors are observed. By allowing the subjects to interact with the stimulus animals and experience the consequences of such a choice, we are able to determine the sexual nature of the partner preference. Therefore, the use of both a No Contact phase and a Contact phase, immediately after one another or separated by a week [15], provides information about both the motivational state of the subject with and without the influence of mating stimulation and copulatory behavior.

6 The Procedure

Before beginning data collection, subjects are typically familiarized with the chamber in 2–3 separate, 10- to 15-min sessions. During these acclimation sessions, the areas where the stimulus animals will be placed are empty. Subjects that are familiar with the apparatus will be more likely to investigate the stimulus animals present during an experiment rather than be influenced by the novelty of the apparatus [1]. Stimulus animals should also be acclimated to the apparatus, but at different times from the subjects. Any necessary training of the stimulus animals, such as tapping the stimulus animals on the nose if they attempt to leave their designated location, can be accomplished during acclimation to the apparatus.

Prior to the start of each test, the floor of the chamber is covered with fresh bedding and then the subject and stimulus animals are placed in designated areas of the chamber for 5–15 min. Across many labs, as published in the literature, tests are typically 10 min long [7, 38], with a minimum of 5 min, which is rarely used [20]. Some labs test for slightly longer durations (i.e., 15 min; [4, 29, 32–34, 47], but rarely over 30 min [48]). Sexually receptive female rats show a robust preference for the sexual stimulus animal in the first 20 min of a partner preference test that restricts physical Contact, but then cease to differentiate between the stimulus animals [10], indicating that a test of 10–15 min captures sexual motivation effectively. Although tests can be 10 min in duration, in some experiments with female subjects if the male stimulus ejaculates before the test is over, he can then be replaced with another male stimulus so that mating is possible for the female subject during the entire test period. This procedure is similar to what is commonly used in tests of paced mating behavior when capturing contact-return latency is critical. Tests are typically conducted during the start of the active phase of the light/dark cycle (2–3 h after lights go off) under dim red light. This is ideal for testing naturally cycling female rats, who often go into behavioral estrus within the first hours of transition to darkness [49]. After each partner preference test, the chambers are emptied of bedding and cleaned to remove the scents of the previous animals.

7 The Dependent Measurements

A number of dependent measures can be recorded during both phases of the partner preference test (No Contact and Contact), no matter the size or shape of the chamber or the nature of the stimulus animals. As mentioned above, each apparatus includes designated locations for measuring the subject's proximity to each stimulus animal (i.e., when the subject is in the designated zone it is considered *with* the stimulus animal). Time spent with each stimulus animal, time spent alone, and number of visits made to each stimulus are common measures and the basis of most transformed measures.

Raw time data are commonly transformed into preference scores: a sexual preference score (time spent with the sexual stimulus divided by the sum of time spent with both stimulus animals) and a social preference score (time spent with either stimulus divided by the total test duration) [15]. Another transformation can be calculated as the subtraction of the time with the social stimulus from the time with the sexual stimulus, with a positive score representing a sexual preference [29, 47, 50, 51]. Relative to raw time, transformed data tend to be more homogeneous and allow for comparisons across experiments, with different conditions, different test durations, or even different chambers. Conclusions drawn from either raw data or transformed data are usually similar. Increases in time spent with a sexual stimulus and corresponding increases in preference score for the sexual stimulus can be assumed to indicate high sexual motivation (i.e., the incentive value of the sexual stimulus is greater than the non-sexual, social stimulus). On the other hand, a decrease in time spent with a sexual stimulus and corresponding decrease in preference score for a sexual stimulus can be assumed to indicate low sexual motivation (i.e., the incentive value of the sexual stimulus is less than or equal to the non-sexual, social stimulus). Indications of low sexual motivation can be interpreted as avoidance behavior, especially when combined with greater time spent with the non-sexual social stimulus. However, only when you use raw data can you see if there is avoidance of physical contact with both stimulus animals or avoidance specifically of the sexual stimulus. Using either raw data or preference scores, comparisons can be made between control conditions (e.g., vehicle) and manipulated conditions (e.g., drug). Raw data and preference scores can both be informative and add value to an analysis of partner preference results.

Visits to each stimulus animals are another important dependent measure. Approaching a reward is indicative of interest in obtaining that reward [2]; therefore, visits to the stimulus animals can reflect motivation; more visits to the sexual stimulus relative to the social stimulus is consistent with heightened sexual motivation.

Visits are also a measure of locomotor behavior, but may not be the best measure of general locomotion because sexual behavior and sexual cues can influence locomotion [52]. An open-field test, where the subject is alone, is a better measure of general locomotion. However, when manipulating hormones, experiences, or neurotransmitter levels, having a measure of visits is helpful in screening overall changes in locomotor behavior and may inform conclusions about sexual motivation. For instance, if a rat is hyperactive, an equal but high number of visits might be made indiscriminately to both stimulus animals resulting in no significant preference. Following up with an open-field test could confirm hyperactivity and limit conclusions about motivation.

When physical contact is possible, even more dependent measures can be recorded. For example, the timing and type (mounts, intromissions, ejaculations) of sexual stimulation displayed and/or received can be recorded. From these observations, a number of additional variables can be calculated. For male subjects, frequency of copulatory behaviors, intervals between behaviors, duration of behaviors, frequency, and nature of mount bouts and time-outs [53], and post-ejaculatory interval can be calculated. For female subjects, lordosis can be observed in response to each sexual stimulation received, from which lordosis quotients can be calculated. Using the type and timing of sexual stimulations received by female subjects, measures of paced mating behavior can be calculated (i.e., the likelihood of leaving the male and the time to return to the male after receiving a particular type of sexual stimulation). In addition, pre-copulatory behavior and proceptive behaviors can be observed. These behaviors act to bring the subject and the sexual stimulus in close contact and stimulate copulatory behavior [54]. In male rats, pursuit and anogenital investigation serve to indicate interest in the sexual stimulus [5]. In female rats, ear wiggling and hopping and darting serve to indicate interest, soliciting the attention of the male stimulus, and subsequently the receipt of sexual stimulation (Erskine 1989). From the multitude of observed and computed dependent variables, conclusions about choice behavior, sexual motivation, locomotion, and copulatory behavior can be made.

8 Partner Preference in the Broader Context of Sexual Behavior

The study of sexual behavior is intended to better understand the physiological, neural, and hormonal aspects of sexual interactions. Hypoactive sexual desire disorder affects both men and women, at 15% and 30%, respectively [55]; the factors that underlie fluctuations in sexual motivation are an ongoing area of study. Partner preference is one tool that can be used to illuminate the role that motivation plays in sexual behavior. Linking findings about motivation with display of sexual behavior is particularly informative.

For example, hormones are critical for both robust sexual motivation and copulation. In male rats, castration decreases sexual motivation and the number of intromissions and ejaculations achieved during mating, but testosterone treatment restores both aspects of sexual behavior [1, 5]. After ovariectomy, lordosis cannot be elicited in female rats and interest in the sexual stimulus in partner preference tests is reduced. Robust sexual behavior and sexual motivation can be elicited in female rats after sequential administration of EB + P in ovariectomized females. A main sex difference in the role of gonadal hormones in sexual motivation and sexual behavior is that the temporal effects occur within hours or days for female rats [3], but on the order of weeks for male rats [5]. Nonetheless, a main takeaway is that sexual motivation and copulatory behavior in male and female rats vary similarly with the removal or re-introduction of gonadal hormones.

The study of female sexual behavior advanced beyond simply observing the presence or absence of the lordosis reflex by testing female rats in an apparatus that allows them to control proximity to the male rat, thereby enabling the display of paced mating behavior, a series of approaches and withdrawals by which the female temporally controls receipt of sexual stimulation [54]. Among the dependent measures obtained from tests of paced mating behavior are (a) percentage of exits, the rate of withdrawal from the male following a mount, intromission, or ejaculation; (b) contact-return latency, the time elapsed from receipt of a sexual stimulation to the return to the male; and (c) time with the male, frequently expressed as a percentage of the entire test [56, 57]. Erskine [58] proposed that shorter contact-return latencies reflect increased sexual motivation. To test that idea, we had female rats receive a 10 min partner preference test with limited physical contact (No Contact phase) followed immediately by a paced mating test until 15 intromissions were received (Meerts & Guarraci, unpublished). Regression analysis showed a significant negative correlation between time spent with male stimulus during the partner preference test and contact-return latencies after intromission during the paced mating test, as well as a trend toward a significant negative correlation between time spent with male stimulus during the partner preference test and contact-return latencies after ejaculation during the paced mating test. See Fig. 4. Similarly, administration of caffeine leads to shorter contact-return latencies after ejaculation during paced mating and more visits to the male stimulus during partner preference tests [59], whereas ketamine shortens contact-return latencies after intromission and increases time spent with a male stimulus during partner preference tests [37, 60]. In some cases, time spent with the male stimulus is measured only during the paced mating test (and no partner preference test is conducted). During paced mating tests, more time spent with the male stimulus is associated with shorter contact-return latencies after intromission in sexually

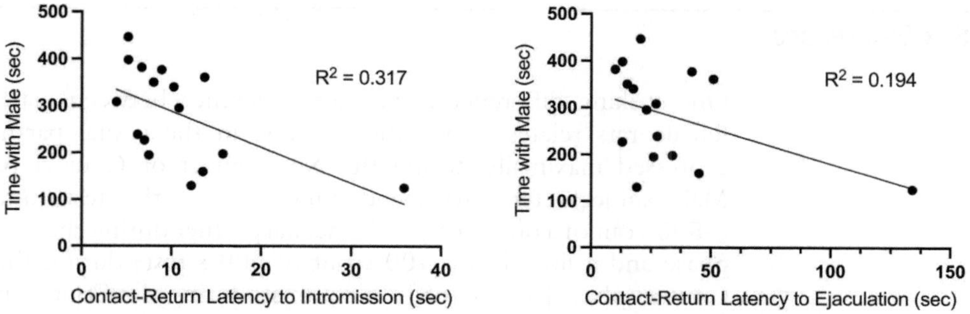

Fig. 4 The relationship between time with the male stimulus (*y*-axis) and contact-return latency (*x*-axis) to intromission (left) or ejaculation (right) is shown. Time with male stimulus was obtained during a 10-min No Contact partner preference test and measures of paced mating behavior were obtained during a separate paced mating test to 15 intromissions on the same day ($n = 16$). As time with male increases, contact-return latency to intromission decreases significantly [$F(1, 15) = 6.5$, $p = .02$); $R^2 = 0.317$]. As time with male increases, contact-return latency to ejaculation shows a trend toward decreasing [$F(1, 15) - 3.4$, $p = 0.09$; $R^2 = 0.194$]

experienced female rats [56]. Collectively, these findings support the idea that different approaches to measuring sexual motivation (partner preference and paced mating behavior) provide information about sexual behavior and also inform each other.

In male rats, experience with a sexually receptive female promotes more efficient mating characterized by shorter latencies to mount, intromit, and ejaculate as well as achieving ejaculation after fewer intromissions [61]. Whereas, sexually naïve male rats fail to exhibit a preference for a sexual stimulus relative to a social stimulus [10, 11]. Exposure to female rat odor increases expression of c-fos in sexually experienced male rats in a number of brain regions linked to male sexual behavior, such as the medial amygdala, the granule layer of the accessory olfactory bulb, and the bed nucleus of the stria terminalis [62]. Ejaculation produces more c-fos in the medial preoptic area of experienced relative to naïve male rats [63]. A subset of otherwise healthy male rats do not mate even after repeated exposure to a receptive female (i.e., non-copulators) [64] or take a long time to ejaculate (i.e., sexually sluggish male rats) [65]. Non-copulators do not exhibit a preference for a sexual stimulus in either No Contact or Contact partner preference tests [14, 62]; however, sexually sluggish male rats, like typically copulating experienced male rats, prefer the sexual stimulus. Apparently, sexual stimulation, even at slow rates, alters male rat brains to promote sexual motivation a finding that relied on partner preference testing.

9 Sex Differences

One striking difference in partner preference between male and female rats relates to whether interest in the sexual partner is expressed maximally during the No Contact or Contact phase. Male subjects tend to spend almost all of the test duration (~500 s out of a 600 s test) with a sexual partner during the Contact phase and a little less (~400 s out of 600 s test) during the No Contact phase [37, 66–68]. In contrast, during the Contact phase when mating is possible, female subjects pace sexual contact at their preferred interval [15]. Because the female subject retreats from the sexual stimulus as she paces, she only spends about 200 s out of the 600 s test with a sexual partner during the Contact phase [37]. Interestingly, when physical contact is limited and no pacing is necessary, female subjects spend a bit more time near the sexual partner (~300 s out of a 600 s test) [37, 41]. Thus, maximal preference for the sexual partner is expressed during the Contact phase for male subjects and during the No Contact phase for female subjects. Moreover, during both phases of the partner preference test (No Contact and Contact), female subjects do not spend nearly as much time with a sexual partner as their male counterparts.

This sex difference in maximal sexual preference may be due to a variation in the incentive value of either stimulus animal. The incentive value of the non-sexual social stimulus may be higher for female rats than for male rats; or the incentive value of the sexual partner may be lower for female rats than for male rats. On the other hand, the sex difference may merely reflect an artifact of female rats pacing sexual contact, which requires them to exit the male compartment [54]. Studies in which the stimulus animals are tethered provide the female subject with the opportunity to be in the incentive zone but just out of reach of the male rat, thus controlling the receipt of sexual stimulation but still considered to be choosing the sexual stimulus [19, 20, 29, 31, 48]. Preference for the sexual stimulus by female subjects appears to be higher or comparable in tests with tethered stimulus animals relative to No Contact [19, 29], whereas preference for the sexual stimulus is higher in the No Contact phase relative to Contact phase of the partner preference test with untethered stimulus animals [15].

In our standard chambers used during the Contact phase of the partner preference test, as soon as the female chooses the male, she can be mounted. We hypothesized that when the female can choose the male but also be able to pace sexual stimulation within the outer compartment, time with the male would be higher than in the standard apparatus. To test this hypothesis, we compared the partner preference for either a sexual or social partner when physical contact was limited and then when physical contact was unrestricted. However, during the Contact phase we added an area just outside of the stimulus animals that could count as being *with* the stimulus but without physical contact for an additional condition. See Fig. 3c. We compared time spent with a sexual stimulus using this arrangement of the chamber to our standard partner preference test, when access to the stimulus animals was unrestricted. See Fig. 3b. If the sex difference is due to an artifact of pacing, then preference for a sexual stimulus should be higher during the Contact phase when an additional antechamber is available to the female subject than during tests without the antechamber. Specifically, if preference for a sexual stimulus is similar to levels observed in male subjects (~500 s out 600 s test), then it is likely that sex differences are an artifact and merely the result of pacing. However, if the female subjects continue to spend less time near the sexual stimulus even with this added antechamber, then it is possible that there are sex differences in the incentive value of the sexual stimulus and/or the non-sexual social stimulus. That is, the sexual stimulus may be preferred more by male rats than female rats; or the non-sexual social stimulus may be preferred more by female rats than male rats.

10 Experiment

10.1 Methods

10.1.1 Animals

Female rats ($n = 16$) bred from adult Long-Evans rats purchased from Envigo (Indianapolis, IN) were used as subjects. Additional male and female rats were purchased from Envigo and used as stimulus animals. Rats were group housed with same-sex cage mates in hanging polycarbonate cages. Aspen wood shavings were used for bedding and a red plastic tube was placed in each cage for enrichment. Food and water were available ad libitum. Temperature and humidity in the animal colony were controlled and monitored daily. The lights in the colony were maintained on a reversed 12:12-h light–dark cycle (with lights off at 10:00 a.m.). Mating tests occurred during the dark phase of the light–dark cycle (starting at 1:00 p.m.), under dim red light. The subjects were weighed weekly. This experiment was approved by the Southwestern University IACUC and followed NIH guidelines.

10.1.2 Surgery

Female subjects and female stimulus animals were bilaterally ovariectomized (OVX) under sodium pentobarbital (50 mg/kg intraperitoneal (i.p.); Henry Schein, Melville, NY) anesthesia [69, 70]. Each female rat was pretreated with atropine sulfate (2.5 mg; Sigma-Aldrich Chemical Company, St. Louis, MO) to reduce respiratory distress and given 10 days to recover following surgery before behavioral testing.

10.1.3 Hormone Treatment

Female subjects and female stimulus animals were administered 10 μg of estradiol benzoate (EB) 48 h and 1 mg of progesterone (P) 4 h prior to behavior testing. Hormones (Sigma-Aldrich Chemical Company, St. Louis, MO) were dissolved in sesame oil and injected subcutaneously in the flank. These doses of EB and P have been shown to produce high levels of sexual receptivity in OVX rats [71].

10.2 Behavioral Procedures

10.2.1 Apparatus

Each partner preference chamber consisted of a Plexiglas arena (91 cm long × 31 cm high × 37 cm wide) and the floor was covered with aspen wood shavings. For the Standard condition, two clear Plexiglas dividers were used to divide the chamber into three equally sized compartments. See Fig. 3b. For the Antechamber condition, each outer compartment was bisected by an additional clear divider. See Fig. 3c. All clear dividers had a 5-cm hole in each of the bottom two corners.

10.2.2 Acclimation

Once female subjects were randomly assigned to the Standard condition or the Antechamber condition ($n = 8$/condition), then each subject was acclimated to the appropriate partner preference chamber apparatus arrangement twice prior to testing. At the start, a female subject was placed into the center compartment and allowed to move freely between the three (Standard condition) or five (Antechamber condition) compartments for 15 min.

Each stimulus animal was also acclimated to the partner preference chambers at least twice prior to testing. At the start of the 15-min acclimation period, a female stimulus animal was placed in one of the outer side compartments and a male stimulus animal was placed in the other outer side compartment. If either stimulus animal attempted to exit through the holes in the clear dividers, they were tapped lightly on the nose to train them to remain in the designated compartment.

10.2.3 Partner Preference Test

Partner preference was tested in two 10-min phases; phase 1: the No Contact phase followed immediately by phase 2: the Contact phase. Female subjects in both the Standard and Antechamber conditions were tested with the same No Contact procedure. During the No Contact phase, the side compartments were further divided in half with wire mesh dividers (30 cm high × 25 cm wide). These wire mesh dividers restricted physical contact between the subject and stimulus animals, but allowed for the exchange of olfactory, auditory, and visual cues (see Fig. 3a). At the start, a female subject was placed into the center compartment of the partner preference chamber and confined there with two solid opaque Plexiglas dividers, which covered the holes in the clear Plexiglas dividers, thereby blocking entry into the two outer side compartments. The outermost side compartments of each chamber held a female or male stimulus animal; the location of which was randomly assigned. Phase 1 started when the opaque dividers were removed and the female subject was allowed to move into either of the side compartments, in close proximity to either stimulus animal. After 10 min, the solid opaque dividers were replaced and the female subject was returned to the center compartment.

For phase 2, the wire mesh dividers were removed from the outer compartment for the Standard condition thereby granting the rats in the Standard condition full access to either stimulus animal through the holes of the clear Plexiglas dividers. For the Antechamber condition, the wire mesh dividers were replaced with clear Plexiglas dividers with holes in the bottom corners so the female subject could enter either the area just outside of the stimulus or travel through holes in the second clear divider and have full access to the stimulus animal. See Fig. 3c. After 10 min, the opaque dividers were replaced and all of the animals were returned to their home cages. The partner preference chambers and dividers were cleaned with Windex and fresh bedding was added before the next subject was tested.

10.2.4 Behaviors Recorded and Calculated

During both phases of the partner preference test, we recorded the timing of entries into and exits from the center compartment into each of the side compartments. A female subject was considered to have entered a compartment when all four paws passed through one of the holes in the clear Plexiglas dividers and into an adjacent compartment. Time spent in each compartment was calculated. For tests with the extra set of dividers (providing an antechamber), entries through only the first set of dividers contributed to the time with the stimulus. Additional behaviors were recorded throughout phase 2 of the partner preference test including, frequency of solicitation behaviors (i.e., hops and darts, ear wiggling), the type and timing of sexual stimulations (i.e., mounts, intromissions, ejaculations), and sexual receptivity, as measured by the lordosis response (LR) to the receipt of sexual stimulation. Each LR was rated on a 4-point scale (0–3), and a lordosis quotient (LQ) was calculated as the percent of LRs ≥ 2 [72, 73]. Paced mating behaviors were also calculated; however, entries and exits from the antechamber to the male compartment were used for these calculations in the Antechamber condition. Percentage of exits was calculated as the likelihood that a subject left the male's compartment following the receipt of each type of sexual stimulation (i.e., mount, intromission, ejaculation). Contact-return latency was calculated as the time (in seconds) that elapsed between receiving each type of sexual stimulation (i.e., mount, intromission, ejaculation), leaving the male's compartment, and re-entering the male's compartment.

Results

We calculated a 2 (Standard vs. Antechamber condition) × 2 (No Contact vs Contact phase) × 2 (male vs female stimulus) repeated measure analysis of variance on each dependent variable (time with stimulus animals and visits to stimulus animals) measured. The analysis of time with the stimulus animals yielded a significant main effect of sex on time spent with the stimulus animals ($F(1,14) = 7.46$ $p < .02$, $\eta_p^2 = 0.35$). Independent of the testing condition (Standard or Antechamber) or phase of the partner preference (No Contact or Contact), female subjects spent more time with the male stimulus than the female stimulus. See Fig. 5 TOP. The three-way ANOVA approached statistical

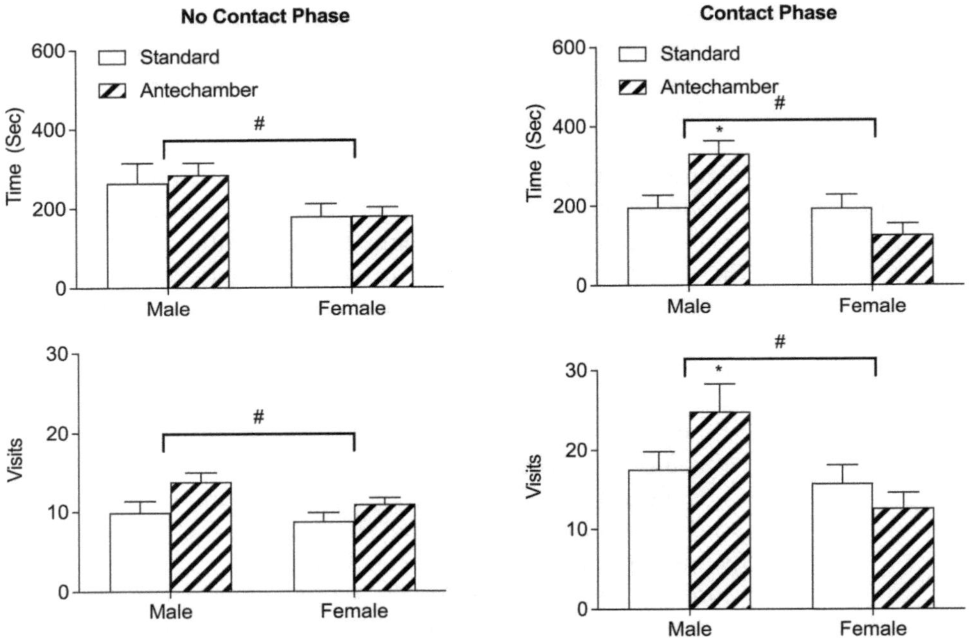

Fig. 5 Interest in a male stimulus during both phases of the partner preference test with restricted physical contact (i.e., the No Contact phase) and unrestricted physical contact (i.e., the Contact phase) is measured with time spent with the male stimulus and visits to the male stimulus animal relative to interactions with the female stimulus. TOP LEFT: Female rats spent significantly more time with the male stimulus than the female stimulus, independent of Testing condition (Standard or Antechamber) or Phase (No Contact or Contact) of the partner preference test. However, female rats tested in the Antechamber condition spent significant more time with the male stimulus than female rats tested in the Standard condition (Standard condition: $n = 8$; Antechamber condition: $n = 8$). BOTTOM LEFT: Female rats made significantly more visits to the stimulus animals during the Contact phase and during the No Contact phase of the partner preference test, independent of testing condition (Standard or Antechamber). However, female rats tested in the Antechamber condition made significantly more visits to the male stimulus than female rats tested in the Standard condition (Standard condition: $n = 8$; Antechamber condition: $n = 8$). Data are presented as means (± SEM) during the entire 10-min test; $n =$ number of subjects in each condition. An asterisk (*) indicates a significant pairwise comparison with testing conditions

significance ($F(1,14) = 3.61$ $p = .078$, $\eta_p^2 = 0.21$). Post hoc Tukey's HSD tests were used to analyze pairwise comparisons. Assignment to testing condition (Standard or Antechamber) had no effect on time spent with either stimulus animal during the No Contact phase of the partner preference test. However, during the Contact phase of the partner preference test, female subjects tested with an antechamber spent significantly more time with the male stimulus than female subjects tested in the standard apparatus ($p < 0.05$). See Fig. 5 TOP.

The analysis of visits to the stimulus animals also yielded a significant main effect of sex on visits to the stimulus animals ($F(1,14) = 8.58$ $p < 0.02$, $\eta_p^2 = 0.38$). Independent of the testing condition (Standard or Antechamber) or phase of the partner preference (No Contact or Contact), female subjects made significantly more visits to the male stimulus than the female stimulus. See Fig. 5 BOTTOM. In addition, there was a significant main effect of phase of the partner preference test (No Contact vs. Contact) ($F(1,14) = 49.55$ $p < .001$, $\eta_p^2 = 0738$). Independent of the testing condition (Standard or Antechamber) female subjects made more visits to either stimulus animal during the Contact phase of the partner preference test than during the No Contact phase. The three-way ANOVA approached statistical significance ($F(1,14) = 3.37$ $p = .078$, $\eta_p^2 = 0.19$). Post hoc Tukey's HSD tests were used to analyze pairwise comparisons. Similar to our results with time spent with the stimulus animals, testing condition had no effect on visits to either stimulus animal during the No Contact phase of the partner preference test. However, during the contact phase of the partner preference test, female subjects tested with an antechamber made significantly more visits to the male stimulus than female subjects tested in the standard apparatus ($p < 0.05$). See Fig. 5 BOTTOM. There were no differences between test conditions on any other measures (contact-return latency, percentage of exits, LQ; all *p's* > 0.05) *see* Table 1.

Table 1
Measures of paced mating behavior during Contact phase

	Contact-return latency		**Percentage of exits**	
	Mount	**Intromission**	**Mount**	**Intromission**
Antechamber	8.6 ± 0.6	26.1 ± 6.1	80.0 ± 20.0	97.9 ± 2.1
Standard	33.2 ± 21.3	34.2 ± 10.2	69.4 ± 16.3	84.2 ± 9.0

Means ± SEM are shown

11 Conclusions

In most partner preference tests, female rats spend equal amounts of time with a sexual stimulus and a social stimulus when physical contact is unrestricted, whereas male rats spend almost all of the test with the sexual partner. We now have evidence to suggest that this sex difference is partly artifactual. We confirmed that the lack of a preference for a sexual partner when physical contact is unrestricted is a consequence of the female rat pacing sexual stimulation. When we provide female rats with an area inside of the male's chamber to escape to during the contact phase of the test, their preference for the male becomes robust, similar to the preference observed when physical contact is limited. Furthermore, measures of paced mating behavior did not differ between the Standard and Antechamber conditions indicating that providing an antechamber does not alter female mating behavior. Nevertheless, female rats, even with the antechamber, still spend less time with the sexual stimulus than what is typically observed in male rats when physical contact is unrestricted. A likely explanation is that the social stimulus for a male subject is an unfamiliar male rat but male rats are frequently aggressive with strange unfamiliar, male conspecifics [74–76], whereas female rats are not aggressive and may even mount another female conspecific [77, 78]. Although there are still differences between male and female rats in terms of maximal preference for a sexual partner, the striking distinction when physical contact is unrestricted vs. when contact is limited is diminished. Finally, female rats tested with the antechamber made more visits to the male stimulus than females tested in the standard chamber, suggesting an increase in preference for the male stimulus. Future studies could further explore the nature of this increased preference as well as the continued presence of sex differences.

We have described the various forms that the partner preference test can take across labs, using different chambers, and different stimulus animals, testing either male or female subjects. We have summarized why certain procedures are critical (e.g., sexual vigor, familiarity, timing, restricted/unrestricted physical contact). We have also highlighted some of the numerous research questions we have asked using the partner preference test. Finally, we have investigated a potential explanation for sex differences commonly observed, finding that female preference for a sexual partner is artifactually reduced by pacing and not a difference in incentive value of the sexual partner. Although the stimulus animals may differ (e.g., gonadally intact, hormone-primed and gonadectomized), chambers may come in different sizes and shapes (e.g., squares, ovals, rectangles), access to stimulus animals may be restricted (with mesh dividers or teethers), and different dependent measures used (e.g., raw time, preference scores), all versions of the partner preference test share a common elemental feature: all offer the subject a choice. It is this choice (between a sexual partner and a non-sexual social partner) that can be used as an indirect measure of sexual motivation and continue to be useful in advancing our understanding of the hormonal and neural basis for sexual motivation as modeled in the rat.

Acknowledgments

We would like to acknowledge Lauran Muskara (BA, BS, Scientific Illustrator) for creating scientific illustrations for the manuscript.

References

1. Agmo A, Turi AL, Ellingsen E, Kaspersen H (2004) Preclinical models of sexual desire: conceptual and behavioral analyses. Pharmacol Biochem Behav 78:379–404
2. Paredes RG (2009) Evaluating the neurobiology of sexual reward. ILAR J 50:15–27
3. Meyerson BJ, Lindstrom L (1973) Sexual motivation in the neonatally androgen-treated female rat. In: Lissak K (ed) Hormones and brain function. Plenum Press, New York, p 443–8
4. Vega Matuszczyk J, Larsson K (1993) Sexual orientation and sexual motivation of the adult male rat. Physiol Behav 53:747–750
5. Pfaff DW, Joels M (eds) (2017) Hormones, brain and behavior, vol 1. Elsevier, Kidlington, Oxford UK
6. Agmo A (2003) Lack of opioid or dopaminergic effects on unconditioned sexual incentive motivation in male rats. Behav Neurosci 117: 55–68
7. Agmo A, Pfaff DW (1999) Research on the neurobiology of sexual behavior at the turn of the millennium. Behav Brain Res 105:1–4
8. Vega Matuszczyk J, Shree Appa R, Larsson K (1994) Age-dependent variations in the sexual preference of male rats. Physiol Behav 55:827–830
9. Hosokawa N, Chiba A (2005) Effects of sexual experience on conspecific odor preference and estrous odor-induced activation of the vomeronasal projection pathway and the nucleus accumbens in male rats. Brain Res 1066:101–108
10. Carr WJ, Loeb LS, Dissinger ML (1965) Responses of rats to sex odors. J Comp Physiol Psychol 59:370–377
11. Stern JJ (1970) Responses of male rats to sex odors. Physiol Behav 5:519–524
12. Olvera-Hernandez S, Hernandez A, Reyes R, Fernandez-Guasti A (2019) Establishment of partner preference in male rats: effect of prenatal letrozole and sexual experience. Horm Behav 109:56–63
13. Garcia-Cardenas N, Olvera-Hernandez S, Gomez-Quintanar BN, Fernandez-Guasti A (2015) Male rats with same sex preference show high experimental anxiety and lack of anxiogenic-like effect of fluoxetine in the plus maze test. Pharmacol Biochem Behav 135: 128–135
14. Portillo W, Paredes RG (2003) Sexual and olfactory preference in noncopulating male rats. Physiol Behav 80:155–162
15. Clark AS, Kelton MC, Guarraci FA, Clyons EQ (2004) Hormonal status and test condition, but not sexual experience, modulate partner preference in female rats. Horm Behav 45: 314–323
16. Haensel SM, Mos J, Olivier B, Slob AK (1991) Sex behavior of male and female Wistar rats affected by the serotonin agonist 8-OH-DPAT. Pharmacol Biochem Behav 40:221–228
17. Snoeren EM et al (2011) A new female rat animal model for hypoactive sexual desire disorder; behavioral and pharmacological evidence. J Sex Med 8:44–56
18. Agmo A (1999) Sexual motivation--an inquiry into events determining the occurrence of sexual behavior. Behav Brain Res 105:129–150
19. Rivas FJ, Mir D (1990) Effects of nucleus accumbens lesion on female rat sexual receptivity and proceptivity in a partner preference paradigm. Behav Brain Res 41:239–249
20. Avitsur R, Yirmiya R (1999) The partner preference paradigm: a method to study sexual motivation and performance of female rats. Brain Res Protocol 3:320–325

21. Wee BE, Francis TJ, Lee CY, Lee JM, Dohanich GP (1995) Mate preference and avoidance in female rats following treatment with scopolamine. Physiol Behav 58:97–100
22. Coria-Avila GA et al (2006) Conditioned partner preference in female rats for strain of male. Physiol Behav 88:529–537
23. Ismail N, Jones SL, Graham MD, Sylvester S, Pfaus JG (2011) Partner preference for strain of female in Long-Evans male rats. Physiol Behav 102:285–290
24. Holley A, Shalev S, Bellevue S, Pfaus JG (2014) Conditioned mate-guarding behavior in the female rat. Physiol Behav 131:136–141
25. Meerts SH, Clark AS (2006) Stimulus animal characteristics do not modulate the expression of partner preference by female rats. Physiol Behav 30:623–626
26. Xiao K, Kondo Y, Sakuma Y (2004) Sex-specific effects of gonadal steroids on conspecific odor preference in the rat. Horm Behav 46:356–361
27. Adkins-Regan E, Mansukhani V, Thompson R, Yang S (1997) Organizational actions of sex hormones on sexual partner preference. Brain Res Bull 44:497–502
28. Vasey PL (2002) Same-sex sexual partner preference in hormonally and neurologically unmanipulated animals. Annu Rev Sex Res 13:141–179
29. Slob AK, de Klerk LW, Brand T (1987) Homosexual and heterosexual partner preference in ovariectomized female rats: effects of testosterone, estradiol and mating experience. Physiol Behav 41:571–576
30. Landauer MR, Wiese RE, Carr WJ (1977) Responses of sexually experienced and naive male rats to cues from receptive vs. nonreceptive females. Anim Learn Behav 5:398–402
31. Williams GW, Goldman J, McGinnis MY, Possidente B, Lumia AR (1991) Effects of ovarian hormones on sexual receptivity, proceptivity, and motivation in olfactory bulbectomized female rats. Physiol Behav 50:751–755
32. Edwards DA, Walter B, Liang P (1996) Hypothalamic and olfactory control of sexual behavior and partner preference in male rats. Physiol Behav 60:1347–1354
33. Edwards DA, Pfeifle JK (1983) Hormonal control of receptivity, proceptivity and sexual motivation. Physiol Behav 30:437–443
34. Edwards DA, Einhorn LC (1986) Preoptic and midbrain control of sexual motivation. Physiol Behav 37:329–335
35. Erskine MS (1985) Effects of paced coital stimulation on estrus duration in intact cycling rats and ovariectomized and ovariectomized-adrenalectomized hormone-primed rats. Behav Neurosci 99:151–161
36. Marshall GE, Guarraci FA, Meerts SH II. (2020) Antidepressants and sexual behavior: acute fluoxetine, but not ketamine, disrupts paced mating behavior in sexually experienced female rats. Pharmacol Biochem Behav 199: 173040
37. Guarraci FA et al (2020) I. Antidepressants and sexual behavior: weekly ketamine injections increase sexual behavior initially in female and male rats. Pharmacol Biochem Behav 199: 173039
38. Winland C et al (2011) Methamphetamine enhances sexual behavior in female rats. Pharmacol Biochem Behav 98:575–582
39. Matuszczyk JV, Larsson K (1994) Experience modulates the influence of gonadal hormones on sexual orientation of male rats. Physiol Behav 55:527–531
40. Vega-Matuszczyk J, Hillegaart V, Larsson K, Ahlenius S (1993) Effects of exposure to an estrous female on forebrain monoaminergic neurotransmission in the non-copulating male rat. Brain Res 630:82–87
41. Blaustein JD, Erskine MS (2002) Feminine sexual behavior: cellular integration of hormonal and afferent information in the rodent forebrain. In: Pfaff DW (ed) Hormones, brain and behavior, vol 1. Academic Press, New York, pp 139–214
42. Memos NK, Vela R, Tabone C, Guarraci FA (2014) Endocannabinoid influence on partner preference in female rats. Pharmacol Biochem Behav 124:380–388
43. Meerts SH, Park JH, Sekhawat R (2016) Sexual experience modulates partner preference and mPOA nitric oxide synthase in female rats. Behav Neurosci 130:490–499
44. Clark AS, Meerts SH, Guarraci FA (2009) Zaprinast, a phosphodiesterase type-5 inhibitor, alters paced mating behavior in female rats. Physiol Behav 96:289–293
45. Henley CL, Nunez AA, Clemens LG (2011) Hormones of choice: the neuroendocrinology of partner preference in animals. Front Neuroendocrinol 32:146–154
46. Henley CL, Nunez AA, Clemens LG (2009) Estrogen treatment during development alters adult partner preference and reproductive behavior in female laboratory rats. Horm Behav 55:68–75
47. Woodson JC, Balleine BW, Gorski RA (2002) Sexual experience interacts with steroid exposure to shape the partner preferences of rats. Horm Behav 42:148–157

48. Gilman DP, Westbrook WH (1978) Mating preference and sexual reinforcement in female rats. Physiol Behav 20:11–14
49. Goldman JM, Murr AS, Cooper RL (2007) The rodent estrous cycle: characterization of vaginal cytology and its utility in toxicological studies. Birth Defects Res B Dev Reprod Toxicol 80:84–97
50. Brand T, Slob AK (1988) Peripubertal castration of male rats, adult open field ambulation and partner preference behavior. Behav Brain Res 30:111–117
51. Brand T, Kroonen J, Mos J, Slob AK (1991) Adult partner preference and sexual behavior of male rats affected by perinatal endocrine manipulations. Horm Behav 25:323–341
52. Ortiz-Pulido R et al (2011) Sexual behavior and locomotion induced by sexual cues in male rats following lesion of Lobules VIa and VII of the cerebellar vermis. Physiol Behav 103:330–335
53. Huijgens PT, Guarraci FA, Olivier JDA, Snoeren EMS (2021) Male rat sexual behavior: insights from inter-copulatory intervals. Behav Process 190:104458
54. Erskine MS (1989) Solicitation behavior in the estrous female rat: a review. Horm Behav 23: 473–502
55. Rosen RC (2000) Prevalence and risk factors of sexual dysfunction in men and women. Curr Psychiatry Rep 2:189–195
56. Meerts SH, Schairer RS, Farry-Thorn ME, Johnson EG, Strnad HK (2014) Previous sexual experience alters the display of paced mating behavior in female rats. Horm Behav 65: 497–504
57. Guarraci FA, Megroz AB, Clark AS (2004) Paced mating behavior in the female rat following lesions of three regions responsive to vaginocervical stimulation. Brain Res 999:40–52
58. Erskine MS (1992) Pelvic and pudendal nerves influence the display of paced mating behavior in response to estrogen and progesterone in the female rat. Behav Neurosci 106:690–697
59. Guarraci FA, Benson A (2005) "Coffee, tea and me": moderate doses of caffeine affect sexual behavior in female rats. Pharmacol Biochem Behav 82:522–530
60. Guarraci FA et al (2018) The effects of ketamine on sexual behavior, anxiety, and locomotion in female rats. Pharmacol Biochem Behav 165:36–44
61. Hull EM (2011) Sex, drugs and gluttony: how the brain controls motivated behaviors. Physiol Behav 104:173–177
62. Portillo W, Paredes RG (2004) Sexual incentive motivation, olfactory preference, and activation of the vomeronasal projection pathway by sexually relevant cues in non-copulating and naive male rats. Horm Behav 46:330–340
63. Lumley LA, Hull EM (1999) Effects of a D1 antagonist and of sexual experience on copulation-induced Fos-like immunoreactivity in the medial preoptic nucleus. Brain Res 829: 55–68
64. Whalen RE, Beach FA, Kuehn RE (1961) Effects of exogenous androgen on sexually responsive and unresponsive male rats. Endocrinology 69:373–380
65. De Gasperin-Estrada GP, Camacho FJ, Paredes RG (2008) Olfactory discrimination and incentive value of non copulating and sexually sluggish male rats. Physiol Behav 93:742–747
66. Davis LK, Bolton JL, Hanson H, Guarraci FA (2020) Modified limited bedding and nesting is a model of early-life stress that affects reproductive physiology and behavior in female and male Long-Evans rats. Physiol Behav 224: 113037
67. Guarraci FA et al (2021) Daily GnRH agonist treatment delays the development of reproductive physiology and behavior in male rats. Horm Behav 132:104982
68. Ali M et al (2020) Neonatal exposure to genistein affects reproductive physiology and behavior in female and male Long-Evans rats. Behav Pharmacol 31:610–621
69. Kirkpatrick ME (2009) The student surgeon: a behavioral neuroendocrinology laboratory exercise in rats. J Undergrad Neurosci Educ 7:A80–A84
70. Khajuria DK, Razdan R, Mahapatra DR (2012) Description of a new method of ovariectomy in female rats. Rev Bras Reumatol 52:462–470
71. Zipse LR, Brandling-Bennett EM, Clark AS (2000) Paced mating behavior in the naturally cycling and the hormone-treated female rat. Physiol Behav 70:205–209
72. Hardy DF, DeBold JF (1971) Effects of mounts without intromission upon the behavior of female rats during the onset of estrogen-induced heat. Physiol Behav 7:643–645
73. Hardy DF, DeBold JF (1972) Effects of coital stimulation upon behavior of the female rat. J Comp Physiol Psychol 78:400–408
74. Albert DJ, Dyson EM, Walsh ML (1987) Competitive behavior in male rats: aggression and success enhanced by medial hypothalamic lesions as well as by testosterone implants. Physiol Behav 40:695–701
75. Albert DJ, Dyson EM, Walsh ML (1987) Intermale social aggression: reinstatement in castrated rats by implants of testosterone

propionate in the medial hypothalamus. Physiol Behav 39:555–560

76. Albert DJ, Dyson EM, Walsh ML, Gorzalka BB (1987) Intermale social aggression in rats: suppression by medial hypothalamic lesions independently of enhanced defensiveness or decreased testicular testosterone. Physiol Behav 39:693–698

77. Fang J, Clemens LG (1999) Vaginocervical stimulation inhibits female-female mounting in laboratory rats. Physiol Behav 67:75–79

78. Fang J, Clemens LG (1999) Contextual determinants of female-female mounting in laboratory rats. Anim Behav 57:545–555

Chapter 11

Copulation in Rats: Analysis of Behavioral and Seminal Parameters

Rosa Angélica Lucio, María Reyna Fuentes-Morales, and Alonso Fernández-Guasti

Abstract

Male copulation in mammals consists of motor and genital components. Two distinct and complementary aspects of copulation are reviewed: masculine sexual behavior and the seminal parameters of the ejaculate. In the first part, we describe in detail how the rat's masculine sexual behavior must be recorded and analyzed, including sexual training tests to have a homogenous population and to identify animals with endogenous phenotypes of copulation (rapid, intermediate, or sluggish). In the second section, we thoroughly explain how to obtain the rat's ejaculate and how to analyze the different parameters. In the final part, we give an example of how these two areas of research may be reported by comparing the seminal parameters in male rats with different endogenous phenotypes of copulation.

Key words Copulatory parameters, Copulatory patterns, Sexually naïve males, Sexually experienced males, Ejaculation pattern, Copulatory phenotypes, Rapid ejaculators, Sluggish ejaculators, Seminal expulsion, Seminal parameters

1 Introduction

Male copulation in mammals consists of motor and genital-external and genital-internal components [1]. The motor component comprises the skeletal musculature that allows the male to climb, clasp, and mount the female, and execute pelvic thrusting movements against the female's rump. The external genital component includes the penile vascular and perineal musculature responses required for the erection of the penis and its insertion into the vagina. The internal genital component comprises the contractile autonomic and somatic activities of organs implicated in ejaculation, i.e., seminal emission and seminal expulsion [2].

Raúl G. Paredes et al. (eds.), *Animal Models of Reproductive Behavior*, Neuromethods, vol. 200,
https://doi.org/10.1007/978-1-0716-3234-5_11,

2 Part A

2.1 Introduction

2.1.1 Male Copulatory Behavior

Behavior is a complex sequence of skeletal movements that occur in response to internal and/or external stimulus. Since those skeletal movements are stereotyped, they can be identified as motor patterns [3, 4]. The display of certain motor patterns is useful to differentiate one behavior from another. Thus, male copulatory behavior is identified by the execution of three motor patterns (mount, intromission, and ejaculation). These patterns differ among males of different species. For instance, male rabbits display pelvic movements during a mount, and when the thrusting is rhythmic, the penis is inserted into the vagina and seminal expulsion occurs. This is known as effective mounts and last less than 3 s. Mounts that do not end in ejaculation are named ineffective [5]. Men, for example, typically engage in a single intromission with intravaginal pelvic thrusting, and ejaculation occurs in approximately 5 min; occasionally, a man can ejaculate more than once [6]. Male rats execute mounts, and mounts with vaginal insertion, called intromissions. After some mounts and intromissions, the male ejaculate also at around 5 min after beginning the sexual activity [7].

2.1.2 Male Rat Copulatory Motor Patterns

Male rats exhibit a highly stereotyped copulatory behavior described in detail since the middle of the previous century [7]. Before starting copulation, the male pursuits the female and sniffs her anogenital region; these responses correspond to precopulatory behaviors (Fig. 1a, b). Three motor patterns identified as copulatory motor patterns shape copulation. The male rat, as other male mammals, mounts the female dorsally and from the rear placing his abdominal and perineal region on the female rump. Mounting behavior can lead to the display of mount, intromission, and/or ejaculation. Penile erection occurs immediately before the male mounts the female, if the penis is not inserted into the vagina, the motor pattern is finished as a mount, but when penile insertion occurs, the intromission pattern is displayed, and detumescence accompanies each dismount [8].

Mount Pattern

The mount pattern is recognized when the male palpates the female's flanks with his forelimbs to induce the lordosis reflex, at the same time performs anteroposterior pelvic movements [3]. These movements have a thrusting frequency of 19–23 Hz on the female's rump that intensify the lordosis posture [9]. Lordosis is the dorsiflexion of the vertebral column, accompanied by the elevation of the head and rump, and lateral deviation of the tail exposing the vaginal orifice. During mounts, the male does not achieve penile insertion. After a non-intromissive mount, the male dismounts slowly from the female (Fig. 1c). Mounts without

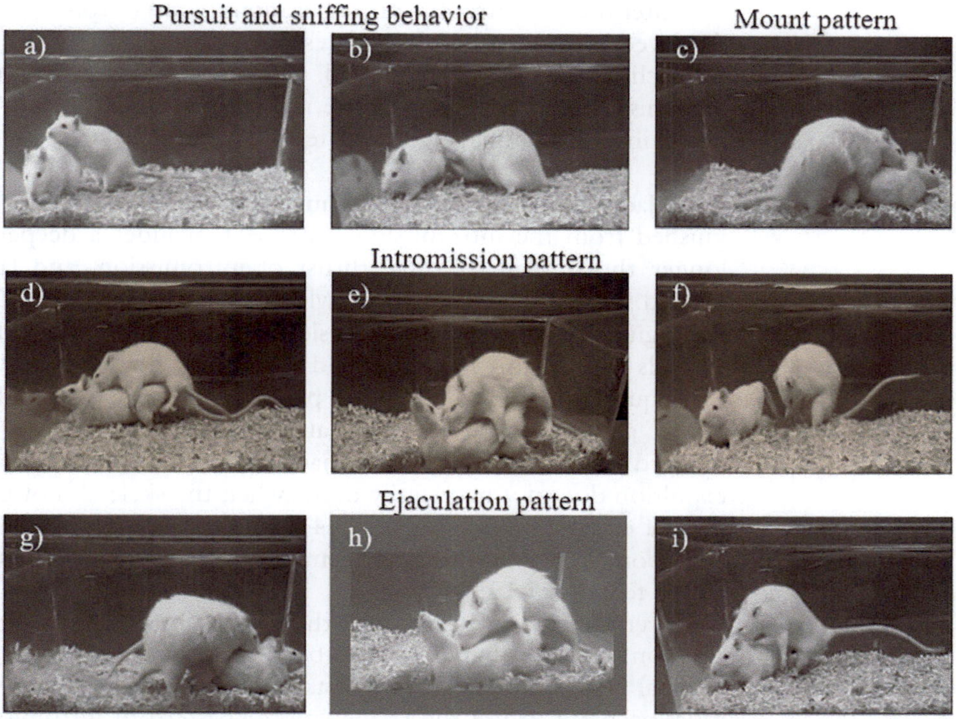

Fig. 1 Photographs of different aspects of masculine sexual behavior in male rats. (**a**, **b**) Precopulatory behaviors, pursuit, and sniffing behavior; (**c**) mount pattern; (**d**–**f**) intromission behavioral pattern; (**g**–**i**) ejaculation behavioral pattern

palpation or without pelvic thrusting are incomplete mounts. After the first mount, other tend to occur with some regularity until the male displays an intromission [3].

Intromission Pattern

The intromission pattern begins like a mount, but in the last pelvic movement, the thrusting is quick and deep when the erect penis is inserted into the vagina. Male rats, as other rodents, present brief penile-vaginal insertions, approximately of 300–400 milliseconds of duration [9]. Dismounting is so abrupt that sometimes causes the male to fall sideways, and genital self-grooming occurs immediately (Fig. 1d–f).

There are treatments that prevent the male from inserting the penis into the vagina, but do not prevent him from executing the characteristic skeletal movements of intromission [10, 11]. It is commonly inferred that penile erection and vaginal insertion occur when the male displays the pattern of intromission. For that reason, penile insertion could be verified using a mirror located at 45° under the copulatory arena [3] or allowing the male to copulate with an inseminated female to dislodge the seminal plug from the vagina [12]. Conveniently, the motor pattern, i.e., the deep pelvic thrusting followed by the abrupt dismount usually is a highly reliable indicator of vaginal insertion [11].

After the first intromission, other mounts and intromissions are displayed before the male ejaculates. Intromissions cause the male to reach the ejaculation threshold. When the stimulation of 6–12 intromissions is accumulated, the male expels semen in the next penile insertion displaying the pattern of ejaculation.

Ejaculation Pattern

The ejaculation pattern is a mount with penile insertion distinguished from the intromission because it includes a deeper and longer thrust than the final thrust of intromission, and lasting around 750–2000 milliseconds, when the ejaculate is placed into the vagina [9]. After seminal expulsion, the male raises his trunk and extends laterally his forelimbs and dismounts slowly (Fig. 1g–i). Subsequently, the male groom his penis and perineal region [13].

In general, when the ejaculation pattern is observed, it is inferred that seminal expulsion had occurred. However, seminal expulsion does not necessarily occur when the skeletal motor pattern of ejaculation is observed [14–16]. Therefore, it is better to corroborate by examining the seminal plug in the vagina of the mated female [3, 12, 17].

Every time an observer uses the terms "intromission and ejaculation"assumes that there has been a penile insertion or/and seminal deposition in the vagina. Unless verified, it would be more accurate to use the terms "motor pattern of intromission" and "motor pattern of ejaculation" and to leave the use of "intromission" and "ejaculation" when the experimenter verifies that indeed a penile insertion and seminal expulsion have occurred together with the corresponding stereotyped movements.

A set of mounts and intromissions that culminates in ejaculation is known as ejaculatory series. Male rats can display many ejaculatory series before reaching sexual exhaustion, ranging from 5 [18] to 18 during a period of 4–6 h with the same or different females [19]. The first ejaculatory series is preceded by about 10 intromissions, while the second and third series may occur after only 3–6 intromissions [8]. Thereafter, the number of intromissions before ejaculation again increases until finally the rat stops copulating or performs repeated mounts without attaining ejaculation. The first ejaculatory series is of long duration compared to the following ones. The second and the third ejaculation latencies are relatively short, and in subsequent ejaculations, this parameter is prolonged [8]. After each ejaculation, the male shows a refractory period being insensitive to sexual stimuli and rather unresponsive to other environmental stimuli. The male lies down, emits ultrasonic vocalizations of 22 kHz [20, 21], and exhibits very little locomotion. The duration of this period of inactivity varies in minutes and is known as the post-ejaculation interval usually defined as the time that elapses between an ejaculation and the intromission of the next ejaculatory series. This interval increases exponentially with each successive ejaculation [22].

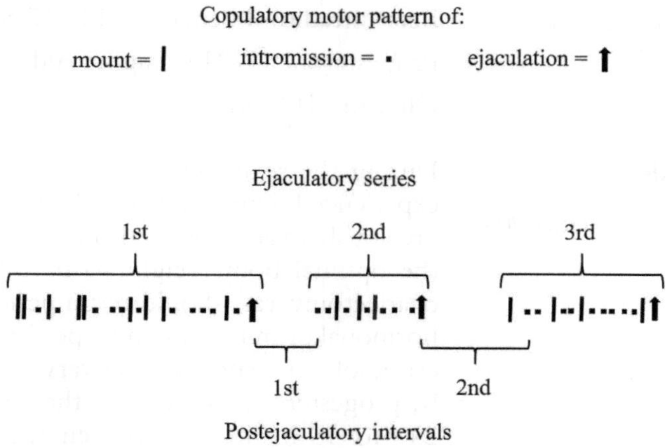

Fig. 2 Schematic representation of the rat copulatory behavioral patterns. Such representation permits an easy visualization of the number of each copulatory pattern per ejaculatory series, their spatial-temporary sequence, the number of ejaculatory series, and their respective postejaculatory intervals

Rat copulatory patterns can be represented graphically using different symbols placed according to the sequence performed during the sexual encounter. Accordingly, it is easy to visualize the number of copulatory patterns per ejaculatory series, as well as the spatial-temporary sequence of each motor pattern. Furthermore, it allows to distinguish the number of ejaculatory series and their respective postejaculatory intervals (Fig. 2).

2.2 Materials

2.2.1 Animals

Male Wistar rats attain sexual maturity between 70 and 100 days old based on sperm production rates and epididymal sperm number [23, 24]. The first mounts occur between 40–45 postnatal day, the first intromissions between 45–75 days old, and the first ejaculations between 48–75 days old [25]. At these ages, male body weights vary from 200–300 grams approximately, i.e., they are young adult rats. Thus, considering sexual maturity, copulation ability and body weight is recommendable to use males of 3 months old at the beginning of the copulatory training. Before starting the copulatory training, females should be ovariectomized and hormonally treated—vide infra: female preparation.

2.2.2 Copulatory Arena, Stopwatch, and Record Sheet

The copulatory arena should be an acrylic cylinder (50 cm diameter × 50 cm height) instead of an acrylic cage. Preferably, the mating arena should be placed in an observation rack or on a table in a separated room adjacent to the vivarium. At least 2 copulatory arenas can be used, each one with a pair of copulating rats. It is highly recommended that rats have a period of copulatory training before being exposed to any manipulation that affects copulatory expression and/or male sexual function. An experienced eye can observe and register two copulatory encounters starting both at the same time using only one stopwatch.

2.2.3 Hormones to Induce Estrus

Beta-estradiol-3 benzoate (E8515 Sigma-Aldrich).

Progesterone (P-0130 Sigma-Aldrich).

Olive oil, 100 mL

2.3 Methods

2.3.1 Female Preparation

During the sexual encounter, it is more practical to use sexually experienced female rats with hormonally induced estrus that were previously ovariectomized. Female Wistar rats of 200 grams have the optimal body weight to be bilaterally ovariectomized. After ovariectomy, rats should rest a period of at least 2 weeks, and the hormonal priming should be performed, i.e., the administration of estradiol-benzoate (in doses varying from 2 to 10 μg/rat) followed by progesterone (in doses in the range of 0.5 to 4 mg/rat) with a 24- to 48-h interval between them. This hormonal priming is needed to sensitize the target organs [26, 27].

One week after the priming, this administration schedule could be repeated. Estradiol-benzoate induces receptive behavior (or lordosis), while progesterone is responsible for producing the proceptive behaviors (hopping, darting, and ear-wiggling that stimulate the male). Females with hormonally induced estrus are receptive during approximately 6 h and can be used for 2–3 copulatory encounters with different males in the same day. At least one week should be elapsed between the administration of estradiol-benzoate and progesterone and the next administration. This hormonal treatment will allow females to show proceptive and receptive behaviors. On the other hand, naturally cycling females may be also used to analyze masculine sexual behavior. Rats in late proestrus, established by a vaginal smear, should be selected.

Preparation of Estradiol-Benzoate (10 μg to Administer 0.1 mL, Subcutaneously)

Weigh 1 mg of estradiol-benzoate on an analytical balance.

Place the weighed estradiol-benzoate in a flask.

Add 10 mL of olive oil to the flask.

Heat (40–50 °C) and dissolve using a magnetic stirrer for 30 min.

Let cool and store in an amber jar at room temperature.

Let the hormone rest for at least 15 days before using it.

Preparation of Progesterone (2 mg to Administer 0.1 mL, Subcutaneously)

Weigh 20 mg of progesterone on an analytical balance.

Place the weighed progesterone in a flask.

Add 1 mL of olive oil to the flask.

Heat (40–50 °C) and dissolve using a magnetic stirrer for 30 min.

Let cool and store in an amber jar at room temperature.

Let the hormone rest for at least 15 days before using it.

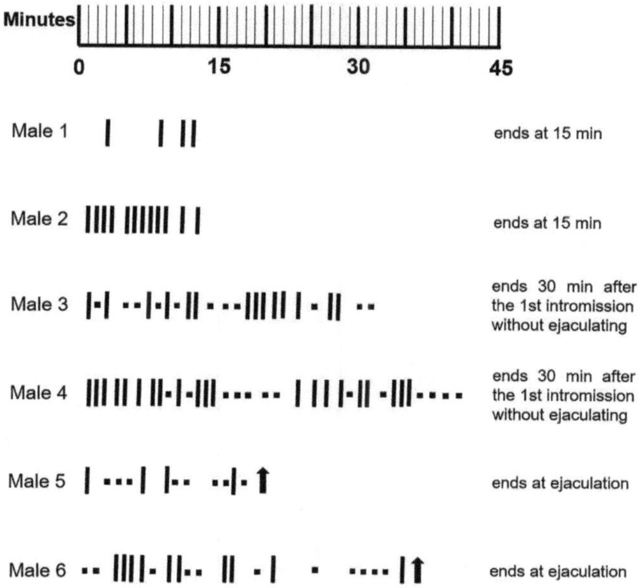

Fig. 3 Schematic representation of different situations where masculine sexual behavior is recorded (see time bar on its upper part): males 1 and 2 only displayed mounts during the first 15 min of the test; males 3 and 4 displayed different number of mounts and intromissions but failed to display the ejaculatory pattern during the 30 min that followed the first intromission; males 5 and 6 displayed the behavioral ejaculation pattern

2.3.2 Duration of Tests for Copulatory Training

Males improve their sexual performance if they are trained sexually [28]. This male rat's population usually is much more homogeneous in its sexual performance than rats selected without sexual training. In addition, in all populations there are non-copulating and sluggish animals [29], which, if undetected, may veil the results of a study. Training tests end if (i) the male does not display copulatory activity or if males only display mounts in 15 min, or (ii) 30 min after the first intromission even if the male fails to display the ejaculatory pattern, (iii) the male displays the ejaculation pattern (Fig. 3).

2.3.3 Copulatory Training Tests Frequency

Copulatory encounters (a single ejaculatory series) should be carried out with intervals of at least two days. Shorter intervals, i.e., every third day or larger intervals, i.e., more than seven days are not recommended because the males display poor copulatory motor patterns, particularly if the males are in the training period. It is very useful to schedule the copulatory tests together with the corresponding hormonal administration and to consider three groups of females for the same group of males.

In the first encounter, usually some males display mounts or mounts and intromissions, but few display the complete sexual behavior repertoire, mounts, intromissions, and ejaculation.

Training calendar

Monday	Tuesday	Wensday	Thursday	Friday	Saturday	Sunday
						E2 to the females Group A
	P4 to the females Group A 1st training test	E2 to the females Group B		P4 to the females Group B 2nd training test		E2 to the females Group C
	P4 to the females Group C 3rd training test	E2 to the females Group A		P4 to the females Group A 4th training test		E2 to the females Group B
	P4 to the females Group B 5th training test	E2 to the females Group C		P4 to the females Group C 6th training test		

Fig. 4 Example of a calendar for training animals for masculine sexual behavior. Observe that females should receive estradiol-benzoate 48 h before progesterone that should be administered 4 h before the tests. Males should rest for a couple of days before being retested

Usually, most males ejaculate from the second test onwards. If males show the ejaculation pattern only in two or three copulatory tests of a total of four, it is preferable to extend the training period to six tests (Fig. 4). In our experience, males must ejaculate in at least four consecutive tests with an interval of 2–4 days between them [30, 31].

Male rats can be considered sexually experienced when they copulate regardless of the length of their ejaculation latencies. However, some researchers are interested in animals that show short or long latencies, and even in those that fail to ejaculate despite copulatory training. In a rat population, there will be male rats with short, intermediate, and long ejaculation latencies and correspond to endogenous ejaculatory phenotypes [32–35]. It should be mentioned that rapid males show short ejaculation latencies from the first, second, or third tests onwards. Sluggish males show complete copulatory activity, i.e., mounts, intromissions, and ejaculation in the last training tests, for example, in the sixth or seventh training. These males ejaculate after long ejaculation latencies and require several sexual trainings. There is undoubtedly a continuum in the length of ejaculatory latencies in a population of rats, so if the interest is to use males of a certain phenotype, particularly those at the extremes, i.e., rapid, and sluggish, it is convenient to test them for at least four tests, once they have presented a stable ejaculatory phenotype.

Sexual training tests should not be considered as control tests. Once males are identified as sexually experienced, they must be summited at least to one control test after one or two weeks after the training. Some sexually experienced males show and maintain the ejaculatory phenotype from the third training test onwards, but other males do not [32]. Thus, there are males with stable and others with unstable ejaculation latencies [35]. Males may be selected accordingly if the study requires the use of male rats with a particular ejaculatory phenotype.

2.3.4 Recording the Copulatory Parameters

Since rats are nocturnal animals, all copulatory tests must be carried out during the dark phase of light–dark cycle inside the vivarium, usually under reversed light/dark cycle (12/12 h) conditions (otherwise the experimenters must come at night to do the observations). Although rats may display copulatory activity at any daytime, it is better to do the observations always at the same time, usually during the second third of the dark phase. A red lightbulb, placed 1 meter away from the copulatory arena, is enough for the observer or the video camera to register the copulatory parameters.

First, a male should be introduced into the copulatory arena, containing a thin layer of sawdust, during 5 min for his habituation inside the copulatory arena. If the male and the female are introduced into the arena at the same time, the male is distracted by sniffing the environment, the sawdust, etc., and that would modify the copulatory test. Once the male is habituated, an estrous female must be introduced, and the stopwatch initiated; at this time, registration begins.

2.3.5 Copulatory Parameters and Other Copulatory Measurements

The male copulatory motor patterns of mount, intromission, and ejaculation are registered and temporarily analyzed as follows, mount latency (ML), intromission latency (IL), and ejaculation latency (EL), as well as the number of mounts (NI) and number of intromissions (NI) preceding ejaculation. Using a stopwatch and a record sheet, these parameters can be viewed sequentially as they occur. At the end of the copulatory session, it is convenient to obtain the data of the registered parameters (Fig. 5). The copulatory parameters allow an interpretation of the copulatory activity.

Mount Latency (ML)

Mount latency (ML) is the interval in seconds from the start of a test (when the female is introduced into the copulatory arena) to the first mount. This parameter is considered a measure of male sexual motivation. Short latencies indicate a higher sexual motivation. In general, sexually naïve males present longer mount latencies than sexually experienced males.

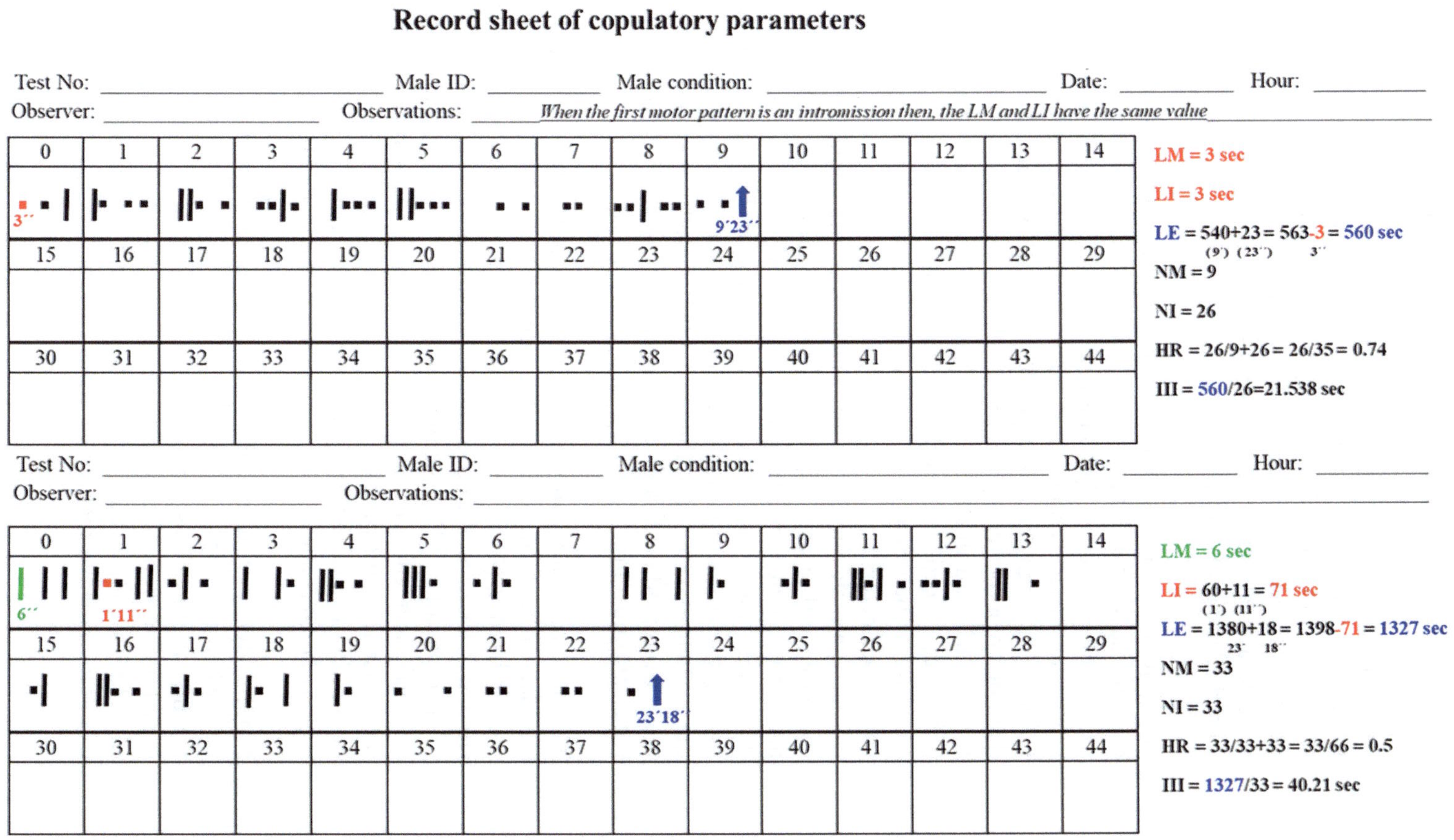

Fig. 5 Examples of record sheets that may be used to register masculine sexual behavior

Intromission Latency (IL)

Intromission latency (IL) is the interval in seconds from the start of a test (when the female is introduced into the copulatory arena) to the first intromission. The intromission latency indicates the time that the male requires to present the first penile erection that allows him to insert the penis into the vagina. Commonly, sexually experienced males have intromission latencies of a few seconds. Furthermore, they could initiate the intercourse with an intromission instead of a mount. In this case, the time of the intromission latency is equal to that of the mount latency. This parameter is also considered a measurement of sexual motivation.

Ejaculation Latency (EL)

Ejaculation latency (EL) is the time in seconds from the first intromission until ejaculation. As males gain sexual experience, they can reduce the duration of the ejaculation latency. However, within sexually experienced males some consistently show short or long ejaculation latencies, being classified as rapid, intermediate, or sluggish ejaculators.

Number of Mounts (NM)

Number of mounts (NM) is the number of this motor pattern (when the male palpates the females' flanks and displays pelvic thrusting) during an ejaculatory series. An increased number of mounts may indicate difficulties in achieving a penile erection.

Number of Intromissions (NI)

Number of intromissions (NI) is the number of this motor pattern (when the male mounts the female and insert the penis into the vagina) during an ejaculatory series. The number of intromissions is considered to reflect the penile sensitivity to reach the ejaculatory threshold. Usually, an experienced male rat shows a greater number of intromissions than the number of mounts. In the first series of copulation, this number is around 10.

Not only the conventional copulatory parameters are needed for the analysis of copulatory behavior, also some mathematical operations are included to obtain the hit rate, and the interintromission interval.

Hit Rate (HR)

Hit rate (HR) also called intromission ratio or copulatory efficiency is an index where the value is between 0 and 1. The formula is $HR = NI/NI + NM$. The hit rate is an indirect measure of erectile potential. When the number of mounts is greater than the number of intromissions, the erectile potential is closer to 0, and when the number of intromissions is greater than the number of mounts, the value is closer to 1.

Interintromission Interval (III)

Interintromission interval (III) is the time elapsed between intromissions expressed in seconds. The formula is: $III = EL/NI$. A shorter III will result in a short EL, and conversely, a longer III will result in a long EL.

It should be noted that all these fine parameters of male sexual behavior must be calculated if the male has achieved ejaculation that usually is considered the top claim of masculine sexual behavior. If the rats do not reach ejaculation, the data must be presented as percentage of males showing mounts, intromissions, and ejaculations. Occasionally, if a given treatment inhibits ejaculation or prolongs the EL, an arbitrary value (for example of 30 min) may be given to be able to calculate the rest of the fine parameters previously described. This procedure should be clearly indicated to avoid confusion.

2.4 Notes

Cylindrical Arenas Instead of Square Arenas: It is advisable to use cylindrical copulatory arenas to facilitate the male to pursuit and mount the female without any impediment. When square copulatory arenas are used, sometimes the female can stay in the corners preventing mounts or intromissions and then altering the values of the copulatory parameters. Copulation in rats, as in many species, has aversive and rewarding components [4]. Therefore, females may show kicking, boxing, or other defensive behaviors toward the male. Females treated with an optimal sequence of ovarian hormones or in natural estrus show few or any rejecting behaviors [36].

Estradiol-Benzoate and Progesterone Instead of Only Estradiol-Benzoate: When males are being sexually trained, they require more sensory stimulation, such as visual stimulation provided by proceptive behaviors. The execution of these behaviors (hopping, darting, and earwiggling) is due to progesterone. For this reason, it is suggested to induce estrus by the administration of estradiol-benzoate followed by progesterone [36]. It should be mentioned that lordosis behavior can also be induced only with repeated doses of estradiol-benzoate, 1.25 mg during 3 consecutive days and tested 24 h later [37]. However, females only will show receptivity but not proceptivity.

Males Control the Pacing of Copulation: Under these conditions, males control the pacing of copulation that is the timing of copulation is under the control of the male. Usually, this pacing is highly rewarding for the male but rather aversive for the female. In natural and seminatural conditions, females pace the rate of the vagino-cervical stimulation they received; in these conditions the aversive properties of mating are reduced [38]. Pacing consists in the intermittent approaches toward and withdrawals from the male [39].

3 Part B

3.1 Introduction

3.1.1 The Ejaculate

The ejaculate corresponds to the expulsion of the seminal fluid deposited in the female reproductive tract, in mammalian males by an intromittent organ, the penis [40, 41]. Nonetheless, the ejaculate can be obtained by masturbation and collected in a container, as occurs in men [42]. Also, by electroejaculation in men with anejaculation, in some clinical cases of spinal cord injury and in farm animals [43].

Seminal fluid is also called semen and is composed of seminal plasma and sperm [44]. Seminal plasma is the collection of secretions from the accessory sexual glands. The variety, number, shape, and size of these glands depend on the species. For example, some marsupials, particularly wallaby, have a prostate and three pairs of bulbourethral glands. In contrast, rodents—and particularly rats—present prostate (three paired lobes), a pair of seminal vesicles, coagulating, and bulbourethral glands [45].

3.1.2 Seminal Emission and Seminal Expulsion, the Two Phases of Ejaculation

Theoretically, ejaculation associated to copulatory behavior refers to the vigorous expulsion of seminal fluid from the proximal urethra to the outside through the urethral meatus [46]. For this, it is required that the components of the seminal fluid that is, sperm (from the epididymal cauda) and seminal plasma (from the accessory sexual glands) come together into the proximal urethra. This process corresponds to seminal emission that also includes bladder neck closure [47]. During the second phase of ejaculation, the seminal fluid located in the proximal urethra (prostatic urethra) travels through the membranous urethra, the distal urethra (penile urethra), and is then impelled through the urethral meatus [47]. Rhythmic pelvic and perineal striated muscle contractions are responsible for the expulsion of seminal fluid. Contractions of the perineal musculature, particularly of the bulbospongiosus, play a major role in seminal expulsion in men [48] and rats [49].

3.1.3 The *Rat*'s Ejaculate

For many years, the male rat has been the laboratory animal used to understand the neural and endocrine mechanisms that allow the execution of male sexual behavior, but also the male sexual functions, i.e., penile erection and ejaculation [50–52]. However, it is striking that little is known about the characteristics of the rat′ s ejaculate.

Only a handful of studies have analyzed the number of sperm in the ejaculate obtained from the inseminated female [19, 30]. No seminal parameter has been evaluated in the semen obtained directly from the male rat.

Obtaining the seminal fluid for macroscopic and microscopic evaluation is called spermatobioscopy. If the sample is obtained from the male is known as direct spermatobioscopy, and it is named indirect spermatobioscopy if obtained from the inseminated female. As far as we know, techniques such as electroejaculation or artificial vaginas have not been developed to obtain semen directly from the male rat. Thus, we obtain the semen from the uterine horns of mated female rats. Although the male deposits the seminal fluid into the vagina, immediately a vaginal plug is formed, and the sperm cells reach the uterine horns. This transcervical sperm transport occurs because the seminal plug is attached strongly to the vaginal walls and cervix [54, 57]. If the seminal plug is removed in less than 5 min, the transcervical sperm transport from the vagina to the uterus horns is interrupted [12]. The postejaculatory interval that usually lasts 5 min after the first ejaculation prevents the male from removing his own plug. However, other males can remove the plug of the male that preceded him when performing 2–3 intromissions. Once the male removed the plug, he can deposit his own semen and promote pregnancy with his gametes.

3.2 Materials

3.2.1 Equipment

Thermo-bath (Felisa, 1–80 °C)

Mixer (Vortex, Type 37600 Touch/On-Off)

Optic microscope (Optiphot-2 Nikon)

Video camera (Digital ELPH, Canon, Power Shot SD900)

Computer (HP Pavilion TV PC)

Analytical balance (Mettler AE 50)

Digital caliper (MyCAL lite)

Differential digital cell counter (Conductronic M220)

Electric razor

Stopwatch

3.2.2 Others

Surgical instruments (operating scissors, scissors blunt-blunt points, tissue forceps, hemostatic forceps, and thin spatula)

Glass Petri dish (9 cm diameter)

Microcentrifuge tubes (1.5 microliters)

Neubauer hemocytometer (Tiefe depth profounder 0,100 mm & 0,0025 mm^2)

Coverslip of the Neubauer hemocytometer

Slides and cover-glass (75 × 25 mm and 22 × 22 mm, respectively)

Silk thread (Suture 33SS7, Gauge3-0)

Strips of pH indicator paper (for viscous substance, range 5.5–9.0)

Micropipettes (10 microliters)

Pipette tips (10 microliters)

Chopsticks

Immersion oil

Syringes 1 mL

3.2.3 Reagents

Sodium pentobarbital (Pfizer)

Xylazine (used for sedation, anesthesia, muscle relaxation; Prozin, Pisa)

Ketamine (used for induction and maintenance of anesthesia; Cheminova)

Sodium chloride crystal (Baker 3624-01)

Yellowish eosin (Hycel CAT 982)

Nigrosine (Hycel CAT 404)

3.3 Methods

3.3.1 Surgical Procedure to Obtain Uterine Content of Mated Females

Once the ejaculation behavioral pattern is observed, the male rat is removed from the copulatory arena, while the inseminated female remains there for at least 5 min (to allow transcervical sperm transport from the vagina to the uterine horns). Then, the female rat is removed from the mating arena and taken to the laboratory to be anesthetized with ketamine (15 mg/kg) and xylazine (1 mg/kg; intraperitoneally). After shaving the abdominal wall, the female is placed in a supine position. An incision is made over the midline of the abdominal wall, including skin and striated muscles (Fig. 6a). The uterine horns are located and ligated with silk thread at their proximal and distal ends. It is recommended that the ligature threads are approximately of 6 centimeters long to facilitate the manipulation of the uterine horns (Fig. 6b). Once the uterine horns are tied, they are removed from the abdomen and placed in a Petri dish containing 0.09% of saline solution at 36 ± 1 °C (Fig. 6c). This temperature is adequate to keep the sperm in good condition. Lower or higher temperatures would cause changes in the evaluation of the ejaculate, particularly sperm motility and viability. Using dissecting forceps and blunt fine-tipped scissors, the fat tissue and blood vessels—adhered to the uterine external wall—are dissected away. The fatty and vascular tissue must be detached with extreme care to avoid leakage through the uterine wall resulting in loss of semen. Once both uterine horns are free of adjacent tissue (Fig. 6d), they are removed from the Petri dish and dried with absorbing paper. An incision is made at one proximal end to pour by applying light pressure to the content into a microcentrifuge tube. Then, the same is done with the other uterine horn. The entire content of each uterine horn is obtained by squeezing from the cervix to the fallopian end (Fig. 6e). Once the content of the uterine horns has been expelled into the microtube, it is necessary to mix in a shaker for about 20 s, approximately, and then keep in a thermo-bath at 36 ± 1 °C.

3.3.2 Seminal Parameters by Indirect Spermatobioscopy

The macroscopic parameters are semen color, semen pH, and seminal volume. In the handful of studies evaluating rat ejaculate, none of the macroscopic parameters have been assessed, although they are easy to evaluate.

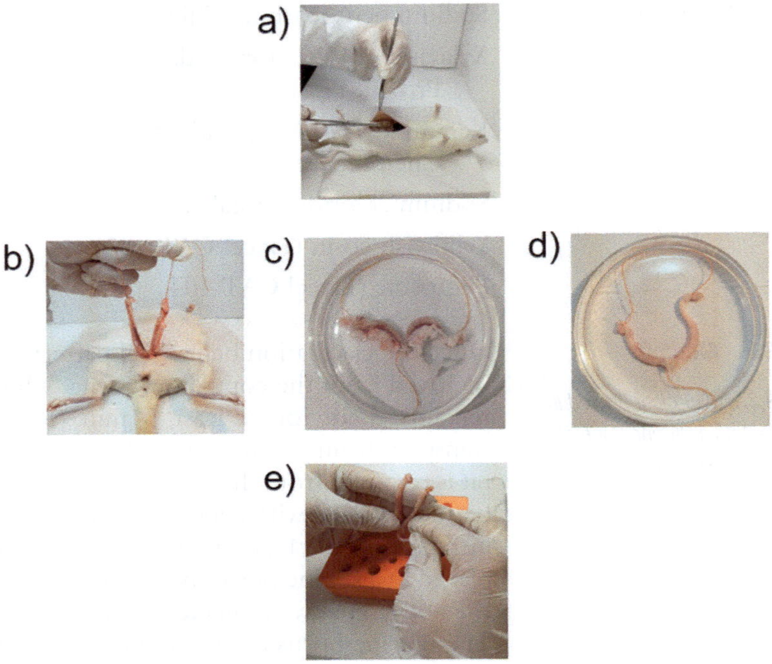

Fig. 6 Removal of the uterine horns to obtain the ejaculate. (**a**) The female in which the male has ejaculated is anesthetized and placed in supine position on the surgical table to make an incision on the midline of the abdominal wall. (**b**) The uterine horns are located and ligated at their proximal and distal ends. (**c**) Both uterine horns are removed from the abdomen and placed in a Petri dish containing saline solution. (**d**) The uterine horns are cleaned; that is, fat and blood vessels are removed. (**e**) The content of each uterine horn is obtained by squeezing from the cervix to the fallopian end

Semen Color

Semen color is the color of the fluid sample contained in the microcentrifuge tube. There are three options by direct observation, whitish, translucent, or bloody. The most common is whitish, indicating the presence of sperm in the sample. The second possibility rarely occurs and means the absence of sperm (Fig. 7a). Finally, if the sample is bloodied, it should be discarded, because it indicates that the uterine horns were not cleaned properly, and the parameters to be evaluated will not be correct.

Semen pH

Semen pH is the alkalinity or acidity of the seminal fluid obtained from the uterine horns. Using a micropipette, take 10 microliters of the collected semen and release it on the strip pH indicator paper for viscous substances (Fig. 7b). After 1 min, determine the pH value by comparing the obtained color of the strip with the standard colors printed on the box.

Because the seminal fluid samples of the male rat are usually obtained from the uterine horns of the inseminated female, the

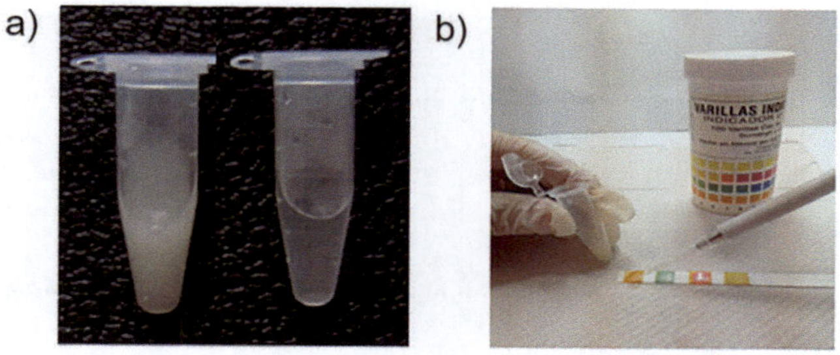

Fig. 7 The macroscopic parameters of semen. (**a**) The normal color of semen is whitish. A transparent sample is abnormal indicating absence or low sperm concentration. (**b**) The pH of the semen is evaluated using pH indicator strips, a 10 μL of semen is placed on each square of the indicator strip, and after 1 min, the strip is compared with the standard colors to determine the pH

seminal volume cannot be quantified. The uterus of an estrous female rat contains luminal fluid secreted by the endometrium and oviduct [58]. Thus, there is a large amount of luminal fluid but can vary among different females, despite all being in behavioral estrous [59]. Different quantities of luminal fluid would affect the quantification of seminal volume.

The microscopic parameters are sperm motility, sperm morphology, sperm viability, and sperm count. Of all of them, the one that has been evaluated in the handful of studies is sperm count [53–55].

Sperm Motility

Sperm motility is the individual movement of each sperm distinguishing three types: progressive motility (spermatozoa with lineal forward displacement), in situ motility (spermatozoa with circular or local displacement), and immobile (spermatozoa without displacement).

Using a micropipette, take 10 microliters of fresh collected semen from the microcentrifuge tube. The content is released on a slide and a cover-glass is placed over the sample to be analyzed using an optical microscope. The slide must be new and non-gelatinized to avoid adhesion.

The observation is carried out from left to right following a horizontal line in the middle of the microscope eyepiece. Another manner is to film the sperm motility using a video camera connected to a computer and placing an acetate on the screen with a drawn horizontal line. Only the sperm cells that touch the horizontal line must be considered (Fig. 8). The film may serve to confirm the data obtained by direct examination from a second blind observer. In different fields, 100 spermatozoa should be observed using the 20× microscope objective and a digital counter. Data are expressed as percentage, i.e., the percentage of spermatozoa with progressive motility, in situ motility, and of immobile.

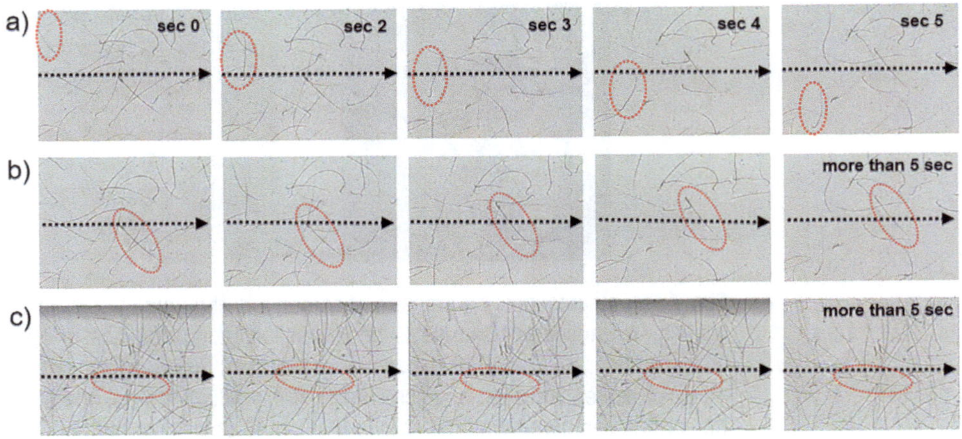

Fig. 8 Real time of frozen images to illustrate sperm motility. The dashed line represents the microscope eyepiece or is drawn on an acetate that may be placed on the computer screen. (**a**) Progressive sperm motility is identified when spermatozoa move with lineal forward displacement; it takes approximately 5 s for the spermatozoa to cross the line. (**b**) In situ motility is when the spermatozoa realize circular or local displacement. (**c**) Immobile spermatozoa remain in the same place. The circles enclose a single spermatozoon to show it movement or immobility

After the analysis, it would be enough to report only progressive sperm motility.

In previous studies, a solution (20% Lique-Nox soap; Alconox; or saline solution) has been injected into the uterine horns and the semen sample obtained by aspiration. This procedure facilitated the sperm count. However, after this procedure ceases sperm motility [53–55].

The sperm morphology and sperm viability are determined using a colorant (eosin-nigrosine). The same stained preparation is used to evaluate both parameters, although separately.

Sperm Morphology

Sperm morphology refers to sperm showing normal (sickle-shaped head and long flagellum) or abnormal characteristics (double head, crooked midpiece, fragmented flagellum, zigzag flagellum, short flagellum; Fig. 9a).

Using a micropipette, take 10 microliters of fresh collected semen and placed them on a slide and add 10 microliters of the colorant. Semen and colorant are mixed using a wooden stick until sample is homogenized. A cover-glass is placed and at least 5 min elapse before the evaluation placing a drop of immersion oil added on the cover-glass. Again, the observation is from left to right, and in different fields, 100 spermatozoa should be observed using the 100× microscope objective and a digital counter. Data are expressed as percentage of normal and of abnormal sperm. It would be enough to report only percentage of normal sperm.

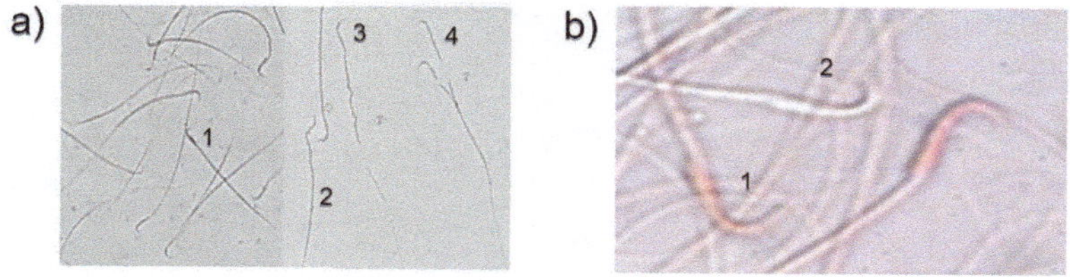

Fig. 9 Sperm morphology and sperm viability. (**a**) Photomicrographs show normal and abnormal spermatozoa. Normal morphology is indicated by a hooked head and long flagellum (1). Abnormal sperm includes a broken neck (2), fragmented flagellum (3), or double head (4). (**b**) Photomicrographs to show died and alive sperm. Eosin-nigrosine-stained sperm indicates died sperm (5), while alive sperm do not stain and remain transparent (6)

Sperm Viability

Sperm viability refers to alive spermatozoa (not stained with the colorant) and dead spermatozoa (stained with colorant; Fig. 9b). In different fields, 100 spermatozoa should be observed using the 100× microscope objective and a digital counter. Data are expressed as percentage of alive and of died sperm. It would be enough to report only percentage of alive sperm.

Sperm Count

Sperm count is the number of spermatozoa inseminated expressed in millions, also called number of sperm present in the uterus [54], or uterine sperm count [53], or number of sperm ejaculated [55]. Sperm count is the seminal parameter considered in the analysis of the rat ejaculate in old studies of reproductive physiology.

Prior to the sperm count, the *sperm density* must be determined. Using a micropipette, take 10 microliters of fresh semen and release it on a slide and place a cover-glass. Then, choose three random fields from the center of the sample using the 20× microscope objective. The image of each field is videotaped using a video camera connected to a computer. Each frozen image is viewed on the screen monitor using an acetate that has 5 zigzag drawn lines (Fig. 10). All the spermatozoa that touch the drawn lines with their head or any part of the flagellum are counted. After counting the sperm in the three fields, the average obtained corresponds to the sperm density. Depending on the mean obtained from the three random fields, the semen dilution is calculated as follows: 1:100 (semen: diluent) corresponds to a mean range of 10–60 spermatozoa; 1:200 to 61–120 spermatozoa; and 1–300 to 121–200 spermatozoa. These dilutions should be considered to express the final sperm count if different mean ranges of spermatozoa are compared. Our laboratory studies indicate that the ejaculate of the rat presents 63 spermatozoa as a minimum value and 120 as a maximum value counted in the three fields. On average 93.33 sperm, consequently the semen dilution most of the time is 1:200. Saline solution is used as diluent. If the semen sample is not diluted, it is impossible to be counted in the Neubauer chamber.

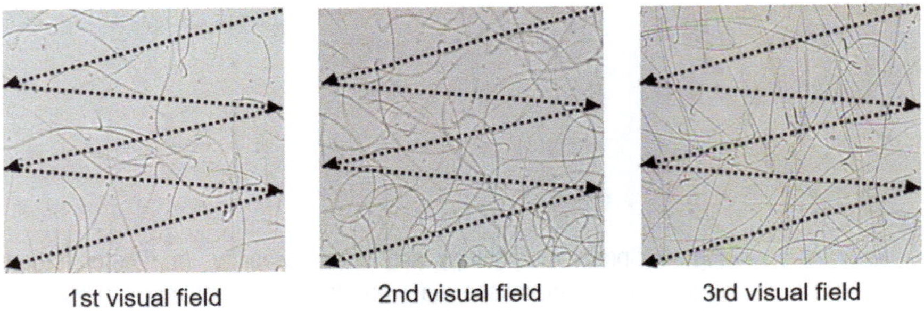

Fig. 10 Sperm density. Randomly select three fields from the center of the sample and observe using the 20× microscope objective. Each frozen image is analyzed on the screen monitor using an acetate with 5 zigzag drawn lines. All spermatozoa that touch the lines with their head or flagellum are counted. The sperm counting in the three fields is averaged to obtain the sperm density

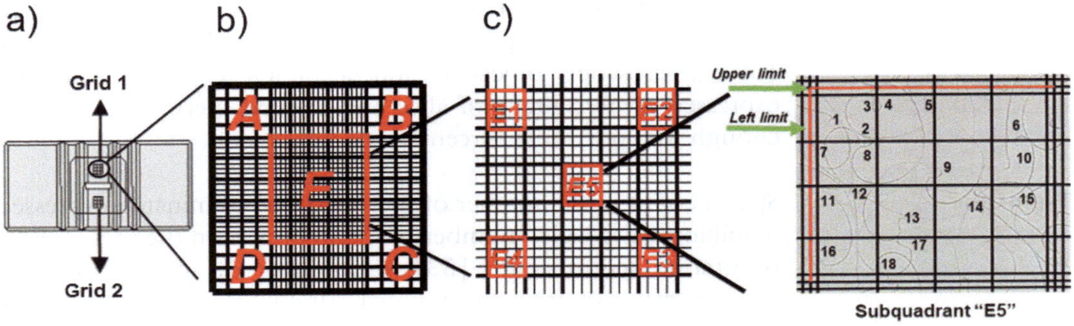

Fig. 11 Sperm count in the Neubauer hemocytometer. From the diluted semen, take a sample of the diluted semen and placed it in the Neubauer hemocytometer. This hemocytometer has two central grooves (**a**), each leading to a grid that consists of nine squares identified with letters starting in the upper left square and going clockwise. The upper squares are identified as A-B, the lower ones as C-D, and the central square as E (**b**). Square E has 25 sub-squares, those in the corner, and that in center are identified using Arabic numbers (**c**). Sperm count is carried out counting only the spermatozoa that are in square E, observing them with the 20× microscope objective. Each sub-square contains 16 smaller squares. The heads of the spermatozoa that are within the 16 small squares must be considered. The number of sperm heads in each sub-square (E1 + E2, + E3, + E4 + E5) is averaged. Counting from left to right is done in both grids and the result is multiplied by 1×10^6

From the diluted semen, take a sample to be placed in a Neubauer hemocytometer to do the sperm count. This hemocytometer has two central grooves (upper and lower), each leading to a grid. Both grids are separated from each other (Fig. 11a). Each grid consists of nine squares; only those in the corners and the central square are identified with letters starting in the upper left square and going clockwise. The upper squares are identified as A-B, the lower ones as C-D, and the central square as E (Fig. 11b). Square E has 25 sub-squares; those in the corner and that in center are identified using Arabic numbers, starting in the upper left square, and following the hands of the clock. Thus, square E contains E1-E5 sub-squares (Fig. 11c).

The Neubauer hemocytometer is covered with a cover slip taking care to cover both grids. Ten microliters of the diluted semen are deposited in the central groove near each grid that will diffuse by capillarity. Then, 8–10 min are allowed to elapse for the sperm to sediment. Sperm count is carried out counting only the spermatozoa that are in square E, observing them with the 20× microscope objective. Each sub-square contains 16 smaller squares. The heads of the spermatozoa that are within the 16 small squares must be considered, as well as the heads that are on the red line that forms the limits (upper and left) of each sub-square, forming an "inverted L." The number of sperm heads in each sub-square (E1 + E2, + E3, + E4 + E5) is added and averaged. Counting is done in both grids and the result is multiplied by 1×10^6. Counting is done from left to right by first counting the heads of the sperm in the first row, then those in the second row, until all four rows are completed (Fig. 11c).

Analysis of all parameters, particularly the microscopic ones, must be done in duplicate to verify that they are correct.

3.3.3 Surgical Procedure to Obtain the Seminal Plug

The seminal plug is obtained surgically from the inseminated female (Fig. 12a). A longitudinal incision is made from the vaginal orifice to the pubis and cut the cartilaginous joint of the pubic symphysis. After, an incision is made in the dorsal vaginal wall to observe the plug attached in its rostral portion to the cervix and in its lateral edges to the vaginal wall. Using a spatula, the plug is detached from the vagina and cervix (Fig. 12b).

3.3.4 Seminal Plug Parameters

Weight

Weight is the mass of the seminal plug expressed in mg. It is determined using an analytical balance (Fig. 13a).

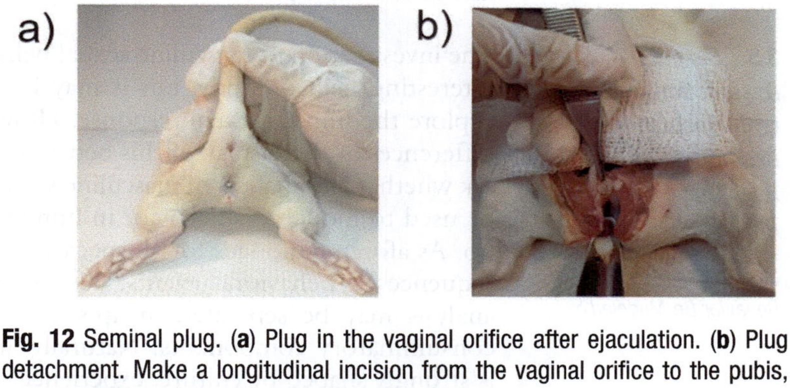

Fig. 12 Seminal plug. (**a**) Plug in the vaginal orifice after ejaculation. (**b**) Plug detachment. Make a longitudinal incision from the vaginal orifice to the pubis, use a spatula to detach the plug from the vagina

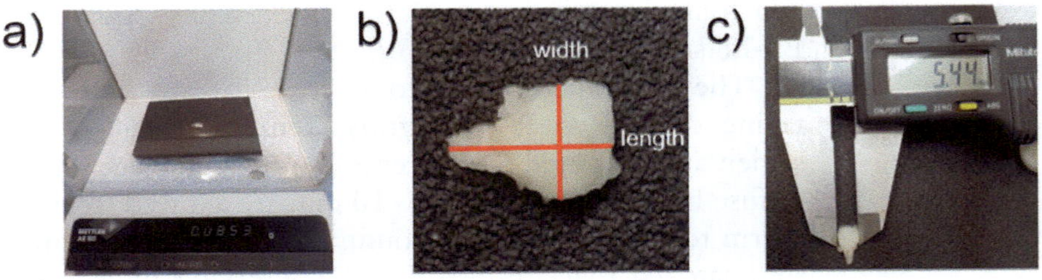

Fig. 13 Parameters of seminal plug. (**a**) The weight of the seminal plug is obtained with an analytical balance. (**b**, **c**) Width and length of the seminal plug calculated using a digital caliper

Size

Size of the seminal plug is the length and width of the seminal plug expressed in centimeters. It is measured using a digital caliper (Fig. 13b, c).

3.4 Notes

Two Incisions Instead of One to Obtain Both Ovaries: When a single medial incision is done to remove the right and the left ovary, many adhesions form around the uterine horns alter sperm motility, in fact just a few spermatozoa present sperm motility. Thus, it is required a right and left incision on the skin and muscles at the level of the flanks. A small incision should be made to locate the ovary without stretching the uterine horns. In addition, it is much better to use ovariectomized females instead of intact ones. Females with natural estrus have many rounds and cornified epithelial cells in the uterine fluid, interfering with the analysis of the microscopic parameters of the seminal fluid.

Removal of Uterine Fat and Blood Vessels: The ability to clean the uterine horns in less than 5 min must be acquired to avoid interfering with sperm motility. At least 2 changes of saline solution must be done to keep the temperature al 36 ± 1 °C, also to avoid interfere with sperm motility.

Temperature of the Thermo-bath: The temperature must be kept constant. Changes in temperature modify sperm motility. Thus, the first microscopic parameter to be analyzed is the sperm motility.

3.5 Sexual Dysfunctions Related with Ejaculation. An Example to Illustrate How These Methods Are Used

3.5.1 Can Man's Sexual Behavior Be Modeled?

The investigation of masculine sexual behavior in the rat may be an interesting field of study, but it may be also used as a model to explore the human sexual response. However, due to the obvious differences in the display of this behavior between species, we may ask whether the analysis of masculine sexual behavior in the rat may be used to model this behavior in humans? The answer is yes and no. As aforementioned, for all species, sexual behavior occurs as a sequence of behavioral events. Also common to all species, its analysis may be separated in appetitive (or precopulatory) and consummatory components. Naturally, in humans, this behavior is strongly shaped by culture, experience, and learning [60]. There are other point links between human and animal sexuality. For example, certain drugs produce erection or reduce sexual arousal in men and animals [61–64], suggesting that certain brain systems are conserved to serve similar functions among species. These findings open the possibility to use animal models to study aspects of human sexual behavior that otherwise could not be experimentally explored in humans, primarily because of ethical reasons. Most of our current understanding of the anatomy and neurobiology of human sexual behavior is based on animal studies using sexually experienced rats that display normal sexual behavior.

3.5.2 Natural Models of Ejaculatory Dysfunctions

It is important to mention that there is a long debate whether premature or delayed ejaculation is pathological states (sexual dysfunctions) or simply the physiological variation normally occurring in men. Premature and delayed ejaculation are considered disorders if they cause distress to the man and low satisfaction to his sexual partner [65, 66]. It should be mentioned that without this criterion the "short" or "long" ejaculation latencies in men cannot be considered as pathological; therefore, no absolute values are given to the length of this parameter to assume that there is an alteration. To date, it is considered that the ejaculatory dysfunctions are the most common forms of sexual dysfunction in men and covers a broad range of disorders including premature and delayed ejaculation, anejaculation, painful ejaculation, and hematospermia [67, 68]. Premature ejaculation is characterized by (i) ejaculation after few penile thrusts, (ii) short ejaculation latencies, and (iii) the occurrence of these symptoms in (nearly) every coitus; it has a prevalence of 20–30% worldwide [69]. On the other hand, delayed ejaculation is characterized by the following symptoms: (i) ejaculation after many penile thrusts, (ii) a long duration of the ejaculation latency, and (iii) the occurrence of these symptoms in nearly every coitus; it has an estimated prevalence of 1–4% [70]. These two disorders may be lifelong or acquired, situational, or persistent [68, 71] and are conditions (physiological or pathological) thought of as related entities on the two ends of a spectrum [72].

Ejaculatory dysfunctions (excluding premature ejaculation) interfere with the delivery of sperm to the female genital tract and are an important and often under-appreciated etiological factor in male sub-or infertility [73]. Interestingly, there is controversy whether men suffering from premature ejaculation have fertility problems [74–76], what seems clear, is that various therapeutic agents used for the treatment of premature or delayed ejaculation produce significant decreases in sperm concentration, motility, and morphology [77, 78].

Various animal models have been proposed to explore the bases of these alterations, including centrally or peripherally acting drugs that reduce or prolong the ejaculation latency or behavioral interventions that result in changes in the ejaculation threshold. For example, the administration of 5-HT1A agonists, such as 8-OH-DPAT, drastically reduces the number of intromissions and shortens the ejaculation latency, effects that are not rewarding for the male [79] and produces a vaginocervical stimulation in the female that fails to make these males more attractive [80]. Conversely, amygdala or bed nucleus of the stria terminalis lesions produce long ejaculation latencies accompanied by a high number of intromissions [81, 82]. Here, we focus our attention on models that spontaneously present different duration of ejaculatory latencies because we believe they resemble the condition closer to what occurs in humans, i.e., in men, the shortening of the ejaculation latency (leading to premature ejaculation) or its prolongation (provoking delayed ejaculation) usually occur by idiopathic reasons, and not by drug administration or brain damage. In line, it was proposed that a worthwhile approach to study ejaculatory dysfunctions is to exploit the individual variability in ejaculatory behavior shown by experimental animals [83]. That is, according to recent reviews [84, 85] we and others consider that this natural model fulfills the construct or homology criterion that implies that the etiological factors are similar between the clinical alteration and the animal's behavior.

In a series of experiments, we asked whether animals with different copulatory phenotypes would show variations in particular seminal parameters of their respective ejaculates. To that end, we tested for sexual behavior a large population of male rats and found that only 7.8% of them were non-copulators, i.e., animals that never showed sexual behavior (in all 6 copulatory tests). Within those that copulate, there were various endogenous phenotypes that are selected according to their ejaculation latency. That is, there was a population of males (71%) that consistently after various training sessions show short ejaculation latencies (named rapid ejaculators), others display long ejaculation latencies (named sluggish, 4%), and a third group show intermediate values (25%) (Table 1). For comparison purposes, we added an extra column showing the behavioral values of another set of animals ($n = 52$) without sexual training, i.e., their values in the first test when they ejaculated. As may be seen from Table 1, the mount (Kruskal–Wallis ANOVA, $H = 31.58$, $p < 0.001$) and intromission latencies (Kruskal–Wallis ANOVA, $H = 48.70$, $p < 0.001$) were much shorter in rapid ejaculating males as compared with the other groups. Naturally, the selecting parameter, ejaculation latency, varied between the groups (Kruskal–Wallis ANOVA, $H = 84.33$, $p < 0.001$) (Table 1), with rapid males showing the shortest and sluggish the longest values. Naïve males displayed relatively long ejaculation latencies that normally shortened after successive trials. The number of mounts (Kruskal–Wallis ANOVA, $H = 62.48$, $p < 0.001$) and of behavioral intromissions (Kruskal–Wallis ANOVA $H = 22.83$, $p < 0.001$) preceding ejaculation were also reduced in rapid ejaculators as compared to the other groups (see Table 1). As expected, animals with short ejaculation latencies copulated

Table 1
Copulatory parameters

	Male rats			
Parameters	Naïve ($n = 52$)	Intermediate ($n = 26$)	Rapid ($n = 76$)	Sluggish ($n = 4$)
Mount latency (s)	28.50 (10.50–117.50)	16 (6–28)	9* (5–14.25)	26.75 (13–40.50)
Intromission latency (s)	89.50 (27.50–251)	21* (6–66)	12*+ (7–20.25)	69 (38–83.50)
Ejaculation latency (s)	775.50 (437–1023)	521 (378–768)	228*+# (160.50–325)	1224 (725–1743)
Number of mounts	12 (7–19.50)	5 (2–7)	2*# (1–5)	17.50 (9–23)
Number of intromissions	11 (8–12)	11.50 (9–14)	8*+# (7–10)	19.50 (15–21)
Hit rate	0.47 (0.38–0.57)	0.69* (0.55.0.83)	0.77* (0.63–0.89)	0.46 (0.44–0.73
Interintromission interval (s)	48.18 (13.74–110.66)	44.32 (35.50–63)	29.37*+# (18.14–35.94)	87.02 (73.02–92.50)

Descriptive statistics. Data are expressed as median and first quartile-third quartile (q1-q3). Kruskal–Wallis ANOVA (see text) followed by Dunn's test *$p < 0.05$ vs. naive; +$p < 0.05$ vs. intermediate; #$p < 0.05$ vs. sluggish

Table 2
Seminal parameters

	Male rats	
Parameters	Intermediate ($n = 21$)	Rapid ($n = 40$)
Progressive sperm motility (%)	70 (55–80.75)	73 (61.75–83)
Sperm morphology (%)	95 (93–97)	94 (92–97)
Sperm viability (%)	86 (81–91)	87 (78–94)
Sperm count (million)	54 (42.06–86.75)	46 (35.12–62.25)
Sperm viability (%)	86 (81–91)	87 (78–94)
Weight of copulatory plug (mg)	95.70 (90.92–117.97)	95.80 (87.5–115.25)
Length of copulatory plug (mm)	10.30 (9.50–11.10)	10.35 (9.35–11.05)
Width of copulatory plug (mm)	5.40 (4.97–5.87)	5.20 (4.85–5.75)

Descriptive statistics. Data are expressed as median and first quartile–third quartile (q1-q3). Mann–Whitney U-test. No differences were found in any parameter between both groups

more efficiently and accordingly their hit rate (Kruskal–Wallis ANOVA, $H = 51.36$, $p < 0.001$) and inter-intromission interval (Kruskal–Wallis ANOVA, $H = 70.78$, $p < 0.001$) differed from the values shown by naïve males. Interestingly, also the hit rate of the males cataloged as intermediate ejaculators was higher than that shown by sexually naïve animals (see Table 1).

Table 2 compared the seminal parameters of males with two different copulating phenotypes: intermediate ($n = 21$) and rapid ($n = 40$) copulators. No statistically significant differences (Mann–Whitney U-test) were found between the various parameters analyzed, supporting the idea that ejaculations coming from animals that have short ejaculation latencies are as fertile as those of intermediates.

Acknowledgments

Authors wish to thank M. Sc. Rebeca Reyes for carefully editing the manuscript.

References

1. Moralí G, Beyer C (1992) Motor aspects of masculine sexual behavior in rats and rabbits. In: Slater PJR, Rosenblatt JS, Beer C, Milinski M (eds) Advances in the study of behavior, vol 21. Academic Press, New York, pp 201–238
2. Moralí G, Soto MAP, Contreras JL et al (2003) Detailed analysis of the male copulatory motor pattern in mammals: hormonal bases. Scan J Psychol 44:79–288. https://doi.org/10.1111/1467-9450.00346
3. Sachs BD, Meisel RL (1988) The physiology of male sexual behavior. In: Knobil E, Neill J (eds) The physiology of reproduction. Raven Press, Ltd, New York
4. Hull EM, Rodríguez-Manzo G (2009) Male sexual behavior. In: Pfaff DW, Arnold AP, Etgen AM, Fahrbach SE, Rubin RT (eds) Hormones, brain and behavior, vol 1. Academic Press, San Diego, pp 5–65
5. Contreras JL, Beyer C (1979) A polygraphic analysis of mounting and ejaculation in the New Zealand white rabbit. Physiol Behav 23: 939–943. https://doi.org/10.1016/0031-9384(79)90204-x
6. Whipple B, Myers BR, Komisaruk BR (1998) Male multiple ejaculatory orgasms: a case study. J Sex Educ Ther 23:157–162. https://doi.org/10.1080/01614576.1998.11074222
7. Larsson K (1956) Conditioning and sexual behavior in the male Albino rat. Almqvist and Wiksell's Stockholm, Stockholm
8. Larsson K, Ahlenius S (1993) Reproductive behavior in behavioral pharmacology. In: van Haaren F (ed) Methods in behavioral pharmacology. Elsevier Science Publishers BV
9. Beyer C, Contreras JL, Moralí G, Larsson K (1981) Effects of castration and sex steroid treatment on the motor copulatory patterns of the rat. Physiol Behav 27:727–730. https://doi.org/10.1016/0031-9384(81)90247-x
10. Lodder J, Zeilmaker GH (1976) Effects of genital deafferentation on mounting and intromission behavior in spayed female rats treated with estrogen. Physiol Behav 16:753–755. https://doi.org/10.1016/0031-9384(76)90246-8
11. Pollak EI, Sachs BD (1976) Penile movements and the sensory control of copulation in the rat. Behav Biol 17:177–186. https://doi.org/10.1016/s0091-6773(76)90427-2
12. Lucio RA, Rodríguez-Piedracruz V, Tlachi-López JL et al (2014) Copulation without seminal expulsion: the consequence of sexual satiation and the Coolidge effect. Andrology 2: 450–457. https://doi.org/10.1111/j.2047-2927.2014.00209.x
13. Sachs BD (1988) The development of grooming and its expression in adult animals. Ann N Y Acad Sci 525:1–17. https://doi.org/10.1111/j.1749-6632.1988.tb38591.x
14. Larsson K, Swedin G (1971) The sexual behavior of male rats after bilateral section of the hypogastric nerve and removal of the accessory genital glands. Physiol Behav 6:251–253. https://doi.org/10.1016/0031-9384(71)90035-7
15. McGlynn JM, Erpino MJ (1974) Effects of vasectomy on the reproductive system and sexual behavior of rats. J Reprod Fertil 40:241–247. https://doi.org/10.1530/JRF.0.0400241
16. Queen K, Dhabuwala CB, Pierrepoint CG (1981) The effect of the removal of the various accessory sex glands on the fertility of male rats. J Reprod Fertil 62:423–426. https://doi.org/10.1530/jrf.0.0620423
17. Lucio RA, Manzo J, Martínez-Gómez M et al (1994) Participation of pelvic nerve branches in male rat copulatory behavior. Physiol Behav 55:241–246. https://doi.org/10.1016/0031-9384(94)90129-5
18. Phillips-Farfán BV, Fernández-Guasti A (2009) Endocrine, neural and pharmacological aspects of sexual satiety in male rats. Neurosci Biobehav Rev 33:442–455. https://doi.org/10.1016/j.neubiorev.2008.11.003
19. Tlachi-López JL, Eguibar JR, Fernández-Guasti A, Lucio RA (2012) Copulation and ejaculation in male rats under sexual satiety and the Coolidge effect. Physiol Behav 106: 626–630. https://doi.org/10.1016/j.physbeh.2012.04.020
20. Fernández-Guasti A, Larsson K, Vega-Sanabria J (1986) Depression of postejaculatory ultrasonic vocalization by (+)-bicuculline. Behav Brain Res 19:35–39. https://doi.org/10.1016/0166-4328(86)90045-8
21. Barfield RJ, Geyer LA (1975) The ultrasonic postejaculatory vocalization and the postejaculatory refractory period of the male rat. J Comp Physiol Psychol 88:723–734. https://doi.org/10.1037/h0076435
22. Rodríguez-Manzo G, Fernández-Guasti A (1994) Reversal of sexual exhaustion by serotonergic and noradrenergic agents. Behav Brain Res 64:127–134. https://doi.org/10.1016/0166-4328(94)90019-1

23. Robb GW, Amann RP, Killian GJ (1978) Daily sperm production and epididymal sperm reserves of pubertal and adult rats. J Reprod Fertil 54:103–107. https://doi.org/10.1530/jrf.0.0540103
24. Campion SN, Carvallo FR, Chapin RE et al (2013) Comparative assessment of the timing of sexual maturation in male Wistar Han and Sprague-Dawley rats. Reprod Toxicol 38:16–24. https://doi.org/10.1016/j.reprotox.2013.02.003
25. Meisel RL, Sachs BD (1994) The physiology of male sexual behavior. In: Knobil E, Neill J (eds) The physiology of reproduction. Raven Press, Ltd, New York
26. Brandling-Bennett EM, Blasberg ME, Clark AS (1999) Paced mating behavior in female rats in response to different hormone priming regimens. Horm Behav 35:144–154. https://doi.org/10.1006/hbeh.1998.1507
27. Corona R, Camacho FJ, García-Horsman P et al (2011) Different doses of estradiol-benzoate induce conditioned place preference after paced mating. Horm Behav 60:264–268. https://doi.org/10.1016/j.yhbeh.2011.05.013
28. Dewsbury DA (1969) Copulatory behavior of rats (Rattus norvegicus) as a function of prior copulatory experience. Anim Behav 17:217–223. https://doi.org/10.1016/0003-3472(69)90004-9
29. De Gasperín-Estrada GP, Camacho FJ, Paredes RG (2008) Olfactory discrimination and incentive value of non-copulating and sexually sluggish male rats. Physiol Behav 93:742–747. https://doi.org/10.1016/j.physbeh.2007.11.027
30. Lucio RA, Tlachi-López JL, Eguibar JR, Agmo A (2013) Sperm count and sperm motility decrease in old rats. Physiol Behav 110-111: 73–79. https://doi.org/10.1016/j.physbeh.2012.12.015
31. Fuentes-Morales MR, Gutiérrez-Ospina G, Fernández-Guasti A et al (2020) Rats ejaculate prematurely and increase the sperm output during competitive mating. Ethol Ecol Evol 32:351–360. https://doi.org/10.1080/03949370.2020.1715488
32. Pattij T, de Jong TR, Uitterdijk A et al (2005) Individual differences in male rat ejaculatory behavior: searching for models to study ejaculation disorders. Eur J Neurosci 22:724–734. https://doi.org/10.1111/j.1460-9568.2005.04252.x
33. Olivier B, Chan JSW, Pattij T et al (2006) Psychopharmacology of male rat sexual behavior: modeling human sexual dysfunctions? Int J Impot Res 18:S14–S23. https://doi.org/10.1038/sj.ijir.3901330
34. Aguilar-Pérez CE, Gómora-Arrati P, Komisaruk BR et al (2020) Threshold for copulation-induced analgesia varies according to the ejaculatory endophenotype in rats. Int J Imp Res. https://doi.org/10.1038/s41443-020-00390-8
35. Lucio RA, Fernández-Guasti A, Fuentes-Morales MR (2020) Ejaculation latency. In: Shakelford TK (ed) The Cambridge handbook of evolutionary perspectives on sexual psychology. Cambridge University Press, Cambridge, UK
36. Blaustein JD, Erskine MS (2002) Feminine sexual behavior: cellular integration of hormonal and afferent information in the rodent forebrain. In: Pfaff DW, Arnold AP, Etgen AM, Fahrbach SE, Rubin RT (eds) Horm. Brain Behav. Academic Press, Amsterdam
37. Fernández-Guasti A, Ahlenius S, Hjorth S, Larsson K (1987) Separation of dopaminergic and serotonergic inhibitory mechanisms in the mediation of estrogen-induced lordosis behaviour in the rat. Pharmacol Biochem Behav 27: 93–98. https://doi.org/10.1016/0091-3057(87)90482-5
38. Paredes RG, Vazquez B (1999) What do females rats like about sex? Paced mating. Behav Brain Res 105:117–127. https://doi.org/10.1016/s0166-4328(99)00087-x
39. Erskine MS (1989) Solicitation behavior in the estrous female rat: a review. Horm Behav 23: 473–502. https://doi.org/10.1016/0018-506x(89)90037-8
40. López-Gatius F (2000) Site of semen deposition in cattle: a review. Theriogenology 53: 1407–1414. https://doi.org/10.1016/S0093-691X(00)00283-1
41. Suarez SS, Pacey AA (2006) Sperm transport in the female reproductive tract. Hum Reprod Update 12:23–37. https://doi.org/10.1093/humupd/dmi047
42. Balaban B, Sakkas D, Gardner DK (2014) Laboratory procedures for human in vitro fertilization. Semin Reprod Med 32:272–282. https://doi.org/10.1055/s-0034-1375179
43. Shaban SF, Seager SW, Lipshultz LI (1988) Clinical electroejaculation. Med Instrum 22: 77–81
44. Poiani A (2006) Complexity of seminal fluid: a review. Behav Ecol Sociobiol 60:289–310. https://doi.org/10.1007/s00265-006-0178-0
45. Setchell BP, Breed WG (2006) Anatomy, vasculature, and innervation of the male

reproductive tract. In: Neill JD (ed) Knobil and Neill's physiology of reproduction, 3rd edn. Elsevier
46. Marberger H (1974) The mechanisms of ejaculation. In: Coutinho EM, Fuchs F (eds) Physiology and genetics of reproduction. Part B. Plenum Press, New York
47. Revenig L, Leung A, Hsiao W (2014) Ejaculatory physiology and pathophysiology: assessment and treatment in male infertility. Transl Androl Urol 3:41–49. https://doi.org/10.3978/j.issn.2223-4683.2014.02.02
48. Gerstenberg TC, Levin RJ, Wagner G (1990) Erection and ejaculation in man-assessment of the electromyographic activity of the bulbocavernosus and ischiocavernosus muscles. Br J Urol 65:395–402. https://doi.org/10.1111/j.1464-410x.1990.tb14764.x
49. Holmes GM, Chapple WD, Leipheimer RE, Sachs BD (1991) Electromyographic analysis of male rat perineal muscles during copulation and reflexive erections. Physiol Behav 49: 1235–1246. https://doi.org/10.1016/0031-9384(91)90357-t
50. Allard J, Truitt WA, McKenna KE, Coolen LM (2005) Spinal cord control of ejaculation. World J Urol 23:119–126. https://doi.org/10.1007/s00345-004-0494-9
51. Carro-Juárez M, Rodríguez-Manzo G (2008) The spinal pattern generator for ejaculation. Brain Res Rev 58:106–120. https://doi.org/10.1016/j.brainresrev.2007.12.002
52. Veening JG, Coolen LM (2014) Neural mechanisms of sexual behavior I the male rat: emphasis on ejaculation-related circuits. Pharmacol Biochem Behav 121:170–183. https://doi.org/10.1016/j.pbb.2013.12.017
53. Austin D, Dewsbury D (1986) Reproductive capacity of male laboratory rats. Physiol Behav 37:627–632. https://doi.org/10.1016/0031-9384(86)90297-0
54. Matthews M, Adler NT (1977) Facilitative and inhibitory influences of reproductive behavior on sperm transport in rats. J Comp Physiol Psychol 91:727–741. https://doi.org/10.1037/h0077364
55. Pound N, Gage MJG (2004) Prudent sperm allocation in Norway rats, Rattus norvegicus: a mammalian model of adaptative ejaculate adjustment. Anim Behav 68:819–823. https://doi.org/10.1016/j.anbehav.2004.02.004
56. Lucio RA, Fernández-Guasti A, Larsson K (2017) Male sexual satiety and the Coolidge effect in rats: relation between behavioral and seminal parameters. In: Komisaruk BR, González-Mariscal G (eds) Behavioral neuroendocrinology. CRC Press, Boca Raton, FL
57. Carballada R, Esponda P (1992) Role of fluid from seminal vesicles and coagulating glands in sperm transport into the uterus and fertility in rats. J Reprod Fertil 95:639–648. https://doi.org/10.1530/jrf.0.0950639
58. Blandau RJ (1945) On the factors involved in sperm transport through the cervix uteri of the albino rat. Am J Anat 77:253–272. https://doi.org/10.1002/aja.1000770205
59. Toner JP, Adler NT (1985) The role of uterine luminal fluid in uterine contractions, sperm transport and fertility of rats. J Reprod Fertil 74:295–302. https://doi.org/10.1530/jrf.0.0740295
60. Pfaus JG, Kippin TE, Coria-Avila G (2003) What can animal models tell us about human sexual response? Annu Rev Sex Res 14:1–63
61. Petrie WM (1985) The sexual effects of antipsychotic, antidepressant, and psychomotor stimulant drugs. In: Segal M (ed) Psychopharmacology of sexual disorders. John Libbey, London
62. Lal S, Laryea E, Thavundayil JX et al (1987) Apomorphine-induced penile tumescence in impotent patients preliminary findings. Prog Neuro-Psychopharmacol Biol Psychiatry 11: 235–242. https://doi.org/10.1016/0278-5846(87)90066-2
63. Melis MR, Argiolas A, Gessa GL (1987) Apomorphine-induced penile erection and yawning: site of action in brain. Brain Res 415:98–104. https://doi.org/10.1016/0006-8993(87)90272-1
64. Pfaus JG, Phillips AG (1991) Role of dopamine in anticipatory and consummatory aspects of sexual behavior in the male rat. Behav Neurosci 105:727–743. https://doi.org/10.1037//0735-7044.105.5.727
65. Althof SE, Abdo CHN, Dean J et al (2010) International Society for Sexual Medicine's guidelines for the diagnosis and treatment of premature ejaculation. J Sex Med 7:2947–2969. https://doi.org/10.1111/j.1743-6109.2010.01975.x
66. American Psychiatric Association (2013) Diagnostic and statistical manual of mental disorders, 5th edition (DSM-5). American Psychiatric Publishing, Washington, DC
67. Wolters JP, Hellstrom WJ (2006) Current concepts in ejaculatory dysfunction. Rev Urol 8 (Suppl 4):S18–S25
68. Gray M, Zillioux J, Khourdaji I, Smith RP (2018) Contemporary management of ejaculatory dysfunction. Transl Androl Urol 7:686–

702. https://doi.org/10.21037/tau.2018.06.20

69. Carson C, Gunn K (2006) Premature ejaculation: definition and prevalence. Int J Impot Res 18(Suppl 1):S5–13. https://doi.org/10.1038/sj.ijir.3901507
70. Chen J (2016) The pathophysiology of delayed ejaculation. Transl Androl Urol 5:549–562. https://doi.org/10.21037/tau.2016.05.03
71. Waldinger MD (2002) The neurobiological approach to premature ejaculation. J Urol 168:2359–2367. https://doi.org/10.1016/S0022-5347(05)64146-8
72. Waldinger MD, Schweitzer DH (2005) Retarded ejaculation in men: an overview of psychological and neurobiological insights. World J Urol 23:76–81. https://doi.org/10.1007/s00345-004-0487-8
73. Barazani Y, Stahl PJ, Nagler HM, Stember DS (2012) Management of ejaculatory disorders in infertile men. Asian J Androl 14:525–529. https://doi.org/10.1038/aja.2012.29
74. Kondoh N (2012) Ejaculatory dysfunction as a cause of infertility. Reprod Med Biol 11:59–64. https://doi.org/10.1007/s12522-011-0108-3
75. Lotti F, Corona G, Castellini G et al (2016) Semen quality impairment is associated with sexual dysfunction according to its severity. Hum Reprod 31:2668–2680. https://doi.org/10.1093/humrep/dew246
76. Yang X, Gao M, Zhang L et al (2018) Central neural correlates during inhibitory control in lifelong premature ejaculation patients. Front Hum Neurosci 12:206. https://doi.org/10.3389/fnhum.2018.00206
77. Koyuncu H, Serefoglu EC, Yencilek E et al (2011) Escitalopram treatment for premature ejaculation has a negative effect on semen parameters. Int J Impot Res 23:257–261. https://doi.org/10.1038/ijir.2011.35
78. Koyuncu H, Serefoglu EC, Ozdemir AT, Hellstrom WJ (2012) Deleterious effects of selective serotonin reuptake inhibitor treatment on semen parameters in patients with lifelong premature ejaculation. Int J Impot Res 24:171–173. https://doi.org/10.1038/ijir.2012.12
79. Camacho FJ, Castro M, Hernández V, Paredes RG (2007) Facilitation of ejaculation induced by 8-OH-DPAT does not produce conditioned place preference in male rats. Behav Neurosci 121:579–585. https://doi.org/10.1037/0735-7044.121.3.579
80. Haensel SM, Mos J, Olivier B, Slob AK (1991) Sex behavior of male and female Wistar rats affected by the serotonin agonist 8-OH-DPAT. Pharmacol Biochem Behav 40:221–228. https://doi.org/10.1016/0091-3057(91)90543-b
81. Claro F, Segovia S, Guilamón A, Del Abril A (1995) Lesions in the medial posterior region of the BST impair sexual behavior in sexually experienced and inexperienced male rats. Brain Res Bull 36:1–10. https://doi.org/10.1016/0361-9230(94)00118-k
82. Harris VS, Sachs BD (1975) Copulatory behavior in male rats following amygdaloid lesions. Brain Res 86:514–518. https://doi.org/10.1016/0006-8993(75)90906-3
83. Waldinger MD, Olivier B (2005) Animal models of premature and retarded ejaculation. World J Urol 23:115–118. https://doi.org/10.1007/s00345-004-0493-x
84. Belzung C, Lemoine M (2011) Criteria of validity for animal models of psychiatric disorders: focus on anxiety disorders and depression. Biol Mood Anxiety Disord 1:9. https://doi.org/10.1186/2045-5380-1-9
85. Treit D, Engin E, McEown K (2010) Animal models of anxiety and anxiolytic drug action. Curr Top Behav Neurosci 2:121–160. https://doi.org/10.1007/7854_2009_17

Chapter 12

Paced Mating Behavior

Zacnité Mier-Quesada, Natalia Robles, and Raúl G. Paredes

Abstract

Sexual behavior is crucial for the survival of many species because through the display of this behavior males and females transmit genetic information assuring that it would be carried into the next generation. On the other hand, for individuals, sexual behavior is not crucial for their survival as is the case for eating and drinking, In fact, in different species, there are males and females who would not display sexual behavior despite they are tested repeatedly. Under appropriate hormonal conditions, males and females are an incentive for members of the opposite sex. An incentive is a stimulus, in this case, a male or a female, which activates an approach behavior. Early studies in rodents assumed that males had an active role, while females were passive mostly responding to the male. In this chapter, we will briefly review different methodologies used in rats to evaluate female sexual behavior demonstrating that they have an active role in sexual behavior. Then, we will describe in detail how pacing behavior is used to evaluate the distinct aspects of female sexual behavior. Finally, we will describe behavioral sex differences when males and females paced the sexual interaction.

Key words Paced mating, Percentage of exits, Return latencies, Sexual incentive, Reward

1 Introduction

Early studies clearly identified the active role that females have during mating, both under laboratory conditions [1] and in semi-natural settings [2, 3]. The first study in which female sexual behavior was categorized and defined was done by Frank Beach (1976) based on numerous studies done on several mammalian species. In that seminal manuscript, he defined three concepts that represent the behavior of several female mammals when they are in estrus: Attractivity "refers to the female's stimulus value in evoking sexual responses by the male." Proceptivity "connotes various reactions by the female toward the male, which constitute her assumption of initiative in establishing or maintaining sexual interaction"; and Receptivity "defined in terms of female responses necessary and sufficient for the male's success in achieving intravaginal ejaculation" [4]. Since then, several research groups have

Raúl G. Paredes et al. (eds.), *Animal Models of Reproductive Behavior*, Neuromethods, vol. 200,
https://doi.org/10.1007/978-1-0716-3234-5_12,

studied, advanced, and sometimes redefined the different components of female sexual behavior. Attractivity is inferred from the behaviors displayed by the male in response to olfactory and/or auditory stimuli displayed by the female. These behaviors include approach and investigation of the female, including her genital region [4–6]. It has also been demonstrated that the female rat is an unconditioned sexual stimulus for the male. That is, cues emitted by the female induce an approach behavior by the male associated with her incentive value [7]. Proceptivity includes behaviors such as approach, where the female gets close to the male; orientation, where the female displays a series of behaviors to place her genital region close to the male nose; and runaway where the female displays hops and darts in zigzag movements exhibiting many times ear wiggling [4, 6]. Receptivity is associated with the display of a posture that facilitates intromissions and ejaculation. In the rat, the lordosis reflex is the arching of the back with an elevation of the pelvis and deviation of the tail [4, 6]. In rodents, the lordosis reflex is induced when the male mounts the female. However, when the female is receptive tactile stimulation of the perineal region can induce lordosis [8] as well as artificial vaginocervical stimulation in anestrous females [9]. Sexual behavior in female rats has also been divided into copulatory, paracopulatory, and progestative behaviors [10]. The copulatory behaviors are those that contribute to transfer sperm from the male to the female and are used instead of the term receptivity, indicating an active role by the female. The paracopulatory behaviors include hopping, darting, ear wiggling, a presenting posture, approach, withdrawn behaviors, and ultrasonic vocalizations, which aroused the male. The progestative behaviors are those that increase the probability that pregnancy can occur during a sexual interaction regulating the timing and frequency of intromissions and ejaculations from the male, the postejaculatory interval, and the pacing of the sexual stimulation, through approach withdrawals behaviors (*see* [10] and references therein).

Of all the behaviors displayed by the female rat during a sexual interaction, the most studied and well defined is undoubtedly the lordosis reflex. The neural circuits controlling the pathways carrying the sensory information that triggers the reflex as well as the motor output that induces lordosis have been clearly identified and described, including some of the neurophysiological and molecular mechanisms [11–13]. On the other hand, several methodologies have been used to study different aspects of proceptivity, paracopulatory behavior or female sexual motivation. In the following section, we will briefly mention some of these methodologies and then describe in detail the limitations and advantages of using the paced mating methodology to study female sexual motivation.

2 Methods to Measure Sexual Motivation

Different methods are used to measure proceptivity, paracopulatory behaviors, or sexual motivation in females. Many of them have already been described in detail and here we will only give an overview of the ones most commonly used. The early methods used include the increasing barrier method, the runway method, and operant behavior method [14]. In the barrier method, the subject is placed in a starting box and must cross to the goal box through an electric grid where a stimulus subject is placed. In each trial, the current can increase, and a higher motivation is inferred when the subject endures a higher current to reach the stimulus subject. In the runway method, the subject is placed in a starting box with a runway that reaches the goal box or the testing arena where a stimulus animal is located. There are different variations in this method. At the end of the runway, there could be a stimulus animal, or the subject could choose between two or more stimulus animals. For example, a stimulus female could reach the goal box and choose to interact with either a sexually experienced or a castrated male [15–17].

In the operant behavior method, subjects are trained to press the lever and obtain the presence of a stimulus animal. In a pioneer and classical study, female rats were trained to press a lever to have access to a sexually experienced stimulus male with whom to mate [1]. In each of these methods, several studies have evaluated different aspects of female sexual motivation that have been extensively described (*see* [14] and references therein). In the present book, other methods are described that can be used to evaluate different aspects of female sexual motivation. "Evaluation of sexual behavior in laboratory vs seminatural conditions" (Chu and Agmo); "Pro-choice: Partner Preference as a Method to Assess Sexual Motivation" (Guarraci and Meerts); "Sexual Incentive Motivation" (Huijgens; Heijkoop and Snoeren); and "Conditioned Place Preference" (Bedos). Another method widely used in the study of sexual behavior and sexual motivation is the partner preference test, as described in this book by Guaraci and Meerts. Basically, the subjects are given the choice to interact with two different stimulus animals that could be tethered or not to allow the interaction in the stimulus animal compartment.

It was initially considered that females had a relatively passive role during mating following the male rhythm during sexual interaction. However, several methods have clearly demonstrated that females can control (pace) the sexual stimulation they receive. As mentioned above, females can press a lever to mate with a sexually experienced male [1, 18]. After a sexual contact, the male was removed until the female pressed the lever again. The latencies to press the lever were shorter after they received a mount than after receiving an intromission, which in turn were shorter than when they received an ejaculation. The authors conclude that sexual contacts are positively reinforcing when they are spaced and females can discriminate and space the stimulation they receive [2]. A similar pattern of spaced sexual stimulation and behavior is observed when females mate with tethered males [19, 20].

Two other methods, bilevel chambers and multiple choice preference tests, are used in which the females pace the sexual interaction which will be briefly described. As their name indicates, the bilevel chambers have two levels that are connected by ramps located at the sides. The main characteristic of these chambers is that they are narrow allowing an optimal side way observation. The males pursue the females through the chamber, and the female can pace the sexual interaction by running from one level to the other. These chambers have been used to study consummatory and appetitive aspects of male [21, 22] and female sexual behavior [23, 24].

The multiple partner preference/choice arena is composed of four cylinders facing to the middle. Each cylinder has a hole in the floor allowing the subject to move freely from one cylinder to the other and to the central security compartment. The stimulus animals cannot leave the cylinder because they are tethered but can engage in socio-sexual interactions [25, 26]. Different stimulus animals are placed in the cylinders to evaluate partner preference in males or females [27]

One important characteristic of the different methods used in which the female paced the sexual interaction is that in these methods, the females display behavioral patterns that are similar to those observed in seminatural conditions [28–30]. The measures of appetitive sexual excitement, solicitation, pacing, hopping, darting, and rejection responses are easily distinguished in these chambers and can be used in conjunction with traditional measures of lordosis.

3 Paced Mating

Although there was clear evidence in studies under natural or seminatural conditions, as well as in laboratory settings demonstrating that females could control or pace the sexual interactions, the first scientist who made systemic observations allowing the female to pace the sexual interaction was Mary Erskine. In her seminal paper, in 1989, she described that "the pacing chamber is equally divided by a removable partition, which contains a small hole through which the female can enter or exit from that half of the cage to which the male is restricted." In this way, the female can move freely from one compartment to the other, pacing the sexual interaction. When the male tries to follow the female, he cannot go through the hole because of his normally bigger size. Important behavioral and physiological changes associated with paced mating were described in the same publication by Erskine [31]. We will briefly mention them here because of the importance they have in reproduction. The exits and enters to the compartment containing the male can occur with or without the concurrent display of hops, darts, and ear wiggling. So, in paced mating tests, these behaviors are not an adequate measure of proceptivity. On the other hand, what is clear in paced mating tests is that the female exits the male compartment depending on the previous stimulation. On average, females will leave the male side more frequently after receiving an ejaculation, than after receiving an intromission than after receiving a mount [31]. Two variables important in the study of paced mating are the latencies and percentage of exits after mounts intromissions and ejaculations. Another set of important observations described by Erskine was that females that paced the sexual interaction had a higher possibility of pseudopregnancy (PSP) and pregnancy, and pacing females required fewer intromissions than non-paced females to induce PSP. These contributions done by Erskine [31] set the ground for studies about different mechanisms associated with paced mating. We will now describe in detail the experimental setup and procedure to perform pacing experiments. The description is based on studies that have used the pacing chamber with one hole, as originally described by Erskine [31].

4 Method

4.1 Pacing Chamber

As indicated, we used a chamber like that described by Erskine [31]. Originally, we made them of wood, but we now use acrylic because they are easy to move, clean, and are not easily impregnated by odors (Fig. 1). The Plexiglas chambers measure 62 × 29 × 42 cm and, in the middle, it has a removable Plexiglas partition with a hole near the bottom, 7 cm in diameter, which allows the female to move back and forward from the male side freely. Since males are usually bigger than the females, when the male tries to follow the female, he cannot pass through the hole.

4.2 Non-Pacing Chambers

They are exactly as the pacing chambers, but the Plexiglas partition is removed. In this case, the animals mate freely and the male, not the female, controls the rate of the sexual stimulation (Fig. 2).

4.3 Subjects

We usually use young sexually inexperienced females around 250 grams, which are maintained under a reversed 12 h light/dark cycle with food and water ad libitum. We house 3 or 4 subjects of the same control or experimental group per cage. Depending on the aim of the experiment, they can be ovariectomized or left intact. When we ovariectomized them, we usually wait 1 week to start the hormonal priming and 2 weeks to begin testing. To induce sexual

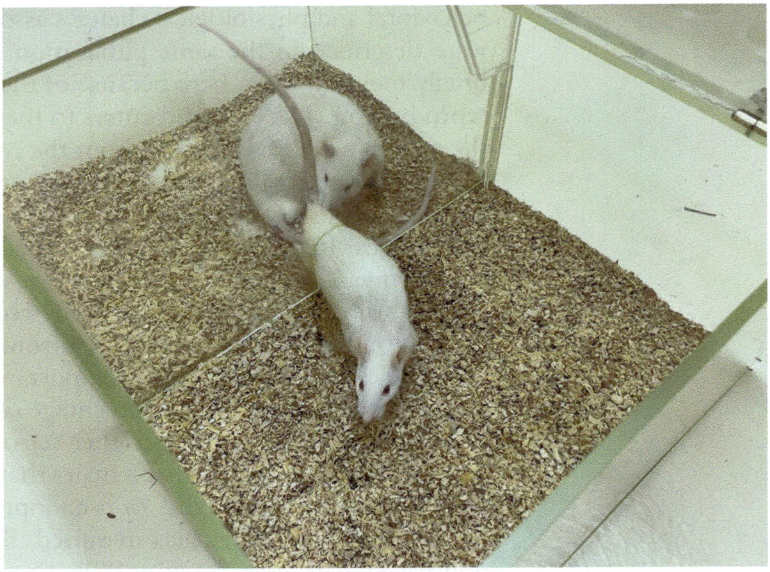

Fig. 1 A paced mating chamber with an acrylic partition with a hole in the bottom that allows the female but not the male to move from one compartment to the other. In this case, the female controls the rate of sexual interaction

Fig. 2 A non-paced mating chamber in which the male controls the rate of sexual interaction

receptivity, subjects are subcutaneously injected with 17-B-estradiol (EB, 25 μg/rat– Sigma, St. Louis, MO, USA) at 48 h and with progesterone (P, 1 mg/rat – Aldrich, St. Louis, MO, USA) 4 h before testing. Both hormones are diluted in corn oil. We used these relatively high doses to ensure that the females are sexually receptive at the time of testing but a lower dose of EB (5, 2.5 or 1.25 μg) plus a 0.5 mg of P reliably induces pacing behavior [32].

4.4 Stimulus Males

Usually, males of the same age as females are bigger and will not be able to follow them through the hole of the pacing chamber. It is convenient to have the males trained in the pacing chamber to acquire sexual experience. In the initial sessions, the male will attempt to follow the female but as they are tested repeatedly, these attempts are reduced.

4.5 Test Duration

The test duration is basically determined by the aim of the study. However, we have evaluated paced mating when females received 10, 15 intromissions or an ejaculation [14]. We have also tested paced mating in 30 min, 1 h tests, or when the females receive several ejaculations [33]. As testing is extended, the latencies to return to the male side increase.

4.6 Procedure

4.6.1 Behavioral Observations

During the sexual behavior tests (paced or non-paced), different parameters are recorded. The most commonly used are the lordosis quotient and the lordosis intensity. The lordosis quotient (LQ) is the number of lordosis divided by the number of mounts + the number of intromissions multiplied by 100. The lordosis intensity (LI) is calculated for each lordosis response and is ranked in intensity between 0 and 3 according to the extent of dorsiflexion observed [34]. To obtain the mean lordosis intensity (MLI), the sum of lordosis score is divided by the number of mounts plus intromissions received by the female. These two variables measure female receptivity.

In the pacing tests, where the females controlled the rate of sexual interaction, the exits from the male compartment after a copulatory event are determined and expressed as a percentage of exits after mounts (% EM), intromissions (% EI), and ejaculations. We also calculate the latencies for the female to re-enter the male's compartment after a mount (mount return latency; MRL), an intromission (intromission return latency; IRL), and an ejaculation (ejaculation return latency; ERL). The percent of exits and the return latencies after a mount, intromission, and ejaculation are the excellent measures of female proceptivity. As already indicated, hops, darts, and ear wiggling are not good indicators of female proceptivity in pacing tests because, in most cases, no differences are observed in these parameters in both testing conditions. In Fig. 3, we present the data of hopping, darting, ear wiggling, and the sum of these events along with the lordosis quotient and the mean lordosis intensity in females tested once a week for four consecutive weeks. They were tested with sexually experienced males for 30 min under non-paced mating conditions followed by 30 min of testing under pacing conditions. In the following week,

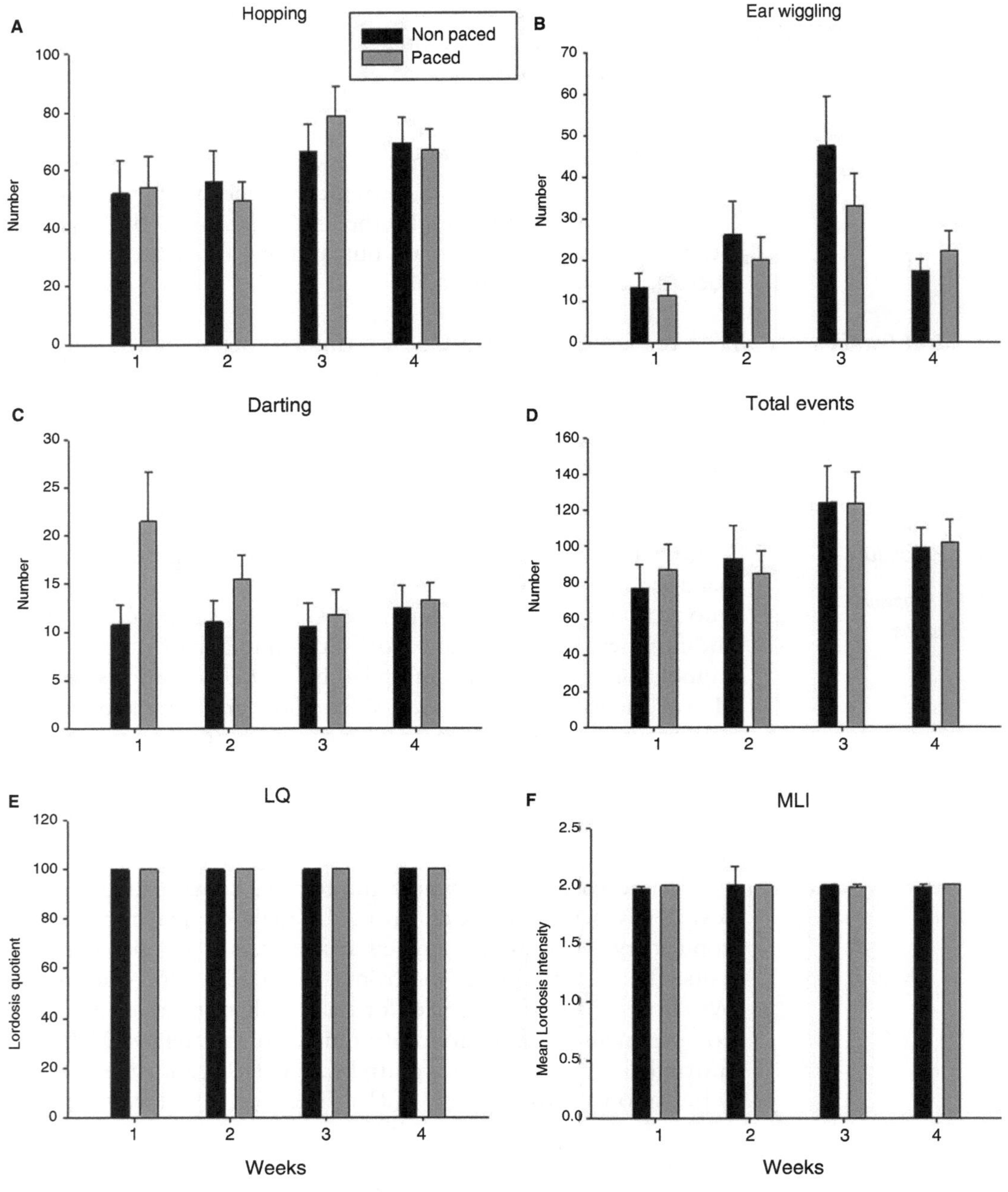

Fig. 3 Ovariectomized hormonally primed females ($N = 8$) were tested for sexual behavior with sexually experienced males once a week for 4 weeks in 1 h tests. In weeks 1 and 3, they mated for 30 min under non-paced mating conditions followed by 30 min of testing under pacing conditions. In the following week, the treatments were reversed. The total number of hooping (**a**), ear wiggling (**b**), and darting responses (**c**), the total number of these events (**d**) as well as the lordosis quotient (**e**) and the lordosis intensity (**f**) were compared with a T test or a Man-Whitney U-test. No significant differences were observed in any of the parameters between groups

Table 1
Paced mating parameters in the females that paced the sexual interaction as described in Fig. 3

Weeks	1	2	3	4
MLI	1.98 ± 0.02	1.83 ± 0.16	2	1.98 ± 0.02
LQ	100	100	100	100
Percentage of exits after				
Mounts	49.2 ± 8.4	65.6 ± 11	45.9 ± 7.4	56.8 ± 7.7
Intromissions	59.6 ± 5.6	65.5 ± 7.2	72.2 ± 4.0	78.5 ± 2.7
Ejaculations	100 ± 0	100 ± 0	100 ± 0	100 ± 0
Return latencies after				
Mounts	15.8 ± 5.8	6.8 ± 3.9	13.1 ± 3.4	9.2 ± 2.5
Intromissions	41.4 ± 15.4	32.8 ± 8.0	43.0 ± 20.2	32.2 ± 9.1
Ejaculations	61.0 ± 16.3	74.5 ± 11.6	103.2 ± 33.2	180.7 ± 73.0

Females were highly receptive as observed in the mean lordosis intensity (MLI) and the lordosis quotient (LQ). The percentage of exits are higher, and the return latencies are longer when the stimulation they received is of higher intensity Data are expressed as mean ± standard error

the treatments were reversed. As can be seen, no differences were observed in these parameters. Moreover, females were highly receptive as indicated by the mean lordosis intensity (MLI) and the lordosis quotient (LQ). As usually observed in paced mating tests, the percentage of exits is higher, and the return latencies are longer when the stimulation they received is of higher intensity. That is, the intensity of the stimulation is higher after an ejaculation than after an intromission than after a mount. These results confirm that in paced mating tests the percent of exits and the return latencies are a better measure of female proceptivity (*See* Table 1).

A trained observer can register three or four females at the same time, but if they are highly sexually active, it might be difficult to keep track and score all the variables properly. Of course, there is also the possibility of recording the sexual behavior tests and analyzing them later.

5 Sex Differences and Similarities in Pacing Behavior

It is well documented that wild rats mate in groups, where several females come into estrous at the same time, and males and females can have sexual contacts with one or several members of the opposite sex [3, 30, 35]. This pattern allows males and females to mate at their own rhythm controlling both sexes their sexual interactions. Classic studies have shown that when males and females pace (control) the sexual interaction, a reward state evaluated by conditioned place preference (CPP) is produced (reviewed in [36]). On the other hand, when males and females mate in a testing environment where the opposite sex controls the sexual interaction, no reward state is observed [37]. A detailed description of how CPP is used to evaluate sexual or other rewards is found in the chapter by Bedos.

We have done several studies using the CPP to unravel important characteristics of paced mating that will be briefly described in the following section. For example, we have shown that males and females when allowed to pace the sexual interaction do not need to ejaculate (in the case of the males) or receive an ejaculation (in the case of the females) to develop a reward state after mating. Males will find sex rewarding by displaying 15 intromissions or 1 ejaculation. However, if they display 5 or 10 intromissions, they will not find sex rewarding. This study demonstrated that males require a minimum amount of stimulation (15 intromissions or 1 ejaculation) to perceive sex as rewarding. This was further confirmed in another study in which males received an injection of the serotonin 1A receptor agonist (8-OH-DPAT) that produces a drastic reduction in the number of intromissions necessary to achieve ejaculation. Males treated with this compound ejaculated in less than five intromissions but did not find sex rewarding, indicating that ejaculation alone is insufficient to induce a reward state and that an optimal amount of sexual stimulation is required [38]. Similar observations have been described in females. Those that received 10 or 15 intromissions or an ejaculation under pacing conditions develop a reward state. On the other hand, those females that received five-paced intromissions or an ejaculation after several intromissions under non-paced conditions did not develop a reward state [14]. Together, these results indicate that, in rats, males and females need to pace the sexual interaction and receive a minimum amount of sexual stimulation, 15 intromissions for the male and 10 for the female for sex to induce a reward state.

One could speculate that when males mate under a condition where the females control the sexual interaction, they would eventually develop a reward state and keep mating. To test this hypothesis, we allowed a group of males to mate once weekly for 10 consecutive weeks in a chamber where the females, not the males, control the sexual interaction. These males continued mating but did not develop a reward state indicating that an estrous female and/or sexual behavior are strong incentives for the males, and they will continue mating even if the rewarding properties of the incentive are reduced [39]. It could also be the case that the CPP is not sensitive enough to detect a reduced reward state when males do not control the rate of sexual stimulation.

In most of the studies that we have done evaluating paced mating in females, we have used ovariectomized subjects to control for possible differences in hormone levels among subjects and, of course, to avoid pregnancy. We have used relatively high doses of the supplemental hormones to ensure that females are receptive, 25 μg/rat of EB, and 1 mg/kg of P. These doses induce high levels of receptive and proceptive behaviors. But as already described above, lower doses of EB (5, 2.5, or 1.25 μg estradiol) plus 0.5 mg of P reliably induce pacing behavior [32]. Females treated with these lower doses that were allowed to pace the sexual interaction developed a reward state. The group treated with 0.625 μg of EB, paced the sexual interaction but showed reduced proceptivity and receptivity without developing a reward state [32]. We have also demonstrated that different ring A-reduced metabolites of progesterone facilitate the reward state following pacing [40].

In another study, we tested if paced mating could induce a reward state in natural cycling females. In the CPP procedure that we follow, subjects have three reinforced and three non-reinforced sessions. That is, females paced the sexual interaction in three different sessions. We could not follow the same procedure in gonadally intact females because they could get pseudo-pregnant or pregnant, changing their hormonal condition. Therefore, we allowed the females to receive three ejaculations in one conditioning session pacing the sexual interaction. Other groups of females mated until they receive three ejaculations without pacing the sexual interaction. Only the groups of females that paced the sexual interaction (receiving one or three ejaculations) developed a reward state [41]. In the same study, we asked if females would develop a reward state if they paced the sexual interaction through a one- or three-hole pacing chamber. We found that it does not make a difference, a reward state is observed in females that mated pacing the sexual interaction in a one- or three-hole chamber [41].

As already indicated, females that mate without pacing the sexual interaction do not find sex rewarding as evaluated by CPP. In our experiments, females mate with a different male in each conditioning session. It could be argued that females do not develop a reward state under non-paced mating conditions because they receive stimulation from various males. To evaluate this possibility, we allowed the females to mate with the same male under non-paced mating conditions in the three conditioning sessions. The results indicate that females do not develop a reward state when they mate with the same male in the non-paced tests [41], further demonstrating the importance of controlling the sexual interaction for the induction of a reward state. As testing is extended in females that do not pace the sexual interaction, they started to show aggressive and avoidance behavior when the males attempt to mount them. However, in extended pacing tests, these behaviors are not observed as females can stay on the side where the male has no access. We have shown that females that mate for 1 h and receive around 25 intromissions controlling the sexual interaction develop a reward state [33].

In other studies, we evaluated if sexual behavior could induce a reward state of the same intensity as a morphine injection. For this purpose, subjects were placed in the non-preferred compartment after achieving an ejaculation; on alternate days, they were placed in the preferred compartment after a morphine injection of 1 mg/kg. In other groups, treatments were reversed. In the case of the males, no group changed their original preference indicating that for males sex can be equally rewarding as a morphine injection of 1 mg/kg [42]. On the other hand, the same treatment in females induces a reward state only in females conditioned for pacing behavior and not for morphine injection indicating that in females, contrary to males, paced mating induces a reward state of higher intensity that an injection of 1 mg/kg of morphine [33].

The reward state induced by sexual behavior is blocked by the administration of the opioid antagonist naloxone in males [43, 44] and females [45], indicating that in both sexes this reward state is mediated by opioids. Interestingly, dopamine antagonists did not block the reward state induced by paced mating in males [43] or females [46]. In a follow-up study, we demonstrated that the infusion of naloxone in the medial preoptic area, or the ventromedial hypothalamus or the medial amygdala (all brain regions important for the expression of sexual behavior) but not in the nucleus accumbens blocked the reward state induced by paced mating [47]. In males, naloxone blocked the reward state induced by sexual behavior when infused in the medial preoptic area but not the nucleus accumbens [48], suggesting that opioids could mediate sex reward in the medial preoptic area in both sexes.

We have done a series of studies investigating if paced mating can induce long-term plastic changes evaluated by the formation of new neurons. A description of these studies is beyond the scope of the present chapter, but we have shown that paced mating in males and females induces neurogenesis in the olfactory bulbs and the hippocampus, for a review (*see* [49] and references therein).

6 Conclusions

The ability to control the rate of sexual stimulation has been observed in seminatural and natural settings and is clear that the females have an active role in the sexual interaction. The pacing method reduces the aversive properties of mating and enhances the appetitive components of the sexual interaction. Paced mating induces physiological and behavioral changes that favor reproduction. It also generates a reward state, ensuring that the behavior will be repeated. Moreover, pacing induces the formation of new neurons in the olfactory bulbs and the hippocampus. Future studies need to address the role of these new neurons. It would also be interesting to determine the possible role of paced mating in other species.

Acknowledgments

Research supported by PAPIIT UNAM grant IN206521.

References

1. Bermant G (1961) Response latencies of female rats during sexual intercourse. Science 133(3466):1771–1773
2. Barnett SA (1975) The rat: a study in behavior. University of Chicago Press, Chicago
3. McClintock MK, Adler NT (1978) The role of the female during copulation in wild and domestic Norway rats (Rattus norvegicus). Behaviour 67(1/2):67–96
4. Beach FA (1976) Sexual attractivity, proceptivity, and receptivity in female mammals. Horm Behav 7(1):105–138
5. Agmo A (1999) Sexual motivation–an inquiry into events determining the occurrence of sexual behavior. Behav Brain Res 105(1):129–150
6. González-Flores O et al (2017) Female sexual behavior in rodents, lagomorphs, and goats. In: Pfaff D, Joëls M (eds) Hormones, brain and behavior. Elsevier, pp 59–82
7. Agmo A (2003) Unconditioned sexual incentive motivation in the male Norway rat (Rattus norvegicus). J Comp Psychol 117(1):3–14
8. Komisaruk BR (1972) Induction of lordosis in ovariectomized rats by stimulation of the vaginal cervix: hormonal and neural interrelationships. UCLA Forum Med Sci 15:127–135
9. Komisaruk BR, Diakow C (1973) Lordosis reflex intensity in rats in relation to the estrous cycle, ovariectomy, estrogen administration and mating behavior. Endocrinology 93(3): 548–557
10. Blaustein JD, Erskine MS (2002) Feminine sexual behavior: cellular integration of hormonal and afferent information in the rodent forebrain. Hormones Brain Behav 1:139–214
11. Pfaff D (2017) How the vertebrate brain regulates behavior. Direct from the lab. Harvard University Press, Cambridge, MA
12. Pfaff DW (1980) Estrogens and brain function: neural analysis of a hormone-controlled mammalian reproductive behavior. Springer, New York
13. Pfaff DW, Gagnidze K, Hunter RG (2018) Molecular endocrinology of female reproductive behavior. Mol Cell Endocrinol 467:14–20
14. Paredes RG, Vazquez B (1999) What do female rats like about sex? Paced mating. Behav Brain Res 105(1):117–127
15. Eliasson M, Meyerson BJ (1975) Sexual preference in female rats during estrous cycle, pregnancy and lactation. Physiol Behav 14(6): 705–710
16. Myerson BJ et al (1973) Sexual motivation in the female rat after testosterone treatment. Physiol Behav 11(4):421–428
17. Paredes RG, Baum MJ (1995) Altered sexual partner preference in male ferrets given excitotoxic lesions of the preoptic area/anterior hypothalamus. J Neurosci 15(10):6619–6630
18. Bermant G, Westbrook WH (1966) Peripheral factors in the regulation of sexual contact by female rats. J Comp Physiol Psychol 61(2): 244–250
19. Broekman M et al (1988) Partner preference behavior of estrous female rats affected by castration of tethered male incentives. Horm Behav 22(3):324–337
20. de Jonge FH, van de Poll NE (1986) On the involvement of progesterone in sexually rewarded choice behavior of the female rat. Physiol Behav 37(1):93–98
21. Pfaus JG, Mendelson SD, Phillips AG (1990) A correlational and factor-analysis of anticipatory and consummatory measures of sexual-behavior in the male-rat. Psychoneuroendocrinology 15(5–6):329–340
22. Pfaus JG, Phillips AG (1991) Role of dopamine in anticipatory and consummatory aspects of sexual-behavior in the male-rat. Behav Neurosci 105(5):727–743
23. Pfaus JG et al (2000) Appetitive and consummatory sexual behaviors of female rats in bilevel chambers II. Patterns of estrus termination following vaginocervical stimulation. Horm Behav 37(1):96–107
24. Pfaus JG, Smith WJ, Coopersmith CB (1999) Appetitive and consummatory sexual behaviors of female rats in bilevel chambers – I. A correlational and factor analysis and the effects of ovarian hormones. Horm Behav 35(3): 224–240
25. Ferreira-Nuno A et al (2010) Copulatory pattern of male rats in a multiple partner choice arena. J Sex Med 7(12):3845–3856
26. Ferreira-Nuno A et al (2005) Sexual behavior of female rats in a multiple-partner preference test. Horm Behav 47(3):290–296

27. Ferrerira-Nuño A et al (2018) Sexual incentive and choice. Curr Sex Health Rep 10:132–141
28. Mcclintock MK, Adler NT (1978) Role of female during copulation in wild and domestic Norway rats (Rattus-Norvegicus). Behaviour 67:67–96
29. Mcclintock MK, Anisko JJ (1982) Group mating among Norway rats. 1. Sex-differences in the pattern and neuroendocrine consequences of copulation. Anim Behav 30(May):398–409
30. Mcclintock MK, Anisko JJ, Adler NT (1982) Group mating among Norway rats .2. The social dynamics of copulation – competition, cooperation, and mate choice. Anim Behav 30 (May):410–425
31. Erskine MS (1989) Solicitation behavior in the estrous female rat: a review. Horm Behav 23(4):473–502
32. Corona R et al (2011) Different doses of estradiol benzoate induce conditioned place preference after paced mating. Horm Behav 60(3): 264–268
33. Arzate DM et al (2011) Extended paced mating tests induces conditioned place preference without affecting sexual arousal. Horm Behav 59(5):674–680
34. Hardy DF, DeBold JF (1972) Effects of coital stimulation upon behavior of the female rat. J Comp Physiol Psychol 78(3):400–408
35. Chu X, Agmo A (2015) Sociosexual behaviors during the transition from non-receptivity to receptivity in rats housed in a seminatural environment. Behav Process 113:24–34
36. Paredes RG (2014) Opioids and sexual reward. Pharmacol Biochem Behav 121:124–131
37. Martinez I, Paredes RG (2001) Only self-paced mating is rewarding in rats of both sexes. Horm Behav 40(4):510–517
38. Camacho FJ et al (2007) Facilitation of ejaculation induced by 8-OH-DPAT does not produce conditioned place preference in male rats. Behav Neurosci 121(3):579–585
39. Camacho F, Sandoval C, Paredes RG (2004) Sexual experience and conditioned place preference in male rats. Pharmacol Biochem Behav 78(3):419–425
40. Gonzalez-Flores O et al (2004) Progestins and place preference conditioning after paced mating. Horm Behav 46(2):151–157
41. Camacho FJ, Garcia-Horsman P, Paredes RG (2009) Hormonal and testing conditions for the induction of conditioned place preference by paced mating. Horm Behav 56(4):410–415
42. Camacho FJ et al (2009) Reward value of intromissions and morphine in male rats evaluated by conditioned place preference. Physiol Behav 98(5):602–607
43. Agmo A, Berenfeld R (1990) Reinforcing properties of ejaculation in the male rat: role of opioids and dopamine. Behav Neurosci 104(1):177–182
44. Mehrara BJ, Baum MJ (1990) Naloxone disrupts the expression but not the acquisition by male rats of a conditioned place preference response for an oestrous female. Psychopharmacology 101(1):118–125
45. Paredes RG, Martinez I (2001) Naloxone blocks place preference conditioning after paced mating in female rats. Behav Neurosci 115(6):1363–1367
46. Garcia Horsman P, Paredes RG (2004) Dopamine antagonists do not block conditioned place preference induced by paced mating behavior in female rats. Behav Neurosci 118(2):356–364
47. Garcia-Horsman SP, Agmo A, Paredes RG (2008) Infusions of naloxone into the medial preoptic area, ventromedial nucleus of the hypothalamus, and amygdala block conditioned place preference induced by paced mating behavior. Horm Behav 54(5): 709–716
48. Agmo A, Gomez M (1993) Sexual reinforcement is blocked by infusion of naloxone into the medial preoptic area. Behav Neurosci 107(5):812–818
49. Bedos M, Portillo W, Paredes RG (2018) Neurogenesis and sexual behavior. Front Neuroendocrinol 51:68–79

Chapter 13

Assessment of Sexual Reward with the Conditioned Place Preference Paradigm

Marie Bedos

Abstract

Sexual behavior, as other naturally motivated behaviors such as eating behavior, involves mesolimbic circuits that belong to the reward system as well as social behavior networks. As a result of the activation of those circuits, individuals normally present an innate motivation to approach those incentives (i.e. potential sexual partner or food) and if they are able to consume such incentives, they will experience a positive affective state. Those mechanisms are crucial for the behavior to be repeated in the future and hence, for the survival of the species. In this context, the conditioned place preference (CPP) paradigm has been a standard behavioral model used to study the rewarding properties of sexual behavior in laboratory models such as rodents. In this chapter, I will provide detailed materials and methods of CPP procedures and specificities about males and females will be addressed. I will also describe the variants in CPP methodology that have been employed to decipher the rewarding properties of sexual behavior and, finally, I will review studies that have elucidated the differences between males and females regarding sexual reward, its neurobiological basis, and the importance of other factors such as sexual experience, steroidal hormones, and sensory cues.

Key words Sexual behavior, Reward, Conditioning, Paced mating, Sensory cues

1 Introduction

Behaviors that are fundamental to obtain basic needs, including food, water, social, and sexual interaction are naturally motivated behaviors. The abovementioned stimuli act as innate positive incentives and therefore activate an approach behavior. Meeting these needs is a requirement for the survival of species and mechanisms involving the reward system guarantee that the behaviors will be repeated [1]. The reward system not only modulates the motivational aspects of those behaviors ("wanting") but also the affective states produced by them ("liking" or rewarding state), as demonstrated by the seminal work of Kent Berridge with food reward [2]. As other naturally motivated behaviors, sexual interaction produces a positive affective (PA) state, which can be measured in laboratory models such as rodents.

Raúl G. Paredes et al. (eds.), *Animal Models of Reproductive Behavior*, Neuromethods, vol. 200,
https://doi.org/10.1007/978-1-0716-3234-5_13,

Several paradigms have been used to assess the rewarding effects of sexual incentives and interactions [3]. Early studies employed operant behavior to demonstrate that the possibility of accessing a sexual partner is a positive reinforcer [4, 5]. Other methods such as partner preference and sexual incentive motivation are frequently used to assess motivational aspects of sexual behavior [6–12]. Both paradigms give an experimental subject the choice between two stimulus animals; however, while subjects can interact physically in the partner preference paradigm, they can only see, hear, and smell stimulus animals in the sexual incentive motivation. It can be assumed from the preference of a subject for a sexual stimulus that the approach behavior is intrinsically rewarding. However, in this chapter, I will focus on the conditioned place preference (CPP) paradigm, which has been the most employed method to evaluate the capacity of a stimulus to induce PA state/ rewarding properties of a stimulus, including sexual behavior. CPP takes advantage of the fact that animals usually develop a preference to remain in a context that was paired with a reward over a context that was not. This experimental paradigm was first reported in a study by Horace Beach (1957), where he showed that rats developed a preference for the compartment where they received a subcutaneous injection of morphine [13]. Almost two decades later, Rossi and Reid (1976) described that the preference of the rats for the compartment where they experienced the rewarding state was due to the PA state induced by the morphine injections [14]. Since then, CPP has been widely employed to provide insights into whether a drug, hormone, or behavior is perceived as rewarding (*see* [15, 16] for detailed reviews on the use of CPP).

In this chapter, I will provide detailed materials and methods of CPP procedures and specificities about males and females will be addressed. I will also describe the variants in CPP methodology that have been employed to decipher the rewarding properties of sexual behavior and, finally, I will review studies that have elucidated the differences between males and females regarding sexual reward, its neurobiological basis, and the importance of other factors such as sexual experience, steroidal hormones, and sensory cues.

2 Materials

2.1 Apparatus

2.1.1 Mating Cages

The mating chamber consists of an acrylic box (40 × 60 × 40 cm) with clean sawdust on the floor. The chamber can be divided into two equal compartments by a partition with a small hole (4 cm diameter) at the bottom (Fig. 1). The hole is big enough to allow the female, but not the male, to go back and forth from the male compartment. This configuration allows the females to control the rate of sexual interaction. When the partition is removed, the males get free and continual access to the female, which allow them to

Fig. 1 Paced mating chamber. The paced mating chamber consists of an acrylic box with clean sawdust on the floor divided in two equal compartments by a partition with a small hole at the bottom. The hole is big enough to allow the female, but not the male, to go back and forth from the male compartment. This configuration allows the females to control the rate of sexual interaction

pace the sexual interaction. Another type of partition has small holes (1 cm in diameter), and only allows olfactory and auditory cues to be exchanged between the subjects without physical contact. The latter can be used as a control or to separate the sensory cues involved in sexual behavior [17].

2.1.2 CPP Apparatus

To test the capacity of sexual behavior to induce a PA state, we use a rectangular apparatus (100 cm long × 30 cm wide × 32 cm high) made of acrylic divided into three compartments (Fig. 2). The device consists of a 20 × 30 × 32 cm compartment painted in grey in the middle that communicates with two lateral compartments (38 × 30 × 32 cm) through sliding doors (9.5 cm wide × 11.5 cm high). One of the lateral compartments is painted white with a removable wrinkled plastic sheet placed on the floor and the contralateral compartment is painted black, has smooth floor, and is moistened with a 2% acetic acid solution immediately before an animal is placed in it. Thus, the lateral compartments offer different cues in terms of color, texture, and odor. The front wall of the middle compartment is made of fine wire mesh that allows the experimenter to observe the subjects. The place preference apparatus and the mating cage are in adjacent rooms illuminated with dim red light.

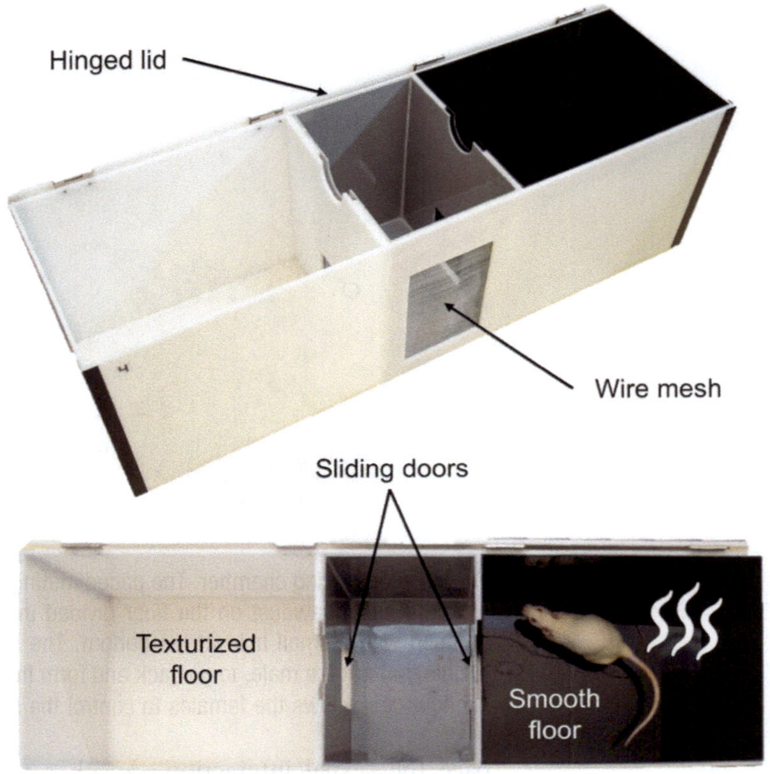

Fig. 2 CPL apparatus. The CPP apparatus consists of a rectangular box made of acrylic divided in three compartments: one in the middle painted grey that communicates with the two lateral compartments through sliding doors. One of the lateral compartments is painted white and has a texturized floor and the contralateral compartment is painted black, has smooth floor and is moistened with a 2% acetic acid solution immediately before an animal is placed into it

2.2 Animal Preparations

2.2.1 Males

If experimental subjects are females, we usually allow males to copulate with an estrous female for three to five sessions before using them as stimuli. We consider them sexually experienced when they can ejaculate at least once in less than 30 min. This training guarantees that males will efficiently copulate with the experimental females, displaying short latencies to mount and intromit [18]. If the paradigm used is paced mating i.e., with the partition that only allows females to pass through, we also train the males to copulate in this configuration. Indeed, untrained males usually try to pass to the female compartment, but they learn not to do so if we prevent them from doing it during training sessions. This training guarantees that females will be able to successfully pace sexual interaction.

If experimental subjects are males, males can be subjected to the abovementioned training or not, depending on the objective of the experiment.

2.2.2 Females

CPL is a conditioning paradigm that implies learning processes. To this aim, reinforcing sessions are alternated with non-reinforcing sessions on consecutive days [19], or, in some cases, one reinforcing and one non-reinforcing sessions are done on the same day [20]. The conditioning must be repeated at least three times. This schedule is incompatible with the estrous cycle of females, which usually presents estrous behavior every 4–5 days. Therefore, it is necessary to ovariectomize females and induce estrous with exogenous steroid hormones every second day to submit them to the protocol. Routinely, we bilaterally ovariectomize females under general anesthesia and let them recover from the surgery for 2 weeks before starting conditioning [19]. A week after the surgery, we start administrating estradiol benzoate (EB) and progesterone to induce estrous behavior [21]. If the females are used as stimuli, we use 25 μg/rat of EB and 1 mg/rat of progesterone via subcutaneous injections, 48 h and 4 h before mating tests, respectively. If the females are the experimental subjects that will be tested with the CPL paradigm, the investigator can decide to use different doses of steroid hormones depending on the objective of the experiment [19].

2.3 Video Analysis

Both pre-and final tests are videotaped and analyzed with the behavioral data acquisition computer program Ethovision 3.0 (Noldus). Briefly, the program enables to define the three areas corresponding to the three compartments and automatically calculates the time spent and the number of entrances in each compartment. The subject is considered in a compartment when more than half of its body is in the area.

3 Methods

3.1 CPP Paradigm

In this section, I will describe the protocol that we have been using in the laboratory (Fig. 3), for the variants in CPP methodology, we invite the interested reader to consult the Notes Subheadings 4.1.1, 4.1.2, and 4.1.3. We usually habituate subjects to handling and -if necessary- to i.p. injections, but we do not expose them to the CPP apparatus before the pre-test. We use a biased design in which the baseline preference of the subjects for each compartment of the apparatus is assessed before the onset of conditioning sessions [16]. During this 10 min pre-test which is videotaped, the lid and sliding doors of the apparatus are open and the subject is free to explore each of the three compartments. This pre-test allows to define each subject's least preferred compartment, which will be paired with the reinforcing event (sexual behavior), and the most preferred compartment, which will be paired with a neutral stimulus (Fig. 3). Moreover, the analysis of the pre-test enables to exclude any animal that spent less than 60 s in one of the

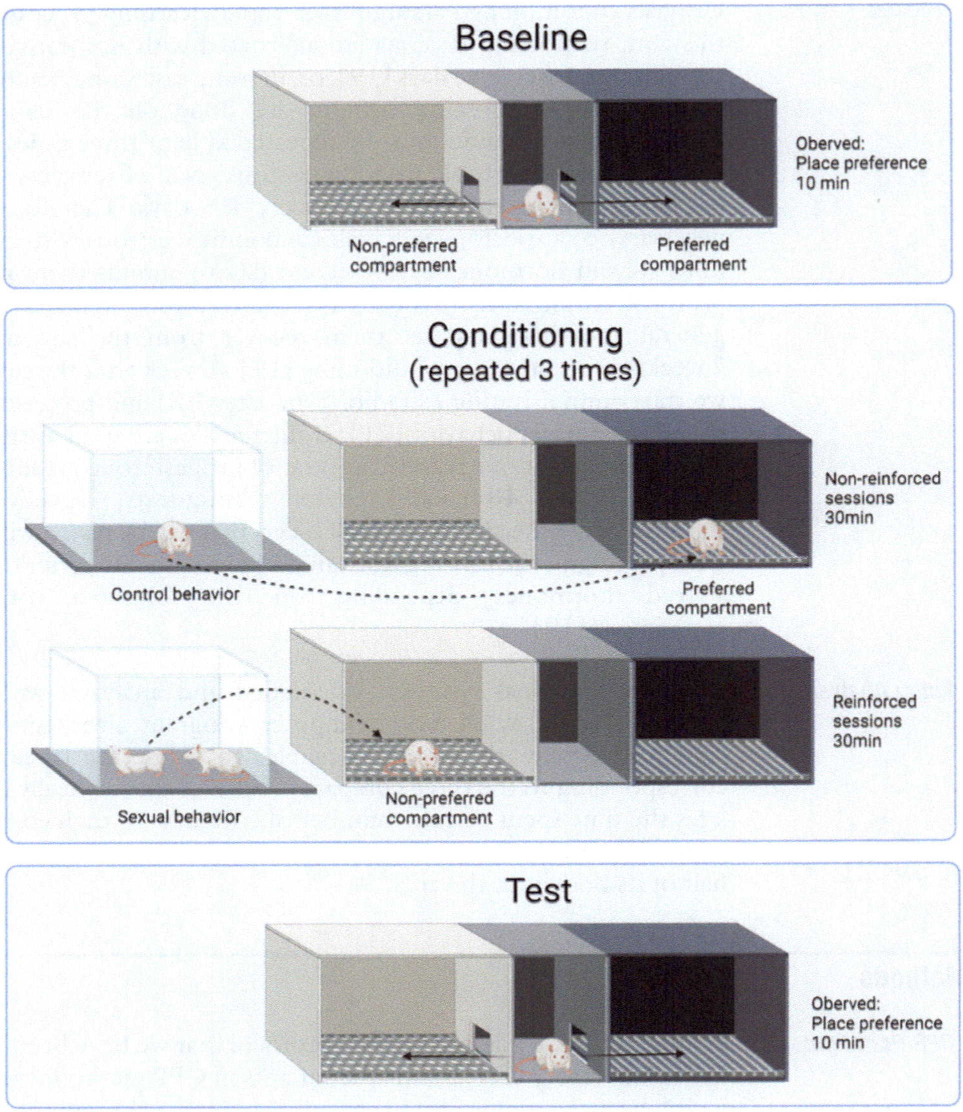

Fig. 3 Experimental procedure for CPP with sexual behavior. The procedure consists of three phases: pretest, conditioning, and final test. Pre- and final tests last 10 min and are videotaped. The analysis of pre-test gives the initial preference, and the final test evaluates the new preference after conditioning. During conditioning, three non-reinforced sessions are alternated with three reinforced sessions, starting with a non-reinforced session where the subject is exposed to a neutral stimulus and then placed in its preferred compartment with the lid closed for 30 min. On alternate days, the subject is allowed to copulate in the mating chamber and immediately placed in the non-preferred compartment with the lid closed for 30 min

compartments and therefore avoid obtaining a biased result because of the low initial preference of the subject for one of the compartments.

The day after the pre-test, we proceed to the conditioning paradigm, which consists of three non-reinforced sessions alternated with three reinforced sessions. We start with a non-reinforced session where the subject is exposed to a neutral stimulus and then placed in its preferred compartment with the lid closed for 30 min. On alternate days, the subject is allowed to copulate in the mating chamber and immediately placed in the non-preferred compartment with the lid closed for 30 min. For each session with sexual behavior, the stimulus subject is placed in the mating cage 2 min before the tested animal is introduced. This sequence of non-reinforced and reinforced sessions is repeated three times. The apparatus is cleaned using a solution of ethanol 70% between each subject. After 6 days of conditioning, we proceed to test the new preference in a 10-min session similar to the pre-test, without any treatment.

Importantly, the neutral stimulus for non-reinforced sessions depends on the behavior or drug tested on reinforced sessions. When sexual behavior is tested, the neutral stimulus consists in moving the subject to the mating chamber for the same period as in reinforced sessions. When the effect of a drug on sexual behavior is evaluated, the subject is injected with the drug's vehicle at the same time interval as when the drug is injected in reinforced sessions. If both behavior and drug are included in the procedure, the subject should be exposed to both neutral stimuli in non-reinforced sessions. Finally, it is mandatory to include a control group in the CPP design, which will be exposed to the same handling as test groups, including the implantation of a cannula if the procedure includes intracerebroventricular injections or exposition to neutral stimuli. If the effect of a drug is tested, the control group should be injected with a vehicle before both reinforced and non-reinforced sessions, at the same time interval as in test groups. No change in place preference is expected in the control groups.

3.2 Experimental Variables

Time in the Reinforced Chamber/Preference Score

Several types of analysis can be made with the results of the CPP. The first one that we usually employ is to compare the time spent in the reinforced chamber (initially least preferred) during the pre-test and the final test. If the experimental subjects spent significantly more time in the initially non-preferred compartment, it can be concluded that the stimuli produced a reward state of sufficient intensity and duration to induce conditioning. This comparison can be expressed as a difference score (the difference between the time spent in the non-reinforced cage and time spent in the reinforced cage) and is expected to decrease significantly after conditioning. The second analysis considers both the time spent in the non-preferred and preferred chambers. The preference score is calculated as [Time spent in the reinforced chamber/(Time spent in the reinforced chamber + Time spent in the non-reinforced chamber)]. The advantage of this analysis that considers times in both chambers is that it excludes a possible effect of the time spent in the grey chamber.

4 Notes

This chapter is intended to introduce the reader to the use of CPP as a methodology to assess reward induced by sexual behavior. Though the principles are similar, choices regarding the type (biased vs. unbiased), number of reinforced and non-reinforced sessions, and analysis vary substantially across studies. In addition, in this section, we discuss how CPP has been helpful in elucidating the differences between males and females regarding sexual reward, its neurobiological basis, and the importance of other factors such as sexual experience, steroidal hormones, and sensory cues.

4.1 Variations in the CPL Paradigm

4.1.1 Conditioning in the CPL Apparatus/in Separate Chambers

It is important to mention that in male rats, CPPs have been established using two different conditioning procedures. In the study of Everitt (1987), male rats were allowed to copulate to ejaculation in the conditioning apparatus, which caused the males to develop a preference for the compartment that was paired with the behavior [22]. In another study, males were allowed to copulate to ejaculation in a separate mating chamber and immediately transferred to the CPP apparatus [23]. Males also developed a clear CPP for the chamber that was paired with sexual behavior in the second procedure. Similarly, it was demonstrated that both procedures induce CPP in female rats [24–26]. By separating sexual behavior from the CPP apparatus, it can be concluded that the change in preference is induced by the PA state induced by copulation and not the behavior itself. As described in the Methods section, we use this methodology, with separate chambers for sexual behavior and conditioning.

4.1.2 Habituation, Duration of Tests, Number of Conditioning Sessions, and Interval Between Conditioning Sessions

In some studies, the subjects are habituated to the CPL apparatus before starting the conditioning [27]. For example, in the study of Domínguez-Salazar et al. (2014), male rats were habituated to the CPP apparatus for 5 min for 2 days before the baseline preference was assessed on the third day [28] and in the study of Hughes et al. (1990), they were habituated during 15 min for 3 days before assessing the baseline preference on the 4th day [29]. In another study in mice, males were allowed for 2–5 days to investigate during 10 min the CPP box before starting the conditioning. This procedure was repeated until the male showed a consistent preference for one of the two compartments [30]. Although some studies habituate subjects before evaluating the initial preference or starting the conditioning, in most investigations that used the CPP paradigm to assess sexual reward, no familiarization with the CPP apparatus is usually given to animals before the pre-test. Therefore, we can think that habituation to the apparatus is not crucial for the procedure to be successful.

There are also differences in the duration of pre- and final tests depending on the studies. As mentioned above, we use 10 min tests but other studies in mice and prairie voles have recorded the initial and final preference during 20 min or 30 min, respectively [31, 32].

CPP is a conditioning paradigm that implies learning processes. To this aim, reinforcing sessions are usually alternated with non-reinforcing sessions on consecutive days, or, in some cases, one reinforcing and one non-reinforcing sessions are carried out on the same day. This schedule facilitates memory formation but is difficult to manage with the estrous cycle of female rodents, which usually present estrous behavior every 4–5 days. Therefore, two types of schemes have been used in ovariectomized female rats. In the first one, the interval between reinforcing and non-reinforcing sessions is 3–4 days, with four or five repetitions of reinforced/non-reinforced sessions and estrous behavior is induced with exogenous hormones before each reinforced session [33–36]. In the second type of scheme, estrous is induced every second day to allow reinforced and non-reinforced sessions on consecutive days [19, 37–39]. In this case, the alternations of reinforced and non-reinforced sessions are usually repeated three times. Similarly, both types of schedules have been used in males [29, 38, 40]. Finally, it should be noted that a different scheme was used in a study with intact female rats, which consisted of a pre-test, one non-reinforced and one reinforced sessions with sexual behavior, and a final test. The reinforced session was carried out the day females were detected in estrous by cytology [41].

4.1.3 Experimental Variables

As mentioned in the methods, most of the CPP results are based on the comparison of the time spent in the reinforced chamber during the baseline and the final test and the preference score, which considers both times in the reinforced and non-reinforced chambers. However, some studies concluded that the subject changed its preference when they observed a significant longer time in the paired versus unpaired side during the final test [42, 43]. Several reviews emphasize the importance of considering the time in the reinforced chamber in CPP analyses [27, 44]. Indeed, only an increase in the time spent in the chamber that was paired with the reinforcer can ensure that the stimulus induced a PA state in the subject.

4.2 Sex

A large number of studies have shown that sexual behavior induces CPP in both male [23, 38, 40] and female rats [19, 21, 38]. However, as described in the next paragraph, the possibility to pace the sexual encounter is crucial for the acquisition of CPP in both sexes. Similarly, the rewarding properties of sexual behavior were demonstrated in both sexes of Syrian hamsters [45–47]. In male mice, it was shown that both intromissions (100 thrusts with intromissions) and one ejaculation induced CPP [30, 32] but to my knowledge, there is no evidence of CPP induced by sexual behavior in female mice. Finally, in other species such as prairie voles that were tested with the CPP paradigm, differences between sexes were described. In fact, mating with one ejaculation and social cohabitation with mating for 6 h induced CPP in males whereas in female voles, mating until one ejaculation, social cohabitation with mating, or exposure to a male without physical interaction for 6 h did not induce CPP [31].

4.3 Paced Mating

Under natural or seminatural conditions, both male and female rats control the sexual interaction because rats usually mate in groups and alternate partners during copulation. However, under conditions of laboratory, when females are mated in a testing arena, they cannot escape from the male resulting in mating being both appetitive and aversive for them. In this context, the possibility for the female to approach or withdraw from the male is crucial and the use of modified mating chambers as described in Subheading 2.1.1 enabled to test animals in seminatural environment. Using the chambers with the partition, several studies demonstrated that sexual behavior only induced a reward state in females that were able to control the sexual interaction. In comparison, females that mated with males that had free access to them (non-paced mating) did not change their place preference [25, 26, 38, 48]. Similarly, it was shown that it is necessary for males to pace the sexual encounter to induce CPP [38, 49]. Interestingly, Mermelstein and Becker (1995) found that the extracellular concentrations of dopamine were higher in the striatum and nucleus accumbens of females that paced the mating compared to females that were not allowed to control the sexual interaction [50]. The latter indicates that the PA state induced by paced mating probably involves the release of dopamine in the reward system. In contrast with these findings, other studies have reported that females can acquire CPP with the sessions of non-paced mating [36, 51]. However, it is important to note that in these studies, the sessions of non-paced mating lasted until the male completed 15 intromissions, whereas in the studies described previously, females were exposed to males for a longer time, which might have been aversive.

4.4 Steroid Hormones

In both males and females, sexual behavior is highly dependent on steroid hormones. Females become sexually receptive during the estrous because of the increase in gonadal estradiol and progesterone and the PA state induced by paced mating requires the females to display a high lordosis quotient. In fact, female rats injected with a lower dose of EB (0.625 μg) and submitted to the CPP paradigm after paced mating sessions did not change their initial preference [19]. In those females, authors reported a decrease in the lordosis quotient. Lordosis is a postural reflex with the dorsiflexion of the vertebral column that is a characteristic of sexually receptive female rodents and lordosis quotient is calculated as the percentage of mounts and intromission that resulted in a lordosis posture. Therefore, their results indicate that a minimum dose of estradiol is necessary to induce full receptivity and produce a reward state [19]. Likewise, female rats primed only with EB did not acquire CPP, even if they were allowed to pace the sexual interaction [25, 26] and another study showed that the administration of progesterone or ring A-reduced metabolites of progesterone such as 5aDHP and 5 h, 3a-Pgl 48 h after an injection of EB-induced CPP after paced mating [52]. Taken together, these studies show that ovarian hormones are crucial for the acquisition of CPP by paced mating likely because they sensitize sensory and motor pathways necessary for sexual behavior and allowing the stimulation that leads to a reward state. However, it was demonstrated that neither estradiol nor progesterone is necessary during the test after the conditioning to observe the effect of the reward state induced by paced mating [35].

Interestingly, female rats ovariectomized before puberty, and therefore, lacking exposure to pubertal ovarian hormones, retained the ability to acquire CPP for paced mating, even if they showed differences in their sexual behavior during paced mating, such as a higher percentage of exits after mounts [53]. In another study, it was found that the neonatal treatment with testosterone propionate did not alter the reward state induced by paced mating, even if females showed increased return latencies with the male after mounts or intromissions [54]. Similarly, the prenatal administration of flutamide, an androgen receptor antagonist, did not affect the capacity of females to experience a reward state after paced mating. However, these females developed CPP with fewer intromissions than controls, indicating that perinatal androgens might modulate the processing of somatosensorial cues that participate in the rewarding properties of mating in females [55].

In males, gonadally intact aromatase knockout mice did not acquire CPP for stimuli associated with an estrous female, indicating that the aromatization of testosterone into estradiol is involved in the rewarding properties of reproductively relevant odors [56]. In Syrian male hamsters, Bell and Sisk (2013) observed that vaginal secretions did not induce CPP in juveniles or gonadectomized adults compared to intact adults or gonadectomized adults with testosterone replacement [57]. Therefore, circulating testosterone is necessary for the expression of the rewarding properties of vaginal secretions in male hamsters. Finally, a recent study demonstrated that male rats treated with degarelix, a GnRH receptor antagonist, reduced the CPP for the environment previously associated with sexual experience [58].

For an exhaustive review of the effects of hormones on CPP and in particular the hormonal aspects of sexual reward, I recommend the interested readers to look at the review of Paredes [44].

4.5 Sexual Experience

It is important to note that the response to sexual reward is dependent on experience and is sexually dimorphic. Indeed, in male rats, sexual behavior induces a PA state from the first time they are allowed to pace mating [59]. In this study, they tested the rewarding values of ejaculation and intromissions depending on prior sexual experience. They found that intromissions induced a clear CPP only in sexually naïve males, whereas ejaculation was rewarding in both sexually naïve and experienced males [59]. Moreover, sexual behavior was still able to induce CPP after male rats were allowed to copulate pacing the mating once a week for 10 consecutive weeks [49]. Interestingly, female hamster vaginal secretions induced CPP in sexually naïve male Syrian hamsters, indicating that vaginal secretions are rewarding before males experience their first sexual encounter [47, 57].

In female rats, it was shown that clitoral stimulation induced a PA state in naïve female rats [33]. However, authors also found that females that experienced five consecutive copulatory sessions did not develop a significant CPP for clitoral stimulation anymore. Therefore, it seems that sexual experience modifies the value of sensorial stimulation in females, affecting their ability to experience a reward state after clitoral stimulation. Interestingly, another study that used the conditioned object preference (COP), a paradigm in which preference for an object is tested after pairing it with sexual behavior, demonstrated that sexual experience is necessary for female rats to show COP for mating after a single training trial since naïve females failed to acquire COP compared to females that had five mating sessions [60].

4.6 Sensorial Stimuli

The characteristics of the mating stimuli that contribute to the CPP have been studied. In this section, I will review the literature that has provided insight into the different sensorial cues that are involved in mating interaction and participated in sexual reward, namely olfactory and genitosensorial stimuli.

4.6.1 Olfactory Stimuli

In mice, the reinforcing properties of sexual pheromones have been mostly studied with the place preference test, a paradigm that differs from the classic CPP consisting of a two-compartment cage where animals are free to visit different odorants. In contrast with CPP, where each subject is placed either in the reinforced or non-reinforced compartment after exposing it to the reinforcer or control environment, instrumental behavior is reinforced in this test. However, the place preference that acquires animals tested with this paradigm is also considered to reflect a reward state as it is assumed that the approach behavior is intrinsically rewarding [3]. Employing this paradigm, it was found that male sexual pheromones constitute a natural reward that can induce conditioned place preference in females [61, 62]. In addition, Agustín-Pavón et al. (2008) demonstrated that naloxone, an antagonist of opioid receptors, did not block the induction of place conditioning when females were reinforced with male pheromones as a rewarding stimulus [63]. Therefore, unlike sexual reward, the conditioning induced by pheromonal cues does not seem to involve opioid signaling in females [63]. Comparable to females, it was demonstrated by the CPP paradigm that bedding from an estrous female induces a reward state in male mice [56]. However, the scent of an estrous female is no longer capable of inducing CPP in male mice that were lesioned in both the main and accessory olfactory bulbs [64]. In another study, they found that aromatase knockout mice did not acquire CPP for the pheromonal cues of females [56]. Thus, it seems that the rewarding properties of female pheromones depend on the aromatization of testosterone into estradiol.

The rewarding properties of reproductively relevant odors have been studied in other species of rodents. For example, female hamster vaginal secretions are rewarding for sexually naïve male Syrian hamsters as demonstrated with the CPP paradigm [47, 57]. However, the exposure to a sexually receptive female without physical contact did not induce CPP in male prairie voles. Similarly, female prairie voles did not change their preference after being reinforced with the exposure to a male without physical interaction [31]. Importantly, Martínez-Ricos et al. (2007) demonstrated that only non-volatile chemicals (pheromones) are responsible for the reward state induced in female mice [61]. In the study of prairie voles, animals were in contact through a plastic screen that allowed them to smell the opposite-sex; however, authors did not mention if the screen had holes that would allow the voles to have close contact, which is necessary for pheromonal detection. Thus, it is possible that animals did not have access to the non-volatile chemicals that innately originate the reward state in other species.

4.6.2 Genitosensorial Stimuli

Studies have demonstrated that artificial vaginocervical stimulation administered to sexually naïve female rats at a rate comparable to that experienced during mating produces a clear CPP [51, 65]. Similarly, Parada et al. (2013) showed that clitoral stimulation induces a PA state in sexually naïve female rats. However, as mentioned before, sexual experience abolished the conditioning induced by clitoral stimulation [33]. Importantly, the place conditioning induced by vaginocervical stimulation was disrupted by lesions in the medial preoptic area [66]. Nonetheless, the transection of the pelvic nerve, which innervates the vagina and cervix had no effect on the acquisition of CPP after paced mating in sexually naïve female rats [67].

In males, different levels of sexual stimulation were tested to evaluate the quantity of stimuli necessary for males to experience a PA state after sexual encounter. Thus, Camacho et al. (2009) found that only males that were allowed to intromit a receptive female 15 times or ejaculate developed CPP whereas males that were allowed 5 or 10 intromissions did not change their place preference [40]. Another study from the same group tested with the CPP paradigm male rats injected with 8-OH-DPAT, a serotonin 1A receptor agonist that facilitates ejaculation (males achieve ejaculation with less intromissions). Interestingly, they found that males that received the drug did not acquire CPP after paced mating. Therefore, reaching ejaculation with fewer intromissions is not enough to induce a reward state in male rats [68].

4.7 Neurotransmitters

4.7.1 Dopamine

The mesolimbic dopamine (DA) pathway is thought to play a primary role in the reward system. However, the first studies testing the involvement of this neurotransmitter in the sexual reward could not demonstrate it. In fact, Ågmo et al. (1990) reported that pimozide, a D2 receptor antagonist, had no effect on the PA state induced by ejaculation [23]. In females, another D2 receptor antagonist called raclopride was tested and authors found no effect on the place conditioning induced by paced mating. In the same study, flupentixol, a non-specific dopaminergic antagonist was also evaluated, and it did not block the CPP induced by paced mating in females [37]. In contrast, a recent study showed that a specific DA D1-like receptor antagonist, Scheme 23390, prevented the acquisition of CPP induced by ejaculation [28]. In male Syrian hamsters, a study showed that the acquisition of CPP for vaginal secretions is dependent on DA signaling since the group of males that was administrated with the DA antagonist haloperidol did not present an increase in their place preference compared to controls that were exposed to vaginal secretions and injected with the vehicle [57]. Finally, in male mice, knock out for the D5 receptor, only the state induced by sexual activity that included ejaculation induced a change in place preference in contrast with WT animals that developed a clear CPP for both intromissions and ejaculations [37]. Overall, it seems that DA signaling is involved in the sexual reward in different species of males but there is yet no evidence of its role in female sexual reward.

4.7.2 Opioids

Interested readers can find a detailed description of the role of opioids in the sexual reward in the review of Paredes [69]. Briefly, several studies employing the opioid antagonist naloxone demonstrated the involvement of opioids in the acquisition and expression of CPP in both male [23, 29, 70–72] and female rats [39, 73]. In prairie voles, the administration of naloxone blocked the rewarding state induced by mating in males but its effect was not tested in females since they did not acquire CPP for mating [31]. Later studies found that the acquisition of CPP with sexual behavior depended on the secretion of opioids in the medial preoptic area, as injections of naloxone in this brain region blocked the place conditioning in both male and female rats [39, 72]. Interestingly, the stereotaxic injection of naloxone in the nucleus accumbens did not block the CPP induced by ejaculation in males [72] or induced by paced mating in females [39] (Fig. 3).

4.7.3 Other Neurotransmitters

Few other neurotransmitters were reported to participate in the sexual reward. First, in male mice, Popik et al. [32] found that the administration of the NMDA receptor channel blocker, memantine, before sexual encounter with a receptive female inhibited the expression of CPP induced by ejaculation. Second, in male rats, it was shown that lesions targeting orexin neurons in the hypothalamus prevented the formation of conditioned place preference for sexual behavior [74]. In contrast, intact sham-treated males or males with partial lesions developed a conditioned place preference for mating. Finally, we have recently found that kisspeptin, a neuropeptide, involved in the control of the hypothalamus-pituitary-gonads axis, and particularly important for the pulsatile secretion of GnRH and steroid positive and negative feedbacks, can induce a PA state. More importantly, the rewarding state induced by paced mating in females was blocked by the systemic injection of p234 penetratin, a kisspepin antagonist (unpublished data). Therefore, we have found a new actor in the mediation of the rewarding properties of mating.

5 Conclusion

The CPP paradigm is a standard behavioral model that has been very useful to assess the rewarding properties of sexual behavior. As highlighted in this chapter, it has been successfully used in different species to evaluate the underpinnings of sexual reward. The test that has been described here along as well as variations in its methodology enabled to unravel the part of the neurobiological basis of the sexual reward and the importance of other factors such as sexual experience, steroidal hormones, and sensory cues. However, the present chapter also reveals that substantial differences exist between males and females regarding sexual reward, and more studies need to be conducted to evaluate more precisely the mechanisms of sexual reward in both sexes.

Acknowledgments

The author thanks Jesús Edgar Hernández Ponce for taking and processing the pictures of Figs. 1 and 2. The research described in this chapter was supported by grants from DGAPA-PAPIIT IA207520 and IA207322.

References

1. Simpson EH, Balsam PD (2016) The behavioral neuroscience of motivation: an overview of concepts, measures, and translational applications. Curr Top Behav Neurosci 27:1–12. https://doi.org/10.1007/7854_2015_402
2. Berridge KC (1996) Food reward: brain substrates of wanting and liking. Neurosci Biobehav Rev 20:1–25. https://doi.org/10.1016/0149-7634(95)00033-b
3. Paredes RG (2009) Evaluating the neurobiology of sexual reward. ILAR J 50:15–27. https://doi.org/10.1093/ilar.50.1.15
4. Bermant G (1961) Response latencies of female rats during sexual intercourse. Science 133:1771–1773. https://doi.org/10.1126/science.133.3466.1771
5. Everitt BJ, Stacey P (1987) Studies of instrumental behavior with sexual reinforcement in male rats (Rattus norvegicus): II. Effects of preoptic area lesions, castration, and testosterone. J Comp Psychol 101:407–419
6. Ågmo A (1999) Sexual motivation—an inquiry into events determining the occurrence of sexual behavior. Behav Brain Res 105:129–150. https://doi.org/10.1016/S0166-4328(99)00088-1
7. Agmo A (2003) Lack of opioid or dopaminergic effects on unconditioned sexual incentive motivation in male rats. Behav Neurosci 117:55–68. https://doi.org/10.1037//0735-7044.117.1.55
8. Agmo A (2003) Unconditioned sexual incentive motivation in the male Norway rat (Rattus norvegicus). J Comp Psychol 117:3–14. https://doi.org/10.1037/0735-7036.117.1.3
9. Portillo W, Paredes RG (2004) Sexual incentive motivation, olfactory preference, and activation of the vomeronasal projection pathway by sexually relevant cues in non-copulating and naive male rats. Horm Behav 46:330–340. https://doi.org/10.1016/j.yhbeh.2004.03.001
10. Romero-Carbente JC, Camacho FJ, Paredes RG (2006) The role of the dorsolateral tegmentum in the control of male sexual behavior: a reevaluation. Behav Brain Res 170:262–270. https://doi.org/10.1016/j.bbr.2006.03.002
11. Ventura-Aquino E, Banos-Araujo J, Fernandez-Guasti A, Paredes RG (2016) An unknown male increases sexual incentive motivation and partner preference: further evidence for the Coolidge effect in female rats. Physiol Behav 158:54–59. https://doi.org/10.1016/j.physbeh.2016.02.026
12. López HH, Olster DH, Ettenberg A (1999) Sexual motivation in the male rat: the role of primary incentives and copulatory experience. Horm Behav 36:176–185. https://doi.org/10.1006/hbeh.1999.1535
13. Beach HD (1957) Morphine addiction in rats. Can J Psychol 11:104–112. https://doi.org/10.1037/h0083703
14. Rossi NA, Reid LD (1976) Affective states associated with morphine injections. Physiol Psychol 4:269–274. https://doi.org/10.3758/BF03332869
15. Tzschentke TM (2007) Measuring reward with the conditioned place preference (CPP) paradigm: update of the last decade. Addict Biol 12:227–462. https://doi.org/10.1111/j.1369-1600.2007.00070.x
16. Prus AJ, James JR, Rosecrans JA (2009) Conditioned place preference. In: Buccafusco JJ (ed). Boca Raton
17. Corona R, Larriva-Sahd J, Paredes RG (2011) Paced-mating increases the number of adult new born cells in the internal cellular (granular) layer of the accessory olfactory bulb. PLoS One 6:e19380. https://doi.org/10.1371/journal.pone.0019380
18. Bedos M, Antaramian A, Gonzalez-Gallardo A, Paredes RG (2019) Paced mating increases the expression of μ opioid receptors in the ventromedial hypothalamus of male rats. Behav Brain Res 359:401–407. https://doi.org/10.1016/j.bbr.2018.11.027
19. Corona R, Camacho FJ, Garcia-Horsman P et al (2011) Different doses of estradiol benzoate induce conditioned place preference after paced mating. Horm Behav 60:264–268. https://doi.org/10.1016/j.yhbeh.2011.05.013
20. Roughan JV, Coulter CA, Flecknell PA et al (2014) The conditioned place preference test for assessing welfare consequences and potential refinements in a mouse bladder cancer model. PLoS One 9:e103362. https://doi.org/10.1371/journal.pone.0103362
21. Arzate DM, Portillo W, Rodriguez C et al (2011) Extended paced mating tests induces conditioned place preference without affecting sexual arousal. Horm Behav 59:674–680. https://doi.org/10.1016/j.yhbeh.2010.08.016
22. Everitt BJ (1990) Sexual motivation: a neural and behavioural analysis of the mechanisms underlying appetitive and copulatory responses of male rats. Neurosci Biobehav Rev 14:217–

232. https://doi.org/10.1016/s0149-7634(05)80222-2

23. Agmo A, Berenfeld R (1990) Reinforcing properties of ejaculation in the male rat: role of opioids and dopamine. Behav Neurosci 104:177–182. https://doi.org/10.1037/0735-7044.104.1.177
24. Oldenburger WP, Everitt BJ, de Jonge FH (1992) Conditioned place preference induced by sexual interaction in female rats. Horm Behav 26:214–228. https://doi.org/10.1016/0018-506x(92)90043-u
25. Paredes RG, Alonso A (1997) Sexual behavior regulated (paced) by the female induces conditioned place preference. Behav Neurosci 111:123–128. https://doi.org/10.1037/0735-7044.111.1.123
26. Paredes RG, Vazquez B (1999) What do female rats like about sex? Paced mating. Behav Brain Res 105:117–127. https://doi.org/10.1016/S0166-4328(99)00087-X
27. Carboni E, Vacca C (2003) Conditioned place preference. A simple method for investigating reinforcing properties in laboratory animals. Methods Mol Med 79:481–498. https://doi.org/10.1385/1-59259-358-5:481
28. Dominguez-Salazar E, Naser HF, Velazquez-Moctezuma J (2014) D1-like antagonist blocks conditioned place preference induced by ejaculation in male rats. Behav Brain Res 269:15–19. https://doi.org/10.1016/j.bbr.2014.04.026
29. Hughes AM, Everitt BJ, Herbert J (1990) Comparative effects of preoptic area infusions of opioid peptides, lesions and castration on sexual behaviour in male rats: studies of instrumental behaviour, conditioned place preference and partner preference. Psychopharmacology 102:243–256. https://doi.org/10.1007/BF02245929
30. Kudwa AE, Dominguez-Salazar E, Cabrera DM et al (2005) Dopamine D5 receptor modulates male and female sexual behavior in mice. Psychopharmacology 180:206–214. https://doi.org/10.1007/s00213-005-2150-5
31. Ulloa M, Portillo W, Díaz NF et al (2018) Mating and social exposure induces an opioid-dependent conditioned place preference in male but not in female prairie voles (Microtus ochrogaster). Horm Behav 97:47–55. https://doi.org/10.1016/j.yhbeh.2017.10.015
32. Popik P, Wrobel M, Rygula R et al (2003) Effects of memantine, an NMDA receptor antagonist, on place preference conditioned with drug and nondrug reinforcers in mice. Behav Pharmacol 14:237–244. https://doi.org/10.1097/00008877-200305000-00008
33. Parada M, Jafari N, Pfaus JG (2013) Sexual experience blocks the ability of clitoral stimulation to induce a conditioned place preference in the rat. Physiol Behav 119:97–102. https://doi.org/10.1016/j.physbeh.2013.06.005
34. Parada M, Chamas L, Censi S et al (2010) Clitoral stimulation induces conditioned place preference and Fos activation in the rat. Horm Behav 57:112–118. https://doi.org/10.1016/j.yhbeh.2009.05.008
35. Parada M, Vargas EB, Kyres M et al (2012) The role of ovarian hormones in sexual reward states of the female rat. Horm Behav 62:442–447. https://doi.org/10.1016/j.yhbeh.2012.07.012
36. Meerts SH, Clark AS (2007) Female rats exhibit a conditioned place preference for non-paced mating. Horm Behav 51:89–94. https://doi.org/10.1016/j.yhbeh.2006.08.007
37. Garcia Horsman P, Paredes RG (2004) Dopamine antagonists do not block conditioned place preference induced by paced mating behavior in female rats. Behav Neurosci 118:356–364. https://doi.org/10.1037/0735-7044.118.2.356
38. Martinez I, Paredes RG (2001) Only self-paced mating is rewarding in rats of both sexes. Horm Behav 40:510–517. https://doi.org/10.1006/hbeh.2001.1712
39. Garcia-Horsman SP, Agmo A, Paredes RG (2008) Infusions of naloxone into the medial preoptic area, ventromedial nucleus of the hypothalamus, and amygdala block conditioned place preference induced by paced mating behavior. Horm Behav 54:709–716. https://doi.org/10.1016/j.yhbeh.2008.07.011
40. Camacho FJ, Portillo W, Quintero-Enriquez-O, Paredes RG (2009) Reward value of intromissions and morphine in male rats evaluated by conditioned place preference. Physiol Behav 98:602–607. https://doi.org/10.1016/j.physbeh.2009.09.012
41. Camacho FJ, Garcia-Horsman P, Paredes RG (2009) Hormonal and testing conditions for the induction of conditioned place preference by paced mating. Horm Behav 56:410–415. https://doi.org/10.1016/j.yhbeh.2009.07.007
42. Packard MG, Cornell AH, Alexander GM (1997) Rewarding affective properties of intra-nucleus accumbens injections of testosterone. Behav Neurosci 111:219–224. https://doi.org/10.1037//0735-7044.111.1.219

43. Packard MG, Schroeder JP, Alexander GM (1998) Expression of testosterone conditioned place preference is blocked by peripheral or intra-accumbens injection of alpha-flupenthixol. Horm Behav 34:39–47. https://doi.org/10.1006/hbeh.1998.1461
44. Paredes RG (2010) Hormones and sexual reward. Vitam Horm 82:241–262. https://doi.org/10.1016/s0083-6729(10)82013-8
45. Meisel RL, Joppa MA, Rowe RK (1996) Dopamine receptor antagonists attenuate conditioned place preference following sexual behavior in female Syrian hamsters. Eur J Pharmacol 309:21–24. https://doi.org/10.1016/0014-2999(96)00389-5
46. Meisel RL, Sachs BD (1994) The physiology of male sexual behavior. Physiol Reprod 2:3–105
47. Bell MR, Meerts SH, Sisk CL (2010) Male Syrian hamsters demonstrate a conditioned place preference for sexual behavior and female chemosensory stimuli. Horm Behav 58:410–414. https://doi.org/10.1016/j.yhbeh.2010.05.017
48. Jenkins WJ, Becker JB (2003) Female rats develop conditioned place preferences for sex at their preferred interval. Horm Behav 43:503–507. https://doi.org/10.1016/S0018-506X(03)00031-X
49. Camacho F, Sandoval C, Paredes RG (2004) Sexual experience and conditioned place preference in male rats. Pharmacol Biochem Behav 78:419–425. https://doi.org/10.1016/j.pbb.2004.04.015
50. Mermelstein PG, Becker JB (1995) Increased extracellular dopamine in the nucleus accumbens and striatum of the female rat during paced copulatory behavior. Behav Neurosci 109:354–365. https://doi.org/10.1037/0735-7044.109.2.354
51. Meerts SH, Clark AS (2009) Artificial vaginocervical stimulation induces a conditioned place preference in female rats. Horm Behav 55:128–132. https://doi.org/10.1016/j.yhbeh.2008.09.003
52. Gonzalez-Flores O, Camacho FJ, Dominguez-Salazar E et al (2004) Progestins and place preference conditioning after paced mating. Horm Behav 46:151–157. https://doi.org/10.1016/j.yhbeh.2004.02.006
53. Meerts SH, Anderson KS, Farry-Thorn ME et al (2017) Prepubertal ovariectomy modulates paced mating behavior but not sexual preference or conditioned place preference for mating in female rats. Physiol Behav 171:142–148. https://doi.org/10.1016/j.physbeh.2017.01.012
54. Gans S, Erskine MS (2003) Effects of neonatal testosterone treatment on pacing behaviors and development of a conditioned place preference. Horm Behav 44:354–364. https://doi.org/10.1016/s0018-506x(03)00157-0
55. Dominguez-Salazar E, Camacho FJ, Paredes RG (2005) Prenatal blockade of androgen receptors reduces the number of intromissions needed to induce conditioned place preference after paced mating in female rats. Pharmacol Biochem Behav 81:871–878. https://doi.org/10.1016/j.pbb.2005.06.011
56. Pierman S, Tirelli E, Douhard Q et al (2006) Male aromatase knockout mice acquire a conditioned place preference for cocaine but not for contact with an estrous female. Behav Brain Res 174:64–69. https://doi.org/10.1016/j.bbr.2006.07.002
57. Bell MR, Sisk CL (2013) Dopamine mediates testosterone-induced social reward in male Syrian hamsters. Endocrinology 154:1225–1234. https://doi.org/10.1210/en.2012-2042
58. Hawley WR, Kapp LE, Dingle CM et al (2021) Sexual motivation and sexual reward in male rats are attenuated by the gonadotropin-releasing hormone receptor antagonist degarelix. J Sex Med 18:240–255. https://doi.org/10.1016/j.jsxm.2020.11.004
59. Tenk CM, Wilson H, Zhang Q et al (2009) Sexual reward in male rats: effects of sexual experience on conditioned place preferences associated with ejaculation and intromissions. Horm Behav 55:93–97. https://doi.org/10.1016/j.yhbeh.2008.08.012
60. Piergies AMH, Hicks MEJ, Schwartz JP, Meerts SH (2019) Sexually experienced, but not naïve, female rats show a conditioned object preference (COP) for mating after a single training trial. Physiol Behav 198:42–47. https://doi.org/10.1016/j.physbeh.2018.09.017
61. Martínez-Ricós J, Agustín-Pavón C, Lanuza E, Martínez-García F (2007) Intraspecific communication through chemical signals in female mice: reinforcing properties of involatile male sexual pheromones. Chem Senses 32:139–148. https://doi.org/10.1093/chemse/bjl039
62. Agustín-Pavón C, Martínez-Ricós J, Martínez-García F, Lanuza E (2007) Effects of dopaminergic drugs on innate pheromone-mediated reward in female mice: a new case of dopamine-independent "liking". Behav Neurosci 121:920–932. https://doi.org/10.1037/0735-7044.121.5.920
63. Agustín-Pavón C, Martínez-Ricós J, Martínez-García F, Lanuza E (2008) Sex versus sweet: opposite effects of opioid drugs on the reward

of sucrose and sexual pheromones. Behav Neurosci 122:416–425. https://doi.org/10.1037/0735-7044.122.2.416
64. Korzan WJ, Freamat M, Johnson AG et al (2013) Either main or accessory olfactory system signaling can mediate the rewarding effects of estrous female chemosignals in sexually naive male mice. Behav Neurosci 127:755–762. https://doi.org/10.1037/a0033945
65. Walker QD, Nelson CJ, Smith D, Kuhn CM (2002) Vaginal lavage attenuates cocaine-stimulated activity and establishes place preference in rats. Pharmacol Biochem Behav 73: 743–752. https://doi.org/10.1016/S0091-3057(02)00883-3
66. Meerts SH, Clark AS (2009) Lesions of the medial preoptic area interfere with the display of a conditioned place preference for vaginocervical stimulation in rats. Behav Neurosci 123:752–757. https://doi.org/10.1037/a0016077
67. Meerts SH, Clark AS (2009) Conditioned place preference for mating is preserved in rats with pelvic nerve transection. Behav Neurosci 123:539–546. https://doi.org/10.1037/a0015267
68. Camacho FJ, Castro M, Hernandez V, Paredes RG (2007) Facilitation of ejaculation induced by 8-OH-DPAT does not produce conditioned place preference in male rats. Behav Neurosci 121:579–585. https://doi.org/10.1037/0735-7044.121.3.579
69. Paredes RG (2014) Opioids and sexual reward. Pharmacol Biochem Behav 121:124–131. https://doi.org/10.1016/j.pbb.2013.11.004
70. Mehrara BJ, Baum MJ (1990) Naloxone disrupts the expression but not the acquisition by male rats of a conditioned place preference response for an oestrous female. Psychopharmacology 101:118–125. https://doi.org/10.1007/BF02253728
71. Miller RL, Baum MJ (1987) Naloxone inhibits mating and conditioned place preference for an estrous female in male rats soon after castration. Pharmacol Biochem Behav 26:781–789. https://doi.org/10.1016/0091-3057(87)90611-3
72. Ågmo A, Gomez M (1993) Sexual reinforcement is blocked by infusion of naloxone into the medial preoptic area. Behav Neurosci 107: 812–818. https://doi.org/10.1037/0735-7044.107.5.812
73. Paredes RG, Martinez I (2001) Naloxone blocks place preference conditioning after paced mating in female rats. Behav Neurosci 115:1363–1367. https://doi.org/10.1037/0735-7044.115.6.1363
74. Di Sebastiano AR, Wilson-Pérez HE, Lehman MN, Coolen LM (2011) Lesions of orexin neurons block conditioned place preference for sexual behavior in male rats. Horm Behav 59:1–8. https://doi.org/10.1016/j.yhbeh.2010.09.006

Chapter 14

Because Sex Matters: The Case of Female Sexual Response

Elisa Ventura-Aquino and Anders Ågmo

Abstract

In recent years, the Food and Drug Administration approved two drugs for the treatment of hypoactive sexual desire disorder in premenopausal women which have been tested in preclinical tests with positive effects on some appetitive sexual behavior in female rats. Some years after their commercial release, the effectiveness of both drugs in women is not clear. In the present chapter, we present some considerations regarding preclinical studies of copulatory behavior, most of them conducted in female rats. Although knowledge of the neurobiological basis of copulatory behavior in the female rat is still growing, the extent to which these findings might be applied to women's sexuality is unknown. On the other hand, sexual response in women is flexible and less hormone-dependent compared with men. Moreover, sociocultural factors deeply influence sexuality in women, and they might contribute to the development of sexual dysfunctions. Additionally, the kind of model that better describes sexual response in women seems to be individual and fluctuating. We briefly mentioned some of the most used methods for measuring genital arousal, putting out the need for developing better strategies to get objective and replicable results. We conclude that the contribution, if any, of animal models of sexual response in female rodents is modest. After the questionable approval and the lack of efficacy of drugs to treat sexual desire issues, it is necessary to realize that the more we understand the female sexual responses, developing better ways to evaluate them, the more we might explain, and eventually treat their dysfunctions.

Key words Female sexual response, Sexual arousal, Female sexual dysfunctions

1 Introduction

The fifth edition of the Diagnostic and Statistical Manual of Mental Disorders (DSM-V) classified female sexual dysfunctions (FSD) into three main categories: (a) sexual interest/arousal disorder, (b) orgasmic disorder, and (c) genitopelvic pain/penetration disorder [1]. The prevalence of each FSD varies depending on instruments applied, the hypoactive sexual desire disorder (HSDD) shows the highest rates (12–20%), followed by the orgasmic disorder (6–12%) and sexual arousal disorder (5–11%) [2]. Additionally, the presence of a FSD also might negatively affect other aspects of sexual response, thus the presence of more than one FDS in the same woman is common.

Raúl G. Paredes et al. (eds.), *Animal Models of Reproductive Behavior*, Neuromethods, vol. 200,
https://doi.org/10.1007/978-1-0716-3234-5_14,

The Food and Drug Administration (FDA) approved flibanserin in 2015 and bremelanotide in 2018 for the treatment of HSDD in premenopausal women. Flibanserin acts on serotonergic and dopaminergic systems and was initially developed for the treatment of depression. Bremelanotide is a melanocortin receptor agonist, originally designed for sunless tanning. Both drugs have positive effects on some appetitive sexual behaviors in sexually experienced and hormone primed female rats. Chronic treatment (14 days of 45 mg/kg/p.o.) of flibanserin and acute bremelatonide (100/200 μg/kg s.c., 5 min before a mating test) enhance solicitation behavior [3].

Some years after their commercial release, actual effects of both these drugs on sexual functions are still a matter of controversy [4]. For example, clinical studies show that 1.75 mg s.c. bremelanotide on demand adds one enjoyable sexual experience every 2 months, whereas the effect of 50–100 mg of flibanserin daily is even more discrete than that. In a meta-analysis by Jaspers et al. [5], they compared flibanserin and placebo effects, and included data from published reports and registries from trials. Authors found that the dropout rate in the flibanserin group was twice that found in the placebo group. The reasons were, among others, dizziness (odds ratio, OR = 4), somnolence (OR = 3.97), and nausea (OR = 2.35) [5]. Post-marketing reports regarding these adverse effects led to modification of the safety labeling of flibanserin, especially concerning its concomitant use with alcohol [4, 6–9]. As showed, the failure to develop efficient drug treatment for female HSDD may be related to a faulty model of the female sexual response (FSR).

In the present chapter, we present some considerations regarding preclinical studies of copulatory behavior, most of them conducted in female rats, to continue mentioning critical aspects of FSR in women described recently, emphasizing peculiarities and sociocultural factors involving FSR. As will be noted, this field is still growing with the need for studying FSR directly in the subjects of interest, i.e., women.

2 Basic Considerations Regarding Copulatory Behavior in the Female Rat

Studies about the mechanisms involved in the dramatic behavioral changes observed between non-sexually receptive and sexually receptive female rats were relevant for extending our understanding of the neuroendocrine and physiological processes underlying these drastical changes. These studies were focused on how sex hormones, such as estrogens, in the central nervous system, induced estrous behavior whereas, in the periphery, they increased the skin's sensitivity of the pudendal nerve field which, in sum, induce and enhance female copulatory behavior [10–12]. During behavioral estrus, rats present two relevant behaviors: receptivity and paracopulatory (formerly proceptive) behaviors. Receptivity manifests itself in a complex neuromuscular reflex, that produces immobilization with flexion of the back, rump elevation, and extension of the neck. The lordosis posture is triggered by tactile stimulation of the flanks and enables penile intromissions into the vagina. Paracopulatory behaviors involve a range of behavioral patterns that females usually display towards the male during estrus, such as hopping, darting, and ear wiggling. A detailed description of sexual behavior in the female rat is found elsewhere [13].

With the unprecedented success of sildenafil (Viagra; Pfizer) for treating erectile difficulties in men, the interest for developing new pharmacological treatments for sexual dysfunctions increased, and many researchers shifted their interest to developing new animal models of sexual dysfunctions, especially regarding erectile and ejaculatory responses [14–16].

For decades, studies about FSB were conducted in small uni-compartmental arenas. Under this condition, a sexually experienced male and a sexually receptive female are placed together., There are a few options, if any, for other interactions but sexual [17, 18]. Research was focused mainly on neuroendocrine mechanisms behind the sexual responses. Since this experimental setup is far from a representative design, i.e., a procedure resembling conditions in the wild, comparisons between different species (e.g., rats and humans) were of limited value [17].

Later, studies concerning other aspects of copulatory behavior in the rat, such as sexual reward and sexual motivation, became available. In this regard, it is worth to mention that robust evidence demonstrating sexual reward in females that paced the mating was published in the late 90s [19–22]. Soon after, the endogenous opioid system was identified as relevant for this effect that also induces neurogenesis in some brain areas, mainly in the accessory olfactory bulb, which is critical in species such as rodents for identifying sexual cues [23–25]. Olfactory cues are critical and required in combination with other sensory stimuli (visual, auditory) to induce a female's approach behavior to a potential mate in rats [26].

The sexual incentive motivation test was proposed as a feasible and specific way to measure sexual motivation in rats [27]. Pharmacological studies used this paradigm to explore the role of neurotransmitters such as serotonin in sexual motivation. For example, the effect of selective serotonin reuptake inhibitors (SSRI) was evaluated, since inhibitory effects on sexual desire have been reported by women taking these drugs for the treatment of depression. Most evidence suggests that increased serotoninergic activity is related to inhibition of approach behaviors in female rats, as a reflection of decreased sexual motivation. Although there is no published data regarding the effect of the current FDA-approved drugs to treat HSDD in procedures especially designed for evaluating sexual motivation, it might be a feasible way to evaluate these drugs in this kind of preclinical tests. Detailed analysis of modeling sexual motivation in rats and humans is found elsewhere [14], and an extensive analysis of that issue goes beyond the scope of the present chapter.

Studies of mating behavior in rats, under seminatural conditions, where a group of males and females are housed together during some consecutive female estrous cycles in a spacious arena, with a burrow system that consists of tunnels and nest boxes connected to an open area. Thus, animals are in contact during transitional periods between receptive and unreceptive status, and conditions enable them to express different behaviors beyond sexual. Contrary to the uni-compartmental condition, the seminatural environment has external validity, i.e., it resembles what occurs in the rats' natural habitat, and results can be generalized to situations outside the specific laboratory setting [17]. Under this condition, the experimental subjects showed a slower rate of copulation and the time spent in mating was less than 1% of the total duration of estrus [14]. Additionally, males and females initiated and maintained sexual interactions equally, not female-dominated as stated previously, and the intensity of paracopulatory behaviors was partially related to female sexual motivation. The results suggest that it is necessary to re-evaluate the previous assumptions regarding the meaning and function of paracopulatory behaviors [28, 29]. Detailed methods for tests are available elsewhere in the present book. Although knowledge of the neurobiological basis of copulatory behavior in the female rat is still growing, the extent to which these findings might be applied to human sexuality is unknown, mainly because we are continually learning about the bases of the human sexual response and its peculiarities.

3 FSR and Some of Its Peculiarities

Our understanding of FSR is growing, and the evidence shows that FSR presents some exceptionalities, which are important for triggering and/or maintaining some FSD. We will discuss some of them in the following sections.

3.1 The Flexibility of Sexual Behavior in Women

Gonadal hormones, mainly estrogens released in the periovulatory period, are necessary to trigger and maintain copulatory behavior in non-primate mammals. However, in some female primates, including women, the role of gonadal hormones is not that determinant. Thus, the sexual response is more flexible and modulated by environmental and non-biological factors. Moreover, human sexual behaviors are shaped by sociocultural conventions. Therefore, humans are used to engage in sexual activities that follow social norms, which change according to social dynamics [30].

From this perspective, the influence of gender roles, stereotypes, and violence on women's sexuality is remarkable, and it is present in the establishment of some sexual dysfunctions, as will be mentioned in the following sections. In this context, biological factors participate marginally in the physiological aspects of sexual response in women, and expectations or previous experiences contribute to the development of women's sexuality. In this regard, studies of sexuality in childhood consistently reported that children were able to respond genitally to sexual stimulation. Whereas masturbation in boys is commonly a discursive topic either negatively or positively, in girls, it is an unspoken issue that usually is discovered by self-exploration [31, 32]. It is possible that whereas boys acquire more experience and knowledge regarding their sexual response as they grow, girls lag in this process, creating a gender gap. Moreover, the importance given to penile–vaginal interaction as the main source of pleasure, in conjunction with the incomprehension about female anatomy and sexual arousal, enlarge this gap. In fact, penile–vaginal intercourse is far less efficient than some other sexual behaviors for leading to orgasm in women. The focus on penetrative sex as the main source of sexual satisfaction might explain, at least partially, the fact that lesbian couples are more likely to experience orgasms than women in heterosexual relationships [33].

Moreover, negative socio-cultural connotations, the lack of information, and weak positive encouragement about FSR increase the probability of developing sexual dysfunctions. Thus, the more we overcome these factors, the more possibilities we have to reduce this gender gap [34]. On the other hand, when humans engage in sex for reasons and consequences beyond recreation, i.e., transactional sex, women's reasons differ from those of men since women tend to show more emotionally based motivations. However, the main reasons for having sex in both sexes are related to recreation [35, 36].

3.2 FSR: Models in Different Shapes and Sizes

Different sequential models of FSR have become available ever since the one proposed by William Masters and Virginia Johnson [37]. Their unprecedented physiological studies performed in actual sexual intercourses undoubtedly established the basis for research in the sexology field. However, to date, no other laboratory has replicated or verified Masters & Johnson's findings, and conclusions derived from them usually are taken as reference in most studies regarding sexual response in the general population. Masters & Johnson's convenience sample required masturbatory experience and the presence of regular coital orgasms [38]. Additionally, they included people with high levels of academic formation and good communicative skills to get a detailed description of their sexual responses [38–40]. Despite the limitations of the sampling procedure, mainly regarding women, Masters and Johnson's model of sexual response in the female, and its later modifications, permitted us to make more precise definitions and classifications of sexual response stages and dysfunctions. Moreover, they established the basis of therapeutic approaches, many of them still used. Around 40 years later, Rosemary Basson proposed a model based on data from questionnaires and interviews with women in long-term relationships. According to Basson, there are four main omissions in previous models:

(a) the lower urge for sexual releasing in women than men,

(b) women reported non-sexual motivations for engaging in sexual activity,

(c) for women, subjective sexual arousal might not be accompanied by genital sexual arousal, or the awareness of that arousal,

(d) in women, the release of sexual tension induced by orgasm might or might not occur.

Basson's model proposes that women might start from neutrality, to decide deliberately to engage in sexual activity, motivated by sexual and non-sexual needs. In this situation, some sexual arousal is present which might be enhanced by sexual stimulation and emotional needs such as closeness, love, and/or affection which enhance the awareness of being sexually aroused. Increased arousal with adequate and sufficient sexual stimulation might lead to orgasm and physical wellbeing. In this theoretical model, desire might precede or accompany arousal, and both elements are reciprocally influenced either positively or negatively by emotional factors, such as intimacy (Fig. 1) [41, 42].

Women's descriptions regarding sexual desire and sexual arousal suggest that these elements are experienced as the same construct, complicating their differentiation each other that fluctuates with sexual experiences. The fluctuation of desire and arousal

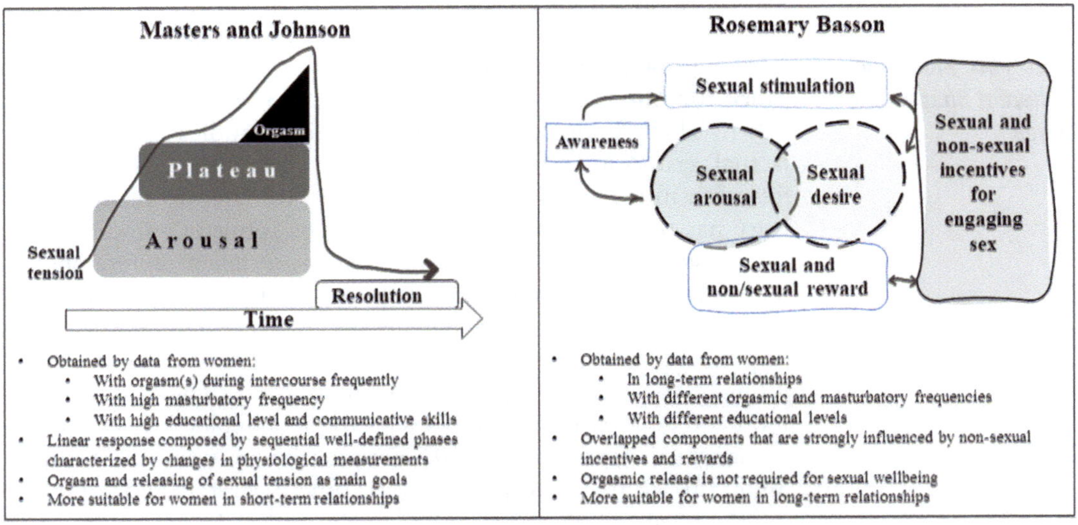

Fig. 1 Comparison between two different models for female sexual responses. (Modified from Masters and Johnson [37] and Basson [41])

is highly variable, depending on relational and sexual outcomes. The results indicated that FSR could not fit into a single model since its variability in the same woman under different situations. Notably, women usually are aware of this variability, but it was not problematic for them. Authors proposed that sexual responses are as variable as women's shapes and sizes [42]. The findings also led us to reconsider how we understand the FSR and if a single model can appropriately represent the variance of sexual responses in women.

3.3 Women's Sexual Arousal: The Bipartite Component and Its Discrepancy

Since sildenafil's release and its tremendous commercial success, the interest in developing analogous drugs in women made sexual arousal the most studied component of the FSR. Although sildenafil has shown inconclusive results and is not approved for female sexual dysfunction treatment, thanks to these studies, our understanding of female sexual arousal is still growing [43, 44].

As aforementioned, women might refer to sexual arousal as the primary mental or psychogenic process. However, experimental studies have reported two different elements that constitute the core of this response: genital arousal (i.e., vasocongestive changes in genitals with engorgement and perceptions of genital pulsing and warmth), and subjective sexual arousal (i.e., feeling 'turned on' during sexual activity) which usually are accompanied by extragenital changes, including skin flushing and nipple erection. Women's descriptions support the interplay between both types of responses during sexual arousal, possibly each one enhances the other, like a positive feedback fashion, facilitating women's realization of being sexually aroused [44, 45]. However, for some women, both components are not related at all. Since the grade of synchrony between genital and subjective sexual arousal is not related to sexual function, its clinical relevance, if any, is unclear [44].

Preclinical experiments using tissue cultures, organ baths, or in vivo tests indicated that peripheral genital arousal in females is mediated by somatic and autonomic responses. Afferent signals from the genitals are conveyed by the pudendal nerve, to the spinal cord, where efferent spinal reflexes induce muscle contraction of the perineum. On the other hand, peripheral autonomic nerves, such as the hypogastric, pelvic, and vagal, facilitate genital blood flow and genital smooth muscle relaxation through parasympathetic and sympathetic influences, which favor vaginal and clitoral engorgement [43, 46]. Electrical stimulation of the pelvic nerve in female rats and rabbits induces hemodynamic changes with vaginal enlargement and increased vaginal wall tension. These responses are also partially mediated by several local mediators, including vasoactive agents such as neuropeptide Y, noradrenaline, vasoactive intestinal polypeptide, nitric oxide synthase (NOS), calcitonin gene-related peptide, and substance P. The role of each one of those in regulating vaginal and clitoral smooth muscle tone remains unclear [43].

Clinical studies in women revealed that the autonomous nervous system is involved in sexual arousal response. The role of generalized sympathetic activity has been demonstrated since (a) noradrenaline plasma levels are higher in women sexually aroused, (b) vaginal arousal induced by erotic films is higher after 20 min of intensive exercise than those presented after a non-exercise test, and (c) this effect was prevented by clonidine, an α-adrenergic receptor blocker [44, 47, 48]. On the other hand, evidence for the regulatory role of parasympathetic is based mainly on women with thoracic spinal injury, who refer getting sexually aroused without vaginal lubrication [49].

Estradiol and testosterone also contribute to genital arousal by regulating the activity of peptides in the vagina, including the vasoactive intestinal polypeptide (VIP) and nitric oxide synthase (NOS). Intravenous and intraepithelial administration of VIP increased vaginal blood flow, measured by heated oxygen electrode, in postmenopausal women, only if they received hormonal supplementation [50, 51]. Although comprehensive VIP mechanisms are not yet elucidated in sexual contexts, results indicate the interaction of endocrine and local factors in vascular vaginal response.

Opposite to genital arousal, which is an autonomic response, subjective arousal involves the experience and feelings of being turned on. Experiments indicate that attentional focus on sexual arousing cues is required to enhance subjective arousal. Thus, distractors reduce subjective arousal in women and they include negative or autocratical thoughts such as situational elements and they are also modeled by psychosocial and cultural inputs [52].

Conventional experiments usually evaluate individual responses. However, growing evidence indicates that it is necessary to broaden our perspective and put on the discussion how personal history, including sexual abuse, and dynamics with the partner are critical for sexual functioning in women [44, 53, 54].

3.4 The Role of Intra and Interpersonal Influences on Women's Sexual Response

Women are historically receptors of different types of violence throughout all their life stages. Up to 85% of women with sexual dysfunctions, mainly of these concerning desire and arousal, also report sexual abuse in childhood, twice as much as those without sexual abuse antecedent. Disruptive changes in sexuality in women with negative sexual experiences include, among others, sexual shame, lack of positive emotions towards sex, impaired attachment security, and negative sexual self-schemas. Those negative associations probably act as distractors that counteract subjective sexual arousal [44]. For this group, psychological therapy is particularly effective [55, 56]. In women veterans, sexual trauma in childhood or during military service was associated with low-sexual satisfaction and especially with sexual pain. Future research is needed to understand this interaction.

Effective communication and the level of emotional intimacy with the partner influence sexual satisfaction in women, especially for those women in long-term relationships. In heterosexual couples, there are some correlations between sexual dysfunctions in the partners. For example, erectile dysfunction is related to low-sexual satisfaction in the female partner, whereas its effective treatment also improves females' sexual life quality [57]. Additionally, a report showed that premature ejaculation might be related to vaginal penetration difficulties in the female partner [58]. The approach considering the couple as a relevant factor in the development, maintenance, and solution of some FSD is being more used [59].

4 Common Methods for Measuring Genital Arousal in Women

Masters and Johnson proposed that besides the obvious anatomical differences, male and FSRs were parallel and similar from a physiological point of view [37].

Currently, researchers continue looking for the gold standard for measuring sexual arousal in women by evaluating different physiological parameters, among others: vaginal blood flow, volume and pressure of genitalia, lubrication, neural responses, and muscular activity. In the meanwhile, there are some available options, each one with strengths and opportunities, summarized in Table 1.

4.1 Vaginal Photoplethysmography

Sexual arousal is followed by vasocongestion of the vaginal capillaries that increase the transudation of plasma to create a lubricative film covering the vaginal walls. The grade of vasocongestive reaction is measured using a photometer with a source of infrared LED light and a phototransistor inside of a tampon-shaped probe, which is inserted into the vagina. Once the LED light is turned on, photons diffuse and reflect on the vaginal walls, which are captured by the phototransistor and transformed into electrical signals. The larger the blood volume in the vagina, the greater the amount of the incident light backscattered to the light detector. Two variables are obtained depending on the signal coupling: vaginal blood volume (VBV) for direct current and vaginal pulse amplitude (VPA) for alternating current. Although VBV and VPP are highly related to blood volume, vaginal photoplethysmography (VPP) is expressed in arbitrary units such as changes in mV or mm of pen deflection [60, 61].

Some advantages of using VVP are: the majority of reports have used VVP for measuring genital arousal and the probe can be placed by the participant. On the other hand, it is still unclear which is the physiological parameter measured by VPP and VPA. Additionally, the need for introducing into the vagina, it might not be feasible in genital pain or vaginismus; after each use, the probe

Table 1
Common methods to measure female sexual arousal in sexual research

Method	Physiological variable	Basic principles	Advantages	Disadvantages
Vaginal photoplethysmography	Genital blood flow	Tampon shape device containing a light source and a photosensitive cell. Light that is backscattered is correlated with vasocongestion	The most extended method for assessing sexual arousal Easily inserted in most of cases by participants Feasible for long and multiple recordings correlates with self-report of sexual arousal	Sterilized is needed after use Not suitable for women with pain or vaginismus Not clear what variable is measured Highly inter-subject variable
Labial thermistor	Temperature changes in the labia minora	A padded clip attached to thermistors that detect temperature of the surface	Minimally invasive Correlates with self-report of sexual arousal	Some participants found it highly aversive Requires room temperature control Movement artifact Affected by body temperatures across the menstrual cycle
Thermography	Temperature changes in the genitals	Thermograms by infrared camera	Non- invasive Correlates with self-report of sexual arousal Objective scale of measures	Expensive devices Requires room temperature
Lubrication measures	pH and amount of lubricant fluid	An electrode placed into the vaginal detect changes in the pH. Components and the amount of vaginal fluid can be measured by filter papers and tampons	Unexpensive Ease of use	Somewhat invasive Scarce use in research Results might be affected by vaginal infection and menstruation

needs to undergo disinfection. Another consideration is the lack of standard units and the high variability across studies, even in the same person.

4.2 Labial Thermistor

Changes in the surface temperature of labia minora are measured by a small thermistor attached to a padded clip and placed on the labia minora. There is a positive correlation between temperature increase and subjective rating of sexual arousal in women, similar to which occurred with vaginal plethysmography. This technique also uses arbitrary units and results show high inter-subject variability but also across the day of the menstrual cycle

4.3 Genital Infrared Thermography

This non-invasive technique also evaluates genital temperature by infrared light camera.

Similarly, with the labial thermistors, infrared thermography correlates with self-reported sexual arousal. Thermography also discriminates between erotic and non-erotic stimuli, making this approach promising to evaluate sexual arousal.

4.4 Lubrication

Vaginal fluid produced during sexual arousal is the result of sequential vasodilation (likely neural VIP-mediated) with vasoconstriction of veins that cause transudation of plasma and interstitial fluid to vaginal walls. Actual mediators of lubrication remain unknown and a report suggests that labia minora produce their lubrication, independent of the vaginal [62]. Although lubrication is considered a physiological response during sexual arousal, there are no reliable ways to measure it. Some reports have focused on vaginal pH changes using a glass pH electrode in the vagina showing that, during clitoral self-stimulation, vaginal pH increased up to 1 unit when the woman reached orgasm. Some preclinical reports have evaluated lubrication fluid by using tampons, but its use is limited in physiological research.

5 Conclusions

As we are still learning from sexual behavior in female rats, we also realize that making extrapolations to humans might not be feasible, mainly because sexual responses in women are so peculiar, variable, and strongly influenced by sociocultural factors. Moreover, our knowledge regarding basic anatomy is still growing nearly 40 years after Masters and Johnson studies: in 2005, the most comprehensive study about the anatomy of the clitoris was reported at last [63]. Thus, the contribution, if any, of animal models of sexual response in female rodents is modest. After the questionable approval and the lack of efficacy of drugs to treat sexual desire issues, it is necessary to realize that the more we understand the FSR, the more we might explain, and eventually treat, their dysfunctions.

Acknowledgments

Funding was provided by Programa de Apoyo a Proyectos de Investigación e Innovación Tecnológica, Universidad Nacional Autónoma de México (No. IA204321)

References

1. Tobia G, IsHak WW (2013) DSM-5 changes in diagnostic criteria of sexual dysfunctions. Reprod Syst Sex Disord. https://doi.org/10.4172/2161-038x.1000122
2. Hayes RD, Dennerstein L, Bennett CM, Fairley CK (2008) What is the "true" prevalence of female sexual dysfunctions and does the way we assess these conditions have an impact? J Sex Med 5:777–787. https://doi.org/10.1111/j.1743-6109.2007.00768.x
3. Pfaus JG, Annette S, Tanya VS et al (2004) Selective facilitation of sexual solicitation in the female rat by a melanocortin receptor agonist. Proc Natl Acad Sci 101:10201–10204. https://doi.org/10.1073/pnas.0400491101
4. Anderson R, Moffatt CE (2018) Ignorance is not bliss: if we don't understand hypoactive sexual desire disorder, how can flibanserin treat it? Commentary. J Sex Med 15:273–283. https://doi.org/10.1016/j.jsxm.2018.01.001
5. Jaspers L, Feys F, Bramer WM et al (2016) Efficacy and safety of flibanserin for the treatment of hypoactive sexual desire disorder in women: a systematic review and meta-analysis. JAMA Intern Med 176:453–462. https://doi.org/10.1001/jamainternmed.2015.8565
6. Mintzes B, Tiefer L, Cosgrove L (2021) Bremelanotide and flibanserin for low sexual desire in women: the fallacy of regulatory precedent. Drug Ther Bull 59(12):185–188. https://doi.org/10.1136/dtb.2021.000020
7. Gelman F, Atrio J (2017) Flibanserin for hypoactive sexual desire disorder: place in therapy. Ther Adv Chronic Dis 8:16–25. https://doi.org/10.1177/2040622316679933
8. Kingsberg SA, Clayton AH, Portman D et al (2021) Failure of a meta-analysis: a commentary on Glen Spielmans's "Re-analyzing phase III bremelanotide trials for 'hypoactive sexual desire disorder in women'". J Sex Res:1–2. https://doi.org/10.1080/00224499.2021.1902926
9. Food and Drug Administration FDA orders important safety labeling changes for Addyi. https://www.fda.gov/news-events/press-announcements/fda-orders-important-safety-labeling-changes-addyi
10. Spiteri T, Musatov S, Ogawa S et al (2010) Estrogen-induced sexual incentive motivation, proceptivity and receptivity depend on a functional estrogen receptor alpha in the ventromedial nucleus of the hypothalamus but not in the amygdala. Neuroendocrinology 91:142–154. https://doi.org/10.1159/000255766
11. Mermelstein PG, Micevych PE (2008) Nervous system physiology regulated by membrane estrogen receptors. Rev Neurosci 19:413–424. https://doi.org/10.1515/revneuro.2008.19.6.413
12. Komisaruk BR, Adler NT, Hutchison J (1972) Genital sensory field: enlargement by estrogen treatment in female rats. Science (80-) 178:1295 LP–1291298
13. Blaustein J, Pfaff D, Arnold A et al (2009) Feminine reproductive behavior and physiology in rodents: integration of hormonal, behavioral and environmental influences. Horm Brain Behav 3:67–107
14. Le Moëne O, Ågmo A (2019) Modeling human sexual motivation in rodents: some caveats. Front Behav Neurosci 13:187. https://doi.org/10.3389/fnbeh.2019.00187
15. Pattij T, Olivier B, Waldinger MD (2005) Animal models of ejaculatory behavior. Curr Pharm Des 11:4069–4077
16. Chung E, De Young L, Brock GB (2011) Investigative models in erectile dysfunction: a state-of-the-art review of current animal models. J Sex Med 8:3291–3305
17. Chu X, Ågmo A (2016) Sociosexual interactions in rats: are they relevant for understanding human sexual behavior? Int J Psychol Res 9:76–95
18. Bergheim D, Chu X, Ågmo A (2015) The function and meaning of female rat paracopulatory (proceptive) behaviors. Behav Process 118:34–41. https://doi.org/10.1016/j.beproc.2015.05.011
19. Erskine MS, Baum MJ (1982) Effects of paced coital stimulation on termination of estrus and brain indoleamine levels in female rats. Pharmacol Biochem Behav 17:857–861
20. Erskine MS, Kornberg E, Cherry JA (1989) Paced copulation in rats: effects of intromission

frequency and duration on luteal activation and estrus length. Physiol Behav 45:33–39

21. Coopersmith C, Candurra C, Erskine MS (1996) Effects of paced mating and intromissive stimulation on feminine sexual behavior and estrus termination in the cycling rat. J Comp Psychol 110:176–186
22. Paredes RG, Alonso A (1997) Sexual behavior regulated (paced) by the female induces conditioned place preference. Behav Neurosci 111:123–128
23. Peretto P, Paredes R (2014) Social cues, adult neurogenesis, and reproductive behavior. In: Mucignat-Caretta C (ed) Neurobiology of chemical communication. CRC Press/Taylor & Francis, Boca Raton, pp 367–387
24. Portillo W, Ortiz G, Paredes RG (2020) Repeated paced mating increases the survival of new neurons in the accessory olfactory bulb. Front Neurosci 14:249
25. Paredes RG (2014) Opioids and sexual reward. Pharmacol Biochem Behav 121:124–131. https://doi.org/10.1016/j.pbb.2013.11.004
26. Ågmo A, Snoeren EMS (2017) A cooperative function for multisensory stimuli in the induction of approach behavior of a potential mate. PLoS One 12:e0174339. https://doi.org/10.1371/journal.pone.0174339
27. Ågmo A, Turi AL, Ellingsen E, Kaspersen H (2004) Preclinical models of sexual desire: conceptual and behavioral analyses. Pharmacol Biochem Behav 78:379–404. https://doi.org/10.1016/j.pbb.2004.04.013
28. Chu X, Guarraci FA, Ågmo A (2015) Sociosexual behaviors and reproductive success of rats (Rattus norvegicus) in a seminatural environment. Physiol Behav 151:46–54. https://doi.org/10.1016/j.physbeh.2015.07.005
29. Chu X, Ågmo A (2016) Studies of sociosexual interactions in rats in an externally valid procedure: are they relevant for understanding human sexual behavior? Int J Psychol Res 9. https://doi.org/10.21500/20112084.2339
30. Simon W, Gagnon JH (2003) Sexual scripts: origins, influences and changes. Qual Sociol 26:491–497. https://doi.org/10.1023/B:QUAS.0000005053.99846.e5
31. Graaf H, Rademakers J (2011) The psychological measurement of childhood sexual development in Western societies: methodological challenges. J Sex Res 48:118–129. https://doi.org/10.1080/00224499.2011.555929
32. Lamb S, Plocha A (2014) Sexuality in childhood. In: APA handbook of sexuality and psychology, Person-based approaches, vol 1. American Psychological Association, Washington, DC, pp 415–432
33. Frederick DA, St John HK, Garcia JR, Lloyd EA (2018) Differences in orgasm frequency among gay, lesbian, bisexual, and heterosexual men and women in a U.S. national sample. Arch Sex Behav 47:273–288. https://doi.org/10.1007/s10508-017-0939-z
34. Laan ETM, Klein V, Werner MA et al (2021) In pursuit of pleasure: a biopsychosocial perspective on sexual pleasure and gender. Int J Sex Health:1–21. https://doi.org/10.1080/19317611.2021.1965689
35. Wyverkens E, Dewitte M, Deschepper E et al (2018) YSEX? A replication study in different age groups. J Sex Med 15:492–501. https://doi.org/10.1016/j.jsxm.2018.02.012
36. Barrada JR, Castro Á, Fernández-del-Río E, Ramos-Villagrasa PJ (2021) Motives to have sex: measurement and correlates with sociodemographic, sexual life, and psychosexual characteristics. Front Psychol 12:2424
37. Masters WH, Johnson VE (1966) Human sexual response. Little, Brown, Boston
38. Levin R (2008) Critically revisiting aspects of the human sexual response cycle of Masters and Johnson: correcting errors and suggesting modifications. Sex Relatsh Ther 23:393–399
39. Tiefer L (1991) Historical, scientific, clinical and feminist criticisms of "The Human Sexual Response Cycle" model. Annu Rev Sex Res 2:1–23
40. Brecher R, Edward M (1966) An analysis of human sexual response. Signet, New York
41. Basson R (2000) The female sexual response: a different model. J Sex Marital Ther 26:51–65. https://doi.org/10.1080/009262300278641
42. Leavitt CE, Leonhardt ND, Busby DM (2019) Different ways to get there: evidence of a variable female sexual response cycle. J Sex Res 56:899–912. https://doi.org/10.1080/00224499.2019.1616278
43. Munarriz R, Kim SW, Kim NN et al (2003) A review of the physiology and pharmacology of peripheral (vaginal and clitoral) female genital arousal in the animal model. J Urol 170:S40–S45. https://doi.org/10.1097/01.ju.0000075352.03144.15
44. Meston CM, Stanton AM (2019) Understanding sexual arousal and subjective–genital arousal desynchrony in women. Nat Rev Urol 16:107–120. https://doi.org/10.1038/s41585-018-0142-6
45. Handy AB, Stanton AM, Meston CM (2019) What does sexual arousal mean to you? Women with and without sexual arousal concerns

describe their experiences. J Sex Res 56:345–355. https://doi.org/10.1080/00224499.2018.1468867

46. Handy AB, Stanton AM, Meston CM (2018) Understanding women's subjective sexual arousal within the laboratory: definition, measurement, and manipulation. Sex Med Rev 6:201–216. https://doi.org/10.1016/j.sxmr.2017.11.001
47. Meston CM, Gorzalka BB (1995) The effects of sympathetic activation on physiological and subjective sexual arousal in women. Behav Res Ther 33:651–664. https://doi.org/10.1016/0005-7967(95)00006-J
48. Meston CM, Gorzalka BB, Wright JM (1997) Inhibition of subjective and physiological sexual arousal in women by clonidine. Psychosom Med 59:399
49. Sipski ML, Rosen RC, Alexander CJ, Gómez-Marín O (2004) Sexual responsiveness in women with spinal cord injuries: differential effects of anxiety-eliciting stimulation. Arch Sex Behav 33:295–302. https://doi.org/10.1023/B:ASEB.0000026629.33441.cf
50. Palle C, Bredkjaer HE, Ottesen B, Fahrenkrug J (1990) Vasoactive intestinal polypeptide and human vaginal blood flow: comparison between transvaginal and intravenous administration. Clin Exp Pharmacol Physiol 17:61–68. https://doi.org/10.1111/j.1440-1681.1990.tb01265.x
51. Palle C, Bredkjœr HE, Fahrenkrug J, Ottesen B (1991) Vasoactive intestinal polypeptide loses its ability to increase vaginal blood flow after menopause. Am J Obstet Gynecol 164:556–558. https://doi.org/10.1016/S0002-9378(11)80019-0
52. Komisaruk BR, Wallman J (1977) Antinociceptive effects of vaginal stimulation in rats: neurophysiological and behavioral studies. Brain Res 137:85–107
53. Meston CM, Kilimnik CD, Freihart BK, Buss DM (2020) Why humans have sex: development and psychometric assessment of a short-form version of the YSEX? Instrument. J Sex Marital Ther 46:141–159. https://doi.org/10.1080/0092623X.2019.1654581
54. Pulverman CS, Meston CM (2020) Sexual dysfunction in women with a history of childhood sexual abuse: the role of sexual shame. Psychol Trauma 12:291–299
55. Pulverman CS, Kilimnik CD, Meston CM (2018) The impact of childhood sexual abuse on women's sexual health: a comprehensive review. Sex Med Rev 6:188–200. https://doi.org/10.1016/j.sxmr.2017.12.002
56. Handy AB, Freihart BK, Meston CM (2020) The relationship between subjective and physiological sexual arousal in women with and without arousal concerns. J Sex Marital Ther 46:447–459. https://doi.org/10.1080/0092623X.2020.1758859
57. Chevret-Méasson M, Lavallée E, Troy S et al (2009) Improvement in quality of sexual life in female partners of men with erectile dysfunction treated with sildenafil citrate: findings of the index of sexual life (ISL) in a couple study. J Sex Med 6:761–769. https://doi.org/10.1111/j.1743-6109.2008.01146.x
58. Bronner G, Kitrey ND, Uziel N et al (2015) Correlation between premature ejaculation and female vaginal penetration difficulties. Int J Impot Res 27:152–156. https://doi.org/10.1038/ijir.2015.3
59. Gregory A (2021) Understanding female sexual dysfunction, its causes and treatments. Br J Nurs 30:S18–S29. https://doi.org/10.12968/bjon.2021.30.18.S18
60. Janssen E (2002) Psychophysiological measurement of sexual arousal. In: Handbook for conducting research on human sexuality. Lawrence Erlbaum Associates Publishers, Mahwah, pp 139–171
61. Hatch JP (1979) Vaginal photoplethysmography: methodological considerations. Arch Sex Behav 8:357–374. https://doi.org/10.1007/BF01541879
62. Levin RJ (2003) The ins and outs of vaginal lubrication. Sex Relatsh Ther 18:509–513. https://doi.org/10.1080/1468199031000l609859
63. O'Connell HE, Sanjeevan KV, Hutson JM (2005) Anatomy of the clitoris. J Urol 174:1189–1195. https://doi.org/10.1097/01.ju.0000173639.38898.cd

Chapter 15

Methods to Assess the Role of Neurogenesis in Reproductive Behaviors of Birds, Rats, and Sheep

Rebeca Corona, Olesya T. Shevchouk, and Ivan E. Gladwyn-Ng

Abstract

Reproduction represents one of the most important biological events for the organism due to its relevance in perpetuating life. It allows the production of offspring with similar characteristics to the progenitor. The behavioral events of reproduction comprise several changes that prepare the organisms and favor the display of specific behaviors. Reproduction starts with the localization and selection of a possible partner, and this specific moment requires the detection of chemosensory cues that guides their attention and behavior. After the adequate partner is selected, the sexual interaction takes place, usually regulated by females. If pregnancy happens, a series of changes and adaptations occurs within the brain that prepares the mother for the future interaction with the offspring. After delivery, interaction with the offspring during early postpartum along with the pregnancy adaptations of the new mother allows the display of a complex set of parental behaviors that facilitate the care and survival of the newborn. During all these reproductive steps, several adaptations occur within the brain that prepare the organism for its current needs and, in some cases, maintain the changes until the next reproductive episode. The aim of the present chapter is to discuss one of the most complex plastic adaptations, namely adult neurogenesis that occurs in the brain and accompanies the different steps of reproduction in life. From partner attraction and selection through sexual interaction to the parental care of the offspring, we selected three different species in which evidence has shown that neurogenesis plays an important role. We will describe how in songbirds, neurons recently incorporated to the high-vocal center are necessary for the female attraction by facilitating a new singing repertoire of the male each reproductive season. In rats, from the first sexual behavior encounter, neurogenesis in the olfactory bulb is stimulated allowing a facilitation of the following interactions. The deployment of maternal behavior in sheep requires an early and highly specialized odor recognition of the offspring by the mother in which newly born olfactory bulb neurons participate. Additionally, in this chapter we overview two of the most used techniques to visualize and study adult neurogenesis, the use of endogenous and exogenous markers revealed by immunostainings and neuronal precursor labeling by electroporation.

Key words Neurogenesis, Markers, Electroporation, Songbirds, Rats, Sheep

1 Introduction

The brain is a remarkable and complex organ that orchestrates most of our functions and behaviors. The organization of the brain network is mainly set during development in the embryonic period,

Raúl G. Paredes et al. (eds.), *Animal Models of Reproductive Behavior*, Neuromethods, vol. 200,
https://doi.org/10.1007/978-1-0716-3234-5_15,

where most of the cells composing the different structures are established and get ready to contend with all the information they must manage during life. For a long time, it was believed that embryonic development was the only period of neurogenesis. In 1912, Allen and collegues started to observe postnatal mitosis in rat brains; however, the scientific community could not confirm these observations due to a lack of techniques to label and identify these new cells [1]. Around the 60s, the groundbreaking work of Altman [2] provided the first evidence of adult neurogenesis in specific regions of the brain, challenging the idea that the central nervous system was static and had no regeneration capacity, the central dogma of neurobiology postulated by Santiago Ramón y Cajal for which he received the Physiology and Medicine Nobel Prize [3]. Around the 80s, Nottebohm and collaborators [4] noted that, in songbirds, the high vocal center (HVC) grew seasonally by the incorporation of new neurons, demonstrating that adult neurogenesis also happened in birds. Although the evidence for adult neurogenesis in birds and mammals accumulated, it was not until the 90s that adult neurogenesis could be conclusively confirmed thanks to advances in immunohistochemistry (IHC) and microscopy and therefore was accepted by the scientific community. This neurogenesis phenomenon was first described in the olfactory bulb (OB) and the hippocampus of an adult mammalian rodent brain [5, 6]. Nowadays, it is widely accepted that mammals and birds are constantly shaping and refining their cerebral network by the processes of plasticity, and that adult neurogenesis is one of the processes that allows the necessary changes in the brain to face behavioral and physiological demands to survive during life.

Interestingly, a large amount of evidence shows that neuronal plasticity, and specifically neurogenesis, is required for critical physiological and reproductive steps differently in males and females. From peripubertal period to adulthood, the neurogenesis process accompanies the adaptative changes needed to respond to a variety of stimuli and learning periods for puberty onset, partner seeking and selection, sexual performance, pregnancy, and parental care [7–11].

1.1 Adult Neurogenesis

Adult neurogenesis is the formation of new functional neurons capable of integrating into existing circuits. Like neurogenesis in the embryonic stage, these new neurons in the adult brain are derived from neuronal precursors located in specific brain neurogenic niches, these are preserved during lifetime. There are now several recognized neurogenic niches in the adult mammal and bird brains like the hippocampus, OB, the HVC and, to a lesser degree, the hypothalamus [12], olfactory tubercles, the anterior olfactory nucleus, the amygdala, the caudate nucleus, the nucleus accumbens, among others [13–16]. In this chapter, we would like to focus in three neurogenic brain regions that are intrinsically regulated and necessary for reproduction in males and females. In mammals, the two most recognized and described neurogenic regions in the adult mammalian brain are the subgranular zone of the dentate gyrus of the hippocampus and the subventricular zone (SVZ)/OB system [17]. In songbirds, new neurons produced in the ventricular zone of the lateral ventricle are incorporated to the HVC, among other regions [18].

Neurogenesis is a highly dynamic and coordinated process that includes cell proliferation, progenitor specification and fate, cell migration, axonal and dendritic growth, differentiation, synaptic integration to the pre-existing circuits, and finally survival of new neurons. The differentiation and fate of progenitor cells is regulated by soluble factors, membrane-associated molecules, and extracellular matrix molecules produced in the progenitor cells and/or in neighboring cells [19]. The differentiation into neurons, the maintainance, and the survival of the progenitor cells depend on the local microenvironment [20].

The production of new neurons occurs mainly in the walls of the lateral ventricles called SVZ. Shortly after birth in the postnatal stage, the rostral extension of the lateral ventricles collapses followed by progressive maturation of the rostral migratory stream (RMS). In this newly formed RMS, neuroblasts form tight chains, aided by glial tubes formed by astrocytes [6, 21]. In the mature RMS, the migrating neuroblasts remain mitotically active, maintaining properties of neuronal progenitors [20, 22]. Neuroblasts generated in the SVZ or along the RMS, migrate through the RMS until reaching the OB, where they disperse and acquire a specific phenotype of interneurons [23, 24]. In fact, it was the discovery of massive migration of newborn neurons via the RMS to the OB [6, 21] that sparked a new level of interest in the phenomenon of adult neurogenesis. The adult rodent SVZ is composed of four main cell types: A, B, C, and ependymal cells [24]. A cells correspond to neuroblasts; they extend throughout the lateral ventricles, from where they begin to migrate through a network of interconnected pathways that converge in the anterior region of the SVZ where the RMS starts. In the RMS, neuroblasts migrate tangentially until reaching the OB where A cells detach from the chains to migrate radially towards the cellular layers [25]. The neuroblast chains are surrounded by astrocytes from the SVZ itself. Some of these astrocytes function as support elements; however, a subpopulation of them are known as B cells [24] and participate in the process of neurogenesis, since they became the progenitor cells [24]. B cells originate from the radial glia, which in the brain of the embryo constitute the neuronal precursors [26]. The A cells form chains which in turn are surrounded by a combination of B cells and rapidly dividing immature cells, called C cells [24]. C cells are also a type of progenitors, however they function as transient amplifying progenitors between B and A cells. Therefore, the process of generating new cells in the SVZ occurs as follows: from B cells (astrocytes) C cells are generated (transient intermediates) and then become A cells (neuroblasts). Once the neuroblasts arrive in the OB, they reach their final position and begin their maturation. The new neurons get incorporated as interneurons of the OB: granule cells and periglomerular cells. After 1 month, approximately half of the newly formed neurons have been eliminated, while the other half have been succesfully integrated into the OB [27, 28].

1.2 Models for Studying the Role of Adult Neurogenesis in Reproduction

In the field of reproduction biology, many important physiological phenomena have been demonstrated in rodent and non-rodent research models. In this chapter, we present a few of the most studied models where adult neurogenesis shows sexually and reproductively differentiated patterns. We will present relevant examples of adult neurogenesis in singing behavior and partner attraction in songbirds, in sexual behavior in rats, and finally in maternal behavior in sheep.

1.2.1 Songbirds

Research on songbird neurogenesis has made a major contribution to our understanding of the functionality of adult neurogenesis. Songbirds include more than 4500 species and possess a unique behavioral characteristic since they have to learn their species-specific vocalizations from other members of the species [29] and use them for complex communication. Importantly, singing behavior is part of the bird's courtship ritual and therefore directly related to reproduction. Some of the brain regions controlling vocalization are highly neurogenic areas [30], allowing us to probe for the role of neurogenesis in singing behavior. Furthermore, many species of songbirds show a stark sexual dimorphism in both the singing behavior and the underlying neuroplasticity [31]. In temperate zone songbird species, especially, only male individuals sing and only during spring and/or early summer. Singing behavior and neurogenesis in regions controlling song are regulated by testosterone levels, as part of the orchestration of the reproductive system that sex hormones perform throughout the body [32]. The sex difference in neurogenesis within song-controlling regions in the brains of songbirds is one of the biggest differences known in all vertebrates, and these differences are well correlated to sex differences in singing behavior. It is an intriguing example of the connection between sex, reproduction, and neurogenesis where adult neurogenesis is more pronounced in one sex than the other.

1.2.2 Rodents

Rats are one of the most studied species for neurogenesis and this includes reproduction-related neurogenesis. In this section, we would like to highlight one of the culminating reproductive behaviors, the sexual interaction. In fact the seminal 1960s papers of Altman showing the discovery of adult neurogenesis in male rats, already mentioned that newborn neurons bind testosterone and are likely involved in sexual processes [2]. Sexual behavior belongs to the reproductive innate behaviors and is highly stereotyped. When females are sexually receptive, they display sexual motivation through solicitation behaviors such as hops and darts, allowing the male to approach and perform different copulatory behaviors including mounting, intromission and ejaculation, and responding to them with sexual receptivity—lordosis. Sexually relevant odors are detected and processed by the olfactory systems, and the OB's are the first of many brain centers to process and organize this information. The OB's are highly dynamic structures because they are constantly adding new granular and periglomerular interneurons, by a neurogenesis process [33]. Evidence shows that adult neurogenesis within the OB participates in the sexual behavior of both sexes [34–37]. The first sexual encounter increases significantly the amount of newly formed cells in the OB, but also sexual experience enhances the survival of these cells [34, 37]. Importantly, the elimination of neurogenesis with a treatment of cytosine arabinoside, while exposing female mice to male pheromones, prevents them from developing a preference for dominant males [38]. Male sexual behavior was also reduced following cytosine arabinoside treatment [39], suggesting intact adult neurogenesis is crucial for sexual behavior learning in both sexes. The full picture of how neurogenesis is differently involved in regulating sexual behavior in males versus females is yet to be discovered; however, the effect of genetic ablation of newly born neurons in the OB of mice was studied in both sexes [40]. Both female and male mutant mice showed an altered olfactory learning response, where the mutants could learn to approach predator odors when previously associated with a reward, while wildtype male and female mice never approached the predator odor, regardless of reward. Male and female mutant mice with ablated OB neurogenesis showed different deficits in sex-specific behaviors. Mutant males displayed reduced male–male aggression and male sexual behaviors toward females, whereas mutant females presented deficits in fertility and maternal behavior. Globally, this study showed that social behaviors are affected in both sexes by the ablation of OB neurogenesis, although a caveat is that females were not tested for sexual or aggressive behavior, while males were not tested for fertility or paternal behavior.

1.2.3 Sheep

A good model for studying the participation of the newly born neurons specifically from the OB in the maternal behavior is the sheep [11, 41, 42]. In this species, olfaction becomes a sensory modality determinant for the establishment of the early mother–infant interaction. Sheep usually give birth to one or two lambs per year and, as a precocial species, the offspring rapidly starts interacting with the mother. In this way, a selective bond between mother and infant is formed in the first few hours after birth. Studies of this model have demonstrated that OB neurogenesis in the mother facilitates the individual recognition of the lamb, allowing the bond to be established and the maternal behavior to be displayed by the mother [11, 41]. Maternally upregulated OB neurogenesis is an example of sex and reproduction interacting to determine the rate and pattern of neurogenesis. This particular form of neurogenesis modulation occurs only in females and only during the perinatal phase following succesful pregnancy and birth of offspring.

2 Methods for Studying Adult Neurogenesis

2.1 Markers of DNA Replication

Given the importance of neurogenesis literature in studying sex differences in reproductive behaviors, we have included an example protocol for labeling newborn neurons in song-controlling regions of the songbird brain, in a female sexual behavior rat model for studying OB and in maternal recognition of the offspring in a sheep model in the Subheadings 2 and 3. Additionally, we included a genetic method called electroporation that is frequently used to study the development of neurons from embryonic stages to post-natally born cells. By using this technique, you can label specific group of cells and trace their lineages in a very elegant and specific manner.

The method of exogeneous markers of DNA replication is based on the use of synthetic nucleoside analogues of thymidine. After administration, they will enter the brain and integrate all the mitotically active cells found in neurogenic niches. If a particular cell is undergoing mitosis at that time, the thymidine analogue will be incorporated into the newly synthesized DNA of the daughter cell, in the place of thymidine. Currently, the most used thymidine analogue is bromodeoxyuridine (BrdU), which replaced the use of radioactive (tritium-) thymidine used in earlier studies [43]. More recently new analogues have been developed such as 5-Chloro-2′-deoxyuridine (CldU), 5-Iodo-2′-deoxyuridine (IdU), and ethynyl deoxyuridine (EdU), that can be used as single tagging or in combination with BrdU or another marker. The thymidine analogue can then be detected ex-vivo by performing IHC to target the analogue with antibodies raised against its epitopes.

Endogenous markers of neurogenesis are proteins that are expressed exclusively or primarily during specific stages of cell division and replication, such as Ki-67, PCNA, Cdk1, and other cyclins/Cdks and PHH3. Ki67 is a marker of proliferation, expressed very specifically by the precursor neurons during all phases of the cell division cycle, but not expressed by quiescent, i.e., non-replicating cells [44]. Another proliferation marker is the proliferating cell nuclear antigen (PCNA) which acts as a scaffolding protein during DNA replication, helping to recruit other proteins necessary for the replication. It is less specific than Ki67 since it is also expressed during DNA repair, chromatin remodeling, and epigenetic mechanisms [45]. Cdk1 is a key regulator of the cell division cycle; by phosphorylating various targets, it helps the dividing cell to progress from one phase of the cycle to the next, therefore its expression in a cell can also be used to infer cell replication [46]. Other cell cycle progression markers including cyclins D, A, and B as well as CDK1, 2 and 4/6, can also be used as neuronal proliferation markers [47]. Phosphohistone H3 (PHH3) is an histone protein that makes up part of the chromatin structure, its expression levels are elevated during mitosis, therefore it is also used as a proliferation marker [48]. To conclude that the newly born cell is destined to become a neuron, cell replication markers must be combined with lineage determination of the new cells, as described in the following section.

The phenotypic identification of recently generated cells is necessary to evaluate if the proliferative process is related to neuronal genesis. Therefore, detecting the co-expression of endogenous/exogenous markers of cell replication (as explained before) with a marker of neuronal phenotype has become a requisite in neurogenesis studies. During the neurogenesis process, the newly generated cells go through a series of critical steps that determine the lineage commitment before acquiring their final and mature neuronal phenotype. These stages include the general steps of proliferation, differentiation, and maturation of the neurons [49]. Within each maturational step, new cells express endogenous markers reflecting the cellular and molecular intrinsic characteristics allowing the identification of the cell phenotype. Even before the new cells have exited the cell cycle, their neuronal lineage commitment is determined. The most common markers that can be used to label neuronal cell commitment early in their development include Nestin, GFAP, Sox2, Pax6 among others. Nestin is an intermediate filament protein expressed in undifferentiated neurons and is used as a progenitor cell marker. It is downregulated once the neuroblast starts to further differentiate, therefore it can be utilized to visualize newly formed neurons in their earliest stages of development. Nestin is also expressed by radial glia cells and astrocytes during prenatal development and disappears gradually in early postnatal ages. Newly formed cells can express the glial fibrillary acidic protein (GFAP), that identifies glial cells, especifically astrocyte-like morphology. Although GFAP is the most commonly used marker of mature astrocytes, a large proportion of the newborn cells within the neurogenic niches in the adult are actually GFAP-positive [50]. Other markers that are transiently expressed on the progenitor cells phenotypically committed to neurons are doublecortin (DCX), the neural cell adhesion molecule (PSA-NCAM) and BIII tubulin (Tuj1) [24, 49, 51, 52]. All of them are cytoskeletal proteins associated to microtubules or to cell membrane and are used to label neurons in early stages. Its staining is found in the cell bodies, dendrites, axons, and axonal terminations of immature neurons. DCX is heavily expressed in migrating neurons and can be detected in the earliest stages of the neuronal maturation. PSA-NCAM is localized mainly in the plasma membrane and is a neuronal cell adhesion molecule that participates in the growth and migration of the committed neurons. Tuj1 is a tubulin involved in neuronal differentiation, it contributes to the building block of the microtubules that are structural components of the cells [49, 53, 54]. Finally, when newly formed cells reach maturity, the expression of specific proteins like calcium-binding proteins (calretinin, calbindin), neurotransmitter enzymes (Tyrosin hydroxylase-TH), and neurotransmitters become evident and can be used as markers [55]. Calbindin is a calcium binding protein that participates in fast calcium buffering and its expression can be observed in cell bodies, dendrites, and spines of the neurons. Calretinin is also a calcium binding protein and is highly expressed in neurons [49, 56]. TH is the rate-limiting enzyme involved in the conversion of tyrosine to dopamine and its been used to label dopamine neurons. The neuronal nuclear protein (NeuN) is commonly used since it is expressed in most of the neuronal cell types throughout the central and peripheral nervous systems. NeuN is frequently used along with BrdU for a clear identification of newly born neurons [49]. Additionally, other markers have been used for a neuronal phenotypic identification. MAP-2 is a cytoskeletal protein expressed in neurons, and specifically the isoform MAP-2b that can be found through life. For a detailed identification of the phenotype of newborn neurons, multiple staining approaches are frequently used, particularly using immunofluorescence labeling due to the possibility of observing colocalization of fluorescent markers by confocal microscopy [17, 23, 49, 57, 58].

2.1.1 Materials

In-Vivo Procedures

1. A minimum of eight animal subjects for each experimental group are suggested;
2. Prepare bromodeoxyuridine (BrdU) in 0.9% NaCl;
3. For perfusion and fixation of the brain, use a perfusion pump and needle to insert inside heart (or inside carotid vein in sheep) and the following buffers: a phosphate-buffered saline (PBS) needs to be prepared for clearing out blood from the brain before perfusion.
   (a) For fixating the brain during perfusion, prepare a solution of 4% paraformaldehyde (PFA). In the specific case of sheep, adding 1% sodium nitrate is recommended.
   (b) After perfusion procedure, scissors and forceps are necessary to extract brain from skull.

Ex-Vivo Procedures

1. After perfusion, the extracted brain is post-fixed in vials with 4% PFA and cryoprotected in a 30% sucrose solution;
2. For cutting the brain sections with a cryostat, the tissue samples are frozen in dry ice and embedded in OCT compound. Sections are usually preserved in eppendorf tubes with anti-freeze solution or glass slides (depending on IHC method).

Immunohistochemistry

The principle of the IHC technique is to visualize the binding of the antibody-antigen by using a chromogen that is attached to a secondary antibody directed to detect the host species of the primary antibody. For the visualization, there are at least two options commonly used, one is the use of peroxidase enzymes to catalyze a color-producing photostable reaction with the biotinylated secondary antibody to be observed in a bright-field microscope. The second option is to visualize the labeling in a fluorescence microscope that possess a filter to excite the fluorophore located in the secondary antibody in a specific wavelength (nm) to produce a fluorescence emission also in a specific wavelength to be detected. Both possibilities are often used in neurogenesis studies, but in general, for visualization of a simple labeling, like the one of BrdU birth-dating, peroxidase staining is preferably used. The bright-field microscopy projects focused light on one plane of the sample and for the quantification of the labeled cells a method of stereology is preferable to obtain a more precise estimation of the cells. This method uses the thickness of each brain section and the number of slices counted to calculate the total number of labeled cells in the area of interest. Immunofluorescence staining is preferable for multiple labeling. The advantage of this approach is the detection of co-expressed labeling due to the use of separate excitation filters and the discrimination of their emission based on spectral properties. In addition, with this second technique, from two to four labels can be detected at a time.

1. IHC in (1) floating sections are performed in baskets and 12-well plate and the tissue is usually 30–40 μm thick; (2) on slides uses glass slides, delimiting pen and slides box and sections usually 10–20 μm thick;
2. Buffers needed: PBS or Tris-buffered saline (TBS) for washing, Triton X-100 or Tween-20 for washing and permeabilization, H_2O_2 for blocking endogenous peroxidase activity, diluted in methanol, blocking serum to decrease background non-specific staining;
3. Primary and secondary antibodies, diluted to a concentration shown to give a strong and specific signal, based on literature and/or in-house optimization tests;

4. For amplification of signal use the avidin-biotin HRP complex;
5. For color revelation of labeled antigen use DAB and H_2O_2;
6. Finally, a coverslip medium is used for preparing slides for microscopy observation.

2.1.2 Methods

Thymidine Analog Administration

1. To study newly born neurons in an adult brain, the use of specific markers is important. The most used method is the BrdU birth-dating marker that needs to be administered to the subject and evaluated afterwards in tissue sections. Usually, BrdU is injected in a dose between 20 and 100 mg/kg, and the number and interval of injections vary according to the species used and the experimental question (*see* **Notes 1–3** for details). According to the stage of development of the new cells, the experimenter wants to detect, BrdU injections can be delivered in different patterns.
   (a) To study cell proliferation, BrdU can be injected twice on the same day, with an interval of 2 h; analysis can be performed under 24 h, or 1 or 2 days after BrdU administration;
   (b) To evaluate neurons in migration, time after injection to evaluate BrdU incorporation could be 1–2 weeks;
   (c) For analyzing cells in differentiating stages or mature cells, 3 weeks to 3 months is enough time to wait after BrdU administration.
2. The day of sacrifice, animals are sedated and intracardially perfused. Detailed videos of whole animal perfusion are available in the literature [59];
3. The brain is removed, post-fixed and sectioned;
3. The free-floating sections are processed by IHC to localize BrdU signal;
4. To determine the nature of the new cells, a double IHC must be performed. The markers that are commonly used for fenotyping the newly born cells are Hu or DCX to identify early differentiated neurons, or NeuN for mature neurons;
5. For clarity, a protocol for a single antigen staining is described (*see* **Note 4**). For double labeling, usually fluorescently labeled antibodies are used, the protocol is similar with some differences and keeping sections as much as possible in the dark after fluorescent antibodies have been applied.
   (i) General protocol: all washes can be done either with PBS or TBS and should always be done three times for 5 min. Washes follow every step except for serum blocking. Every wash after the primary antibody step should use the 0.1%

Tween-20 or Triton X-100 buffer. If sections were cut on slides using non-PFA fixed brains, fixate tissue by applying 4% PFA for 10 min, then wash and apply the delimiting pen around the sections;

(ii) Blocking serum, primary and secondary antibodies should be diluted in buffer with 0.1% Tween-20 or Triton X-100;

(iii) Protocol for both modes of IHC (floating and on slides);

(iv) Block endogeneous peroxidases in 3% H_2O_2 diluted in methanol for 20 min;

(v) Permeabilize in buffer with 0.5% Tween-20 or Triton X-100 for 10 min;

(vi) Block for 1 h with serum from the species corresponding to the secondary antibody host (e.g., for goat anti-rabbit secondaries use normal goat serum for blocking);

(vii) Incubate with primary antibody for one or two overnights at 4 °C. Incubate with secondary antibody for 2 h;

(viii) Amplify with ABC-HRP solution (follow producers instructions for diluting and durations);

(ix) Reveal coloration with a solution of 0.05% DAB and 0.015% H_2O_2, make sure to use buffer without Tween-20 or Triton X-100. Adjust the duration to the antigen studied as well as ambient temperature conditions (the reaction goes faster at higher temperatures)

2.2 Neuronal Precursor Labeling by Electroporation

In utero and postnatal electroporation (IUE and PNE) are rapid and powerful in vivo DNA transfer techniques extensively used to study the molecular and cellular mechanisms underlying the development of many brain regions. These techniques are based upon the introduction of DNA of interest into brain ventricles followed by application of electric current (via either a pair or a trio of electrodes) to direct the entry of genetic material into the cells (typically the neuronal stem cells) lining the brain ventricles.

There are many advantages to these approaches over other methods to study specific steps of brain development in vivo, because they allow the study of neuronal proliferation, neurogenesis, and lineage tracing, in addition to neuronal migration, axonal pathfinding, and synaptic integration [60, 61]. Specifically, electroporation techniques permit the targeting of cell populations that may be spatially and temporally separated. This characteristic eliminates the need for generation of multiple different animal models for studying all the different populations of interest and circumvents problems related to unavailability of specific Cre-recombinase driver rodent models. Additionally, these techniques permit labeling of desired cells with molecular reagents such as fluorescent proteins or channelrhodopsins, which benefit the visualization of functional interactions between those live cell populations ex vivo [61, 62] or manipulating the expression of genes of interest in those cells that facilitate rescuing the embryonic or postnatal lethality due of deletion of gene(s) of interest, and to conduct functional experiments concurrently [63]. In the below section, we describe in greater detail the individual steps necessary to carry out the electroporation of the brain, as well as highlight the technical variations to target different cerebral structures and to investigate distinct steps of neurogenesis.

2.2.1 Materials

Electroporation Preparation

1. DNA plasmid solution: purify plasmid DNA with an Endotoxin-free prep kit, multiple commercial vendors provide different ready to use kits with detailed kit-specific reagents & protocols;
2. Sterile nuclease-free water that is usually available as component of vendor prep-kits
3. Prepare plasmid DNA solution to desired concentrations in sterile nuclease-free water to typical ranges of final concentration of 1 μg/μL (single plasmid) to 2 μg/μL (plasmid mix);
4. Add Fast Green (concentration of 0.05%; Sigma) to visualize injection. Typical range of final volume of DNA plasmid and Fast Green is between 10 and 20 μL;
5. Needles used for the procedure are pulled borosilicate glass micro-capillaries using a micropipette puller. Multiple commercial vendors provide different lengths and gauges of needles, ranging from 1.0 mm O.D. × 0.5 mm I.D. to 1.2 mm O.D. × 0.9 mm I.D. Pre-pulled glass micro-capillaries can also be purchased and are good alternatives if no puller is available;
6. Microloader tips and micropipette.

Surgery

1. To prepare for surgery, autoclave surgical instruments and phosphate buffered saline;
2. Surgical instruments required for this procedure are forceps: graefe forceps, ring forceps, curved forceps, scissors, iris scissors, spring scissors, and a needle holder;
3. A surgical lamp with flexible arm, the recommendation would be a Halogen 20 W or LED equivalent;
4. Use of heating mat/pad for all the surgery, animals need to keep a constant temperature;
5. Surgical and medical protective disposables: masks and sterile gloves, sterile drapes, sterile swabs, cotton buds, and the use of

skin disinfectant (such as Chlorehexidine) and electric razor are basic for this procedure;

6. Conical tubes of 50 mL;
7. Hypodermic syringes from either 1–5 mL are recommended;
8. Use of sutures at the end of the procedure;
9. Analgesics: typically, an opioid such as buprenorphine but NSAID can be considered as alternatives;
10. Anesthetic: typically, flurane-based aesthetics such as isoflurane (alternatives being sevoflurane or desflurane);
11. Heating pads and recovery chamber as well as water bath and beaker for the full recovery of the animal.

Electroporation

The equipment is commercially available by multiple suppliers and may come as a complete kit:

1. Electroporator;
2. Electrodes (paddle-type or wire-type electrodes);
3. Microinjector;
4. Foot control;
5. Capillary holder.

2.2.2 Methods

Electroporation

This protocol describes the material and steps of in utero and post-natal cortical electroporation that has been performed by multiple groups for more than a decade (*see* **Note 5**). For variations, readers may also refer to other published protocols [64–67].

1. Pre-operative handling and preparation of animal for surgery: the operative equipment needed for this step are the following:
   (a) Switch on the water bath (39 °C), heating mat/pads (39 °C), recovery chamber and set them to the appropriate temperature;
   (b) Place sterile PBS in warm water bath (20 mL in a conical tube);
   (c) Connect the electroporator with the electrodes and microinjector with the foot control before switching both on;
   (d) Confirm that the respective settings are set up appropriately for the surgery;
   (e) Place electrodes into a beaker filled with PBS and connect to the electroporator;

(f) Place sterile surgical drape over the heating mat/pads followed by placing sterile surgical instruments and materials on the drape. Using micropipette and micro loader tip, fill the glass capillary with the DNA solution and connect to the capillary holder.

2. For the induction of anesthesia, it is mandatory to refer to the IACUC for regulations, guidelines, and training regarding "Survival Surgery." In addition, analgesia and anesthesia protocols must be adapted to comply with local legislation and institutional guidelines regarding use of these drugs.

(a) Prepare premedication/analgesia: buprenorphine, diluted to 0.03 mg/mL and administered at a final dose 0.1 mg/kg, or carprofen, diluted at concentration 0.5 mg/mL and administered at a final dose 5 mg/kg;

(b) Prepare anesthetic induction chamber according to manufacturer's protocol with appropriate volatile anesthetic;

(c) Connect the anesthesia machine, confirm that the amount of inhalational anesthetic is sufficient for the duration of the surgery, place the anesthetic mask on the surgical table;

(d) Anesthetize the animal in the anesthetic induction chamber with air flow (oxygen of 1–2 L/min (with flurane-based anesthetic at no more than 5% until the animal loses righting reflex, which indicates a state of sedation). Critically, the animal must be closely monitored during induction of anesthesia to avoid lethal overdose;

(e) Administration of pre-operative analgesic: gently remove the animal from the induction chamber and transfer it to the surgical drape in a ventral decubitus position. Rapidly inject the pre-operative analgesic subcutaneously into the interscapular area;

(f) Place a drop of protective moisturizer or gel on each eye of the animal to prevent corneal ulceration during general anesthesia;

(g) Rotate the animal in a dorsal decubitus position, before inserting its head entirely into the anesthetic mask, with air flow at 1 L/min and anesthesia decreased to 2%;

(h) Preparation of aseptic surgical zone on the animal: prepare the surgical zone on the animal by using the electric razor to shear the abdominal fur of the animal, with a region of approximately of 3–4 cm in diameter centered around navel;

(i) Gently stretch the skin of the abdomen during shearing minimize micro skin lesions cuts as dermal irritation has a strong negative impact on post-surgical well-being and can lead to lack of care of newborn pups;

(j) Sterilize the surgical zone on the animal using sterile surgical gauze at least three times with the antiseptic, using a different gauze each time, moving in a circular pattern from navel to the outer part of the sheared region without returning to center.

In Utero Electroporation

The four main procedural stages are: (1) laparotomy to expose the embryos, (2) micro-injection, (3) electroporation, and (4) surgical suture. Turn on the surgical light. At this stage, the mouse should be in dorsal decubitus, on a clean, sterile drape, above the heating pad. Confirm whether the animal is in a deep anesthetized state by pinching the hind paw to indicate the absence of pedal reflex before starting the surgery. Put on mask and sterile surgical gloves and cover the animal with a sterile drape (with a small aperture of between 2 and 3 cm over the abdomen).

1. Laparotomy:
   (a) Using forceps on non-dominant hand to raise the skin above the navel and using scissors on dominant hand to gently perform a mid-sagittal incision through the skin of about 2–2.5 cm centered on the navel. The incision should expose linea alba (white line composed mostly of collagen connective tissue joining the two abdominal muscles);
   (b) Use the forceps to gently separate the adherences between skin and abdominal muscle by placing the tip of the forceps into the wound in close contact with the muscle;
   (c) Gently open the forceps so that its jaws push skin tissue ventrally away from the muscle;
   (d) Again, using forceps on non-dominant hand to raise the umbilicus and using scissors on dominant hand to very carefully perform a very small incision (approximately 2–5 mm).
2. Micro-injection:
   (a) Choose the most accessible embryos and place the ring forceps between two embryos;
   (b) Carefully pull the embryonic chain out of the abdominal cavity to expose one uterine horn up to the ovary;
   (c) From this point on, keep the embryos hydrated with sterile prewarmed PBS;
   (d) Pinch off the tip of the microcapillary needle with forceps;

(e) Start with one of the most lateral embryos. Using the ring forceps with one hand to manipulate the position of one embryo inside the amniotic sac and gently squeeze to stabilize the head between the rings and push it up closer to the uterine wall. Using the other hand to take the capillary holder and insert the microcapillary needle carefully into the desired brain region to target the appropriate ventricle. Researchers may experience that two steps of injection are necessary (2-hit pass) that correspond to the microcapillary needle first piercing the uterine wall, followed by piercing the skin and skull;

(f) Press the foot pedal that is connected to the microinjector to inject the solution. The successful injection can be visualized by filling of the ventricle with the dye (either Fast Green or Fast Red) giving a crescent-like shape. If this is not observed, the injection may be unsuccessful due to either wrong injection angle and/or depth. However, it is not recommended to make more than two injection attempts on any single embryo.

3. Electroporation:
   (a) Place the positively charged anode on the side of DNA injection and negatively charged cathode on the opposite of the embryo's head to target the desired area;
   (b) Press the foot pedal connected to the electroporator and wait until the electroporation has been completed. The embryo may experience slight contractile movements in sync with the electrical pulses. All embryos per uterine horn or in a pregnant mouse can be electroporated with the same DNA construct to avoid confusion when harvesting at either embryonic or post-natal stages, respectively;
   (c) Maintain optimal hydration of the uterus at all times by using the syringe to drip warm sterile PBS over the uterus;
   (d) After either all or selected embryos in one uterine horn have been microinjected and electroporated, use the ring forceps to carefully replace the uterine horn back to its original intra-abdominal location, starting from the extremity attached to the ovary;
   (e) Extract the other uterine horn and repeat procedures for microinjection and electroporation and reinsertion back into the abdomen.

4. Suture and Stitching:
   (a) To close up the abdominal muscle walls, use the forceps and needle holder to perform a typical running suture, make a knot every 2–3 stictches, with spacing between

consecutive stitches approximate the width of the closed forceps tip, repeat for the skin, however, using instead adsorbable sutures. Turn off anesthetic after stitching is completed.

5. Post-Operative Handling and Recovery of Animal:
   (a) Anesthetic Recovery Period: Once the anesthetic is turned off, place the animal in a recovery chamber until it regains consciousness (starting from approximately 1 min, and would be fully awake within 5 min) before transferring it to the home cage placed on a heating pad to provide a warm environment;
   (b) The animal should be monitored regularly until responsive to gentle stimulation, restoration of righting reflex, spontaneous movement and self-cleaning, typically within 15–20 min after termination of anesthesia;
   (c) Post-Operative care: Inspect the stiches daily until the incision is fully healed and monitor general body condition, attitude, and mobility in order to assess pain, suffering, or distress. If needed, analgesics can be administered to minimize pain and discomfort;
   (d) Monitor food and water consumption daily.
6. Tissue Processing:
   (a) Collect either the electroporated embryos or pups or adults at the appropriate age required for the experiment;
   (b) For analyses at embryonic stages, euthanize mother via appropriate institutional guidelines and collect the embryos. Identify the brains of the embryos that have been appropriately electroporated. This can be as visualized either across the skull using a fluorescent binocular microscope prior to dissection or under a fluorescent microscope after dissection. Observe for appropriate intensity and location of the fluorescent signal;
   (c) For analyses at post-natal or adult stages, anesthetize pups or adult mice with intraperitoneal injection of pentobarbital (100 mg/kg; Pentobarbital® sodique, Ceva) and perform transcardial perfusion as explained previously. Identify the brains of the embryos that have been appropriately electroporated. This can be as visualized under a fluorescent microscope after dissection. Observe for appropriate intensity and location of the fluorescent signal;
   (d) Fixation: After dissection and identification of appropriately electroporated brains, fix the brains in 4% PFA overnight and then place in 20% sucrose/PBS overnight at 4 ° C. Embed in brains OCT compound, freeze at −80 °C and section using a cryostat as explained before.

3 Notes

There are several aspects that we would like to discuss regarding the study of neurogenesis in different species and with different objectives. In the present chapter, we included three examples of the study of adult neurogenesis and its relationship with reproductive behaviors. As well, we explained in detail, one genetic method to stain new neurons in an adult brain. Some details to consider for the experimenter are described below.

1. *Songbird Neurogenesis*

   For the experimental design, experimenters should consider that sex, photoperiodic state (for seasonal species), and hormonal status of the individual are powerful regulators of the neurogenesis [68]. Depending on the question being asked, it might be necessary to include both sexes and/or birds in different photoperiodic and hormonal conditions. Photoperiodic manipulations can take many months while circulating sex hormones stabilize approximately 3 weeks after gonadectomy and implantation of 10 mm-long Silastic™ capsules filled with beta-estradiol or testosterone. The following protocol is based on Shevchouk et al. [69]. Photosensitive male canaries were injected with 50 mg/kg BrdU dissolved in NaCl intraperitoneally five times in a single day, with 2 h between the injections. The following day the canaries were implanted with Silastic™ implants filled with testosterone or kept empty and, simultaneously, changed to a photostimulating photoperiod (i.e., long-days). The duration of post-BrdU survival depends on the phase of neurogenesis being studied—shorter durations such as 1, 2 days or under 24 h, can be used to study proliferation and the earliest phase of young neurons detaching from the ventricular zone in preparation for migration. In songbirds durations of about 1 week will show BrdU neurons in migration, many of which will have a bipolar morphology. By 3 weeks post-BrdU injections, a significant number of newborn neurons will be in the differentiating stage, becoming multipolar and forming connections with other neurons, although a proportion will still be migrating. For all durations, on the day of sacrifice the bird is sedated and intracardially perfused (*see* Subheading 2.2.2) and the brain removed from the skull. The brains were post-fixed overnight in PFA and cryopreserved in 30% sucrose until the brain sunk to the bottom of the vial, when they were frozen on dry ice and stored until cryosectioning. For free-floating sections for IHC cut 30 μm sections on the cryostat and store sections in anti-freeze solution in −20 °C until the IHC experiment. To ensure that BrdU signal detected reflects newborn neurons rather than other cell types, IHC experiments should co-localize BrdU with the neuronal marker

Hu (expressed starting from early differentiation) or doublecortin. The latter presents the additional advantage of discriminating between bipolar migrating and multipolar differentiating newborn neurons.

2. *Neurogenesis and Sexual Behavior in Female Rats*

The OB neurogenesis in adult female rats is involved in sexual behavior. We have shown that the first sexual encounter of a female rat significantly increases both cell proliferation in the SVZ and RMS and the incorporation of new cells within the OB [34, 35]. Repeated sexual encounters, providing experience and leading to an improvement of the behavioral efficiency, also augments the survival of the newly incorporated neurons, greatly increasing the percentage of neuronal commitment of the newly incorporated OB cells [37]. Additionally, evidence shows that newly formed OB neurons become part of the neuronal circuits involved in the control of sexual behavior, since their activation, measured by the expression of immediately early gene cFos in the new OB neurons, increases significantly when females are allowed to copulate [35]. Similar results have been published suggesting that OB adult neurogenesis is equally important for sexual behavior in males [70]. In female mice, it has also been shown that OB neurogenesis is determinant for reproductive behaviors, Feierstein et al. [71] demonstrated that the ablation of SVZ neurogenic cells by irradiation impaired the discrimination of a potential male partner in females, implying compromised reproduction.

The following protocol is based on Corona et al. [34] and Unda et al. [36]. Group-housed females are ovariectomized and given 2 weeks to recover from surgery. To induce sexual receptivity, they are injected subcutaneously with female sex hormones diluted in corn oil–estradiol benzoate (25 μg/rat) 48 h and progesterone (1 mg/rat) 4 h before each sexual behavior experience, respectively. Depending on whether sexually naïve or experienced animals are being studied, some sessions of sexual behavior training can be conducted before injecting BrdU. If females are being studied, a paced-mating setup should be used, which implies the use of a mating cage divided by a partition in two compartments, where only the female can navigate between compartments, based on the fact that the partition has a hole at the bottom, too small for the male but big enough for the female to enter. If the male's sexual behavior-induced neurogenesis is being studied, then a single-compartment mating cage can be used. In both cases, the session can be terminated after the male has performed 15 intromissions or 30 min have elapsed. Depending on the research question, a control group of animals should be included that has the same hormonal treatments but no sexual behavior, they should spend the same time in mating cages alone. 100 mg/kg BrdU dissolved in NaCl was injected intraperitoneally 60 min before the sexual behavior test, immediately after the test and 60 min after the test. 15–45 days later, the rats were sedated and perfused intracardially with PBS and then 4% PFA and the brain removed from the skull. If OB neurogenesis is being studied, extra care should be taken not to damage the bulb during removal. The brains were post-fixed for 60 min in PFA and cryopreserved in 30% sucrose until sectioning. For free-floating sections for IHC, cut 30 μm sections on the cryostat and store sections in anti-freeze solution until the IHC experiment.

3. *Neurogenesis and Maternal Behavior in Sheep*

   Motherhood requires physiological changes that support a complex set of behavioral displays from the mother. Especially in ewes, which rely upon an efficient olfactory recognition of their offspring, olfactory neurogenesis becomes a necessary and adaptative plastic change for the mothers' OB. Evidence shows that maternal behavior in ewes promotes an increased response of the newly generated OB neurons [41]. Moreover, by using chemical disruption with cytosine arabinoside of the olfactory neurogenesis ewes showed deficiencies on their behavioral display towards their offpring, strongly suggesting that the OB neurogenesis participates in the maternal behavior in sheep [42].

   Working with a sheep model requires very particular infrastructure, skills, and ethical considerations, and should be done in collaboration with a licensed butcher. The following protocol is based on Corona et al. [41]. BrdU was administered to ewes 1.5 months into their 5-month gestation and 3.5 months before sacrifice—ewes received four intravenous injections within 24 h of 20 mg/kg BrdU dissolved in NaCl. The duration between BrdU injection and sacrifice is optimized for studying young migrating and differentiating neurons, these processes occur more slowly than in rodents [72]. To study maternal neurogenesis in a robust way, tests are done to ensure all ewes are maternal and selective towards their own lamb. To study the function of the BrdU-labeled newborn neurons, the ewes can be exposed to their own lambs versus to other lambs or to other control animals. In this case, the BrdU labeling needs to be combined with a marker of neuronal activation, such as immediate early genes like c-fos, Arc, or egr-1. On the day of sacrifice, the ewes are deeply anesthetized and decapitated by a licensed butcher. The head of the ewes is perfused via carotid arteries with PBS with 1% sodium nitrate and then 4% PFA. The brain is dissected into blocks, post-fixed in PFA for 48 h, and stored in 30% sucrose for at least 2 days until sectioned. 30 μm sections are cut using a freezing microtome.

4. *Limitations in the Use of Markers for DNA Replication*

There are several limitations in these methods that must be considered. As already mentioned, most, if not all, methods can capture events other than neurogenesis, such as DNA repair, etc., and therefore multiple different approaches need to be used to conclude that birth of new cells under conditions and in a particular brain region has been observed. Furthermore, a particular limitation of the endogenous marker approach is that antibodies against these proteins are not always available for all species, especially when it comes to non-rodent experimental models such as sheep, birds, or fish. When using an antibody for the first time, especially in a species other than what the company has validated, it is very important to run tests to verify the specificity of the antibody and the IHC protocol. Some potential methods for this include: (1) co-labeling with other markers of the neurogenesis phase you are studying (e.g., proliferation, differentiation, migration, etc.), (2) verify that the labeling occurs only where you expect to see it and not throughout the brain, (3) if possible test the antibody in brain sections from a knockout/knockdown model to make sure the staining is absent or, in the case of knockdown, reduced, compared to wild-type animals, (4) pre-adsorb with the peptide that the antibody is targeting to make sure this removes the signal, and (5) omit the primary antibody and check that the other reagents are not causing a non-specific signal. Due to these limitations and difficulties, the use of exogeneous markers is a very robust method that can be applied regardless of species. A major disadvantage of using exogeneous markers is the possible toxicity both to the cells that incorporate the thymidine analogues [43, 73, 74] and to the animal in general. Care should be taken to use the lowest possible doses and consider how long the subjects will survive after the thymidine analogue administration. Some analogues only become toxic to the cell after it has undergone further cycles of division, for example, EdU [75, 76]. Combining multiple thymidine analogues, while offering a more detailed view of the neurogenic process, is associated with an even higher risk of toxicity and in some cases dose-specific cross-reactivity of antibodies to multiple analogues (for example, cross-reactivity of BrdU and EdU, [76]). Since DNA synthesis can be initiated independently of mitosis during events such as gene duplication, or apoptosis, to some extent these markers are indicators of only DNA synthesis and not cell division, however, the signal detected when thymidine analogues incorporate during these events is substantially weaker and found throughout the brain, whereas adult neurogenesis occurs only in very specific brain areas. Nevertheless, findings based on exogeneous markers of neurogenesis should generally be corroborated with studies using endogenous markers.

5. *Electroporation*

Critically, all procedures and surgeries must obey the rules for animal care and use in research and must comply with the respective countries' legal framework and institutional guidelines (refer to either IACUC or EU Directives or equivalent). All manipulations must be refined to minimize pain and distress with appropriate pre-operative handling of animals as well as post-operative and analgesia. In addition, strict compliance of surgical sterility avoids post-surgical complications and alleviates the need for antibiotics. It is strongly recommended to seek advice and guidance from local animal welfare and veterinary committees with rigorous records of all pre-operative and post-operative procedures.

There are multiple limitations to in utero and postnatal electroporation, chief of all being invasive procedures that requires the necessary IACUC approval of animal ethics with the appropriate oversight by federal or national LAS organizations (AALAS / FELASA) as well as the training of operators on the surgical techniques and animal health monitoring. In addition, inconsistent techniques or collateral damage of cells by electroporation will impact the repeatability and reproducibility of the findings. Next, the spatio-temporal resolution of cells are dependent on the genetic reagents utilized. For example, non-viral vectors are likely to be maintained episomally in the proliferating cells in the developing stages, making it unclear how long transgene expression continues. This may be overcome by using transposon-mediated gene expression system, including *piggyBac* and *Tol2*, by integrating transgenes into the chromosomal DNA of the targeted cells [77].

References

1. Owji S, Shoja MM (2020) The history of discovery of adult neurogenesis. Clin Anat 33(1): 41–55
2. Altman J, Das GD (1965) Autoradiographic and histological evidence of postnatal hippocampal neurogenesis in rats. J Comp Neurol 124(3):319–335
3. de Castro F, Lopez-Mascaraque L, De Carlos JA (2007) Cajal: lessons on brain development. Brain Res Rev 55(2):481–489
4. Nottebohm F (2002) Neuronal replacement in adult brain. Brain Res Bull 57(6):737–749
5. Snyder JS, Drew MR (2020) Functional neurogenesis over the years. Behav Brain Res 382: 112470
6. Luskin MB (1993) Restricted proliferation and migration of postnatally generated neurons derived from the forebrain subventricular zone. Neuron 11(1):173–189
7. Peretto P, Paredes RG (2014) Social cues, adult neurogenesis, and reproductive behavior. In: Mucignat-Caretta C (ed) Neurobiology of chemical communication. CRC Press, Boca Raton
8. Migaud M et al (2016) Adult neurogenesis and reproductive functions in mammals. Theriogenology 86(1):313–323
9. Trova S et al (2021) Sex steroids and the shaping of the peripubertal brain: the sexual-dimorphic set-up of adult neurogenesis. Int J Mol Sci 22(15):7984
10. Trova S et al (2020) HPG-dependent peripubertal regulation of adult neurogenesis in mice. Front Neuroanat 14:584493

11. Corona R, Levy F (2015) Chemical olfactory signals and parenthood in mammals. Horm Behav 68:77–90
12. Butruille L et al (2018) Seasonal reorganization of hypothalamic neurogenic niche in adult sheep. Brain Struct Funct 223(1):91–109
13. Gould E et al (1999) Neurogenesis in the neocortex of adult primates. Science 286(5439): 548–552
14. Kornack DR, Rakic P (2001) The generation, migration, and differentiation of olfactory neurons in the adult primate brain. Proc Natl Acad Sci U S A 98(8):4752–4757
15. Bernier PJ et al (2002) Newly generated neurons in the amygdala and adjoining cortex of adult primates. Proc Natl Acad Sci U S A 99(17):11464–11469
16. Shapiro LA et al (2009) Subventricular zone-derived, newly generated neurons populate several olfactory and limbic forebrain regions. Epilepsy Behav 14(Suppl 1):74–80
17. Ming GL, Song H (2011) Adult neurogenesis in the mammalian brain: significant answers and significant questions. Neuron 70(4): 687–702
18. Brenowitz EA, Larson TA (2015) Neurogenesis in the adult avian song-control system. Cold Spring Harb Perspect Biol 7(6):a019000
19. Watt FM, Hogan BL (2000) Out of Eden: stem cells and their niches. Science 287(5457):1427–1430
20. Lois C, Alvarez-Buylla A (1993) Proliferating subventricular zone cells in the adult mammalian forebrain can differentiate into neurons and glia. Proc Natl Acad Sci U S A 90(5): 2074–2077
21. Lois C, Alvarez-Buylla A (1994) Long-distance neuronal migration in the adult mammalian brain. Science 264(5162):1145–1148
22. Peretto P et al (1999) The subependymal layer in rodents: a site of structural plasticity and cell migration in the adult mammalian brain. Brain Res Bull 49(4):221–243
23. Alvarez-Buylla A, Garcia-Verdugo JM (2002) Neurogenesis in adult subventricular zone. J Neurosci 22(3):629–634
24. Doetsch F, Garcia-Verdugo JM, Alvarez-Buylla A (1997) Cellular composition and three-dimensional organization of the subventricular germinal zone in the adult mammalian brain. J Neurosci 17(13):5046–5061
25. Carleton A et al (2002) Making scents of olfactory neurogenesis. J Physiol Paris 96(1–2): 115–122
26. Merkle FT, Mirzadeh Z, Alvarez-Buylla A (2007) Mosaic organization of neural stem cells in the adult brain. Science 317(5836): 381–384
27. Petreanu L, Alvarez-Buylla A (2002) Maturation and death of adult-born olfactory bulb granule neurons: role of olfaction. J Neurosci 22(14):6106–6113
28. Winner B et al (2002) Long-term survival and cell death of newly generated neurons in the adult rat olfactory bulb. Eur J Neurosci 16(9): 1681–1689
29. Marler P (1970) Birdsong and speech development: could there be parallels? Am Sci 58(6): 669–673
30. Goldman SA, Nottebohm F (1983) Neuronal production, migration, and differentiation in a vocal control nucleus of the adult female canary brain. Proc Natl Acad Sci U S A 80(8): 2390–2394
31. Arnold AP et al (1986) Sexual dimorphisms in the neural vocal control system in song birds: ontogeny and phylogeny. Brain Behav Evol 28(1–3):22–31
32. Ball GF et al (2020) How does testosterone act to regulate a multifaceted adaptive response? Lessons from studies of the avian song system. J Neuroendocrinol 32(1):e12793
33. Lim DA, Alvarez-Buylla A (2016) The adult ventricular-subventricular zone (V-SVZ) and olfactory bulb (OB) neurogenesis. Cold Spring Harb Perspect Biol 8(5):a018820
34. Corona R, Larriva-Sahd J, Paredes RG (2011) Paced-mating increases the number of adult new born cells in the internal cellular (granular) layer of the accessory olfactory bulb. PLoS One 6(5):e19380
35. Corona R et al (2016) Sexual behavior increases cell proliferation in the rostral migratory stream and promotes the differentiation of the new cells into neurons in the accessory olfactory bulb of female rats. Front Neurosci 10:48
36. Unda NM et al (2016) Sexual stimulation increases the survival of new cells in the accessory olfactory bulb of the male rat. Front Neurosci 10:65
37. Arzate DM et al (2013) Repeated paced mating promotes the arrival of more newborn neurons in the main and accessory olfactory bulbs of adult female rats. Neuroscience 232:151–160
38. Mak GK et al (2007) Male pheromone-stimulated neurogenesis in the adult female brain: possible role in mating behavior. Nat Neurosci 10(8):1003–1011
39. Lau BW et al (2011) Effect of corticosterone and paroxetine on masculine mating behavior:

possible involvement of neurogenesis. J Sex Med 8(5):1390–1403

40. Sakamoto M et al (2011) Continuous neurogenesis in the adult forebrain is required for innate olfactory responses. Proc Natl Acad Sci U S A 108(20):8479–8484
41. Corona R et al (2017) Exposure to young preferentially activates adult-born neurons in the main olfactory bulb of sheep mothers. Brain Struct Funct 222(3):1219–1229
42. Corona R et al (2018) Disruption of adult olfactory neurogenesis induces deficits in maternal behavior in sheep. Behav Brain Res 347:124
43. Taupin P (2007) BrdU immunohistochemistry for studying adult neurogenesis: paradigms, pitfalls, limitations, and validation. Brain Res Rev 53(1):198–214
44. Brown DC, Gatter KC (2002) Ki67 protein: the immaculate deception? Histopathology 40(1):2–11
45. De Biasio A, Blanco FJ (2013) Proliferating cell nuclear antigen structure and interactions: too many partners for one dancer? Adv Protein Chem Struct Biol 91:1–36
46. Jiang Y, Hsieh J (2014) HDAC3 controls gap 2/mitosis progression in adult neural stem/progenitor cells by regulating CDK1 levels. Proc Natl Acad Sci U S A 111(37): 13541–13546
47. Chirivella L et al (2017) Cyclin-dependent kinase 4 regulates adult neural stem cell proliferation and differentiation in response to insulin. Stem Cells 35(12):2403–2416
48. Hans F, Dimitrov S (2001) Histone H3 phosphorylation and cell division. Oncogene 20(24):3021–3027
49. Ming GL, Song H (2005) Adult neurogenesis in the mammalian central nervous system. Annu Rev Neurosci 28:223–250
50. Obernier K et al (2018) Adult neurogenesis is sustained by symmetric self-renewal and differentiation. Cell Stem Cell 22(2):221–234 e8
51. Sawamoto K et al (2011) Cellular composition and organization of the subventricular zone and rostral migratory stream in the adult and neonatal common marmoset brain. J Comp Neurol 519(4):690–713
52. Menezes JR, Luskin MB (1994) Expression of neuron-specific tubulin defines a novel population in the proliferative layers of the developing telencephalon. J Neurosci 14(9):5399–5416
53. Ma DK et al (2009) Adult neural stem cells in the mammalian central nervous system. Cell Res 19(6):672–682
54. Lledo PM, Alonso M, Grubb MS (2006) Adult neurogenesis and functional plasticity in neuronal circuits. Nat Rev Neurosci 7(3):179–193
55. Bagley J et al (2007) Adult neurogenesis and specific replacement of interneuron subtypes in the mouse main olfactory bulb. BMC Neurosci 8:92
56. Whitman MC, Greer CA (2009) Adult neurogenesis and the olfactory system. Prog Neurobiol 89(2):162–175
57. Carleton A et al (2003) Becoming a new neuron in the adult olfactory bulb. Nat Neurosci 6(5):507–518
58. Alvarez-Buylla A, Ling CY, Nottebohm F (1992) High vocal center growth and its relation to neurogenesis, neuronal replacement and song acquisition in juvenile canaries. J Neurobiol 23(4):396–406
59. Gage GJ, Kipke DR, Shain W (2012) Whole animal perfusion fixation for rodents. J Vis Exp (65):3564
60. Tabata H, Nagata K (2016) Decoding the molecular mechanisms of neuronal migration using in utero electroporation. Med Mol Morphol 49(2):63–75
61. Nishimura YV et al (2012) Application of in utero electroporation and live imaging in the analyses of neuronal migration during mouse brain development. Med Mol Morphol 45(1): 1–6
62. Nagahama K et al (2021) Combining electrophysiology and optogenetics for functional screening of pyramidal neurons in the mouse prefrontal cortex. STAR Protoc 2(2):100469
63. Taniguchi Y et al (2012) In utero electroporation as a tool for genetic manipulation in vivo to study psychiatric disorders: from genes to circuits and behaviors. Neuroscientist 18(2): 169–179
64. Hand R, Polleux F (2011) Neurogenin2 regulates the initial axon guidance of cortical pyramidal neurons projecting medially to the corpus callosum. Neural Dev 6:30
65. Pacary E, Guillemot F (2047) In utero electroporation to study mouse brain development. Methods Mol Biol 2020:513–523
66. Saito T (2006) In vivo electroporation in the embryonic mouse central nervous system. Nat Protoc 1(3):1552–1558
67. Shimogori T, Ogawa M (2008) Gene application with in utero electroporation in mouse embryonic brain. Develop Growth Differ 50(6):499–506
68. Balthazart J, Ball GF (2016) Endocrine and social regulation of adult neurogenesis in songbirds. Front Neuroendocrinol 41:3–22

69. Shevchouk OT et al (2017) Testosterone-induced neuroendocrine changes in the medial preoptic area precede song activation and plasticity in song control nuclei of female canaries. Eur J Neurosci 45(7):886–900
70. Portillo W et al (2012) Sexual activity increases the number of newborn cells in the accessory olfactory bulb of male rats. Front Neuroanat 6:25
71. Feierstein CE et al (2010) Disruption of adult neurogenesis in the olfactory bulb affects social interaction but not maternal behavior. Front Behav Neurosci 4:176
72. Brus M, Keller M, Levy F (2013) Temporal features of adult neurogenesis: differences and similarities across mammalian species. Front Neurosci 7:135
73. Breunig JJ et al (2007) Everything that glitters isn't gold: a critical review of postnatal neural precursor analyses. Cell Stem Cell 1(6):612–627
74. Anda S, Boye E, Grallert B (2014) Cell-cycle analyses using thymidine analogues in fission yeast. PLoS One 9(2):e88629
75. Ponti G et al (2013) Cell cycle and lineage progression of neural progenitors in the ventricular-subventricular zones of adult mice. Proc Natl Acad Sci U S A 110(11):E1045–E1054
76. Shevchouk OT et al (2017) Studies of HVC plasticity in adult canaries reveal social effects and sex differences as well as limitations of multiple markers available to assess adult neurogenesis. PLoS One 12(1):e0170938
77. Dumas L et al (2020) In utero electroporation of multiaddressable genome-integrating color (MAGIC) markers to individualize cortical mouse astrocytes. J Vis Exp (159). https://doi.org/10.3791/61110

INDEX

Raul Paredes Guerrero et al. (eds.), *Animal Models of Reproductive Behavior*, Neuromethods, vol. 200, https://doi.org/10.1007/978-1-0716-3234-5,

FSC
www.fsc.org
MIX
Papier aus verantwortungsvollen Quellen
Paper from responsible sources
FSC® C105338

If you have any concerns about our products,
you can contact us on
ProductSafety@springernature.com

In case Publisher is established outside the EU,
the EU authorized representative is:
Springer Nature Customer Service Center GmbH
Europaplatz 3, 69115 Heidelberg, Germany

Printed by Libri Plureos GmbH
in Hamburg, Germany